AIR POWER

Air Power
Promise and Reality

Edited by
Mark K. Wells

Imprint Publications
Chicago
2000

Copyright © 2000 by Imprint Publications, Inc., Chicago, Illinois
All rights reserved. With the exception of brief quotations in a review, no part of
this book may be reproduced in any form or by any means without permission in
writing from the publisher. Statements of fact or opinion made by contributors
in this volume are solely the responsibility of the contributors and do not necessarily
reflect the views of the United States Air Force Academy or any other United States
government agencies.

Cover design and illustration by Chris Hureau

Library of Congress Catalog Card Number 00-110268
ISBN 1-879176-30-0 (Paper)

Military History Symposium Series of the United States Air Force Academy, Vol. 6
Carl W. Reddel, Series Editor

Printed in the United States of America on acid-free paper

*To the combat airmen of all nations,
past, present, and future*

AIR POWER

Contents

Foreword *Ronald R. Fogleman*	xi
Preface *Mark K. Wells*	xiii
Air Power and Warfare: A Historical Overview *Richard J. Overy*	1

I. Air Power and Warfare, 1903–1941

The British Dimension *R. A. Mason*	7
The Continental Experience *Edward Homze*	19
French Military Aeronautics before and during the Great War *Lucien Robineau*	31
The American Dimension *Eugene M. Emme*	49

II. World War II in the Air: National Experiences

The Rise and Fall of the Imperial Japanese Air Forces *Alvin D. Coox*	73
Soviet Air Power in World War II *Kenneth R. Whiting*	85
Higher Command and Leadership in the Luftwaffe, 1935–1945 *Horst Boog*	111
Some Observations on Air Power *Ira C. Eaker*	139

III. Postwar Air Doctrine and Organization, 1945–1953

The Emergence of the Postwar Strategic Air Force, 1945–1953 *John T. Greenwood*	149
American Postwar Air Doctrine and Organization: The Navy Experience *David Alan Rosenberg*	175
The Interaction of Technology and Doctrine in the U.S. Air Force *Robert Perry*	205

IV. The Cold War and Beyond

Korean, Vietnam, and NATO: An Airman's Perspective, from First
Lieutenant to Four-Star General   219
Bryce Poe II

Air Power in the Cold War   237
R. A. Mason

Searching for Victory through Air Power: The Conduct of the Air Wars
in Vietnam   263
Mark Clodfelter

Air Power in the Gulf War: Plans, Execution, and Results   289
Thomas A. Keaney

An Afterword   314
David MacIssac

Contributors   319

Index   323

Foreword

I appreciate the opportunity to rejoin many of my former colleagues in the Air Force Academy's Department of History in the preparation of this volume. It is clear that military affairs over the centuries have been dramatically impacted by a number of significant technological innovations. None has been more profound, nor had a more revolutionary effect, than the introduction of the airplane. Warfare has never been the same since those first flimsy biplanes soared over the deadly trenches of World War I battlefields. Military air power, built by the labor and invigorated by the spirit of scores of visionaries, increasingly dominated twentieth-century warfare. Recent events make it clear that aircraft, precision-guided weapons, and advanced space systems will continue to play a huge role in peacetime as well as war.

During my tenure as Chief of Staff, one of my goals was to prepare the men and women of our nation's Air Force for the future. Meeting the challenges of tomorrow requires vision, commitment, and a clear sense of perspective. This book, along with those I selected in 1996 for the Air Force's professional reading list, helps foster the development of a unifying air and space culture. Decades ago Brigadier General Billy Mitchell argued for Americans to be more "air-minded." His passion, insight, and advocacy for air power—no less relevant to the future of U.S. national security now than it was in the 1920s—runs as a theme throughout this volume. As we transition to the twenty-first century, it is important to reflect on stories of service and sacrifice by airmen of all nations. The collection of essays here presented, each written by a world-class scholar or notable airman, brings into focus almost a hundred years of military aeronautics. This is the history of air power. This is the story of the Aerospace Century.

RONALD R. FOGLEMAN, General, U.S. Air Force
Chief of Staff (1994–97)

Preface

The Department of History published the proceedings of the Air Force Academy's Eighth Military History Symposium, *Air Power and Warfare,* in 1978. Superbly edited by then Col. Alfred F. Hurley and Maj. Robert C. Ehrhart, it has always been one of the more readable and effective proceedings editions. This new volume is designed to focus, synthesize, and update the book to make it even more useful. Accordingly, we asked some of the world's most recognized air power specialists to contribute. This volume preserves the core of the original and presents five completely new essays. It features a powerful introduction by one of the world's top air power historians, Professor Richard J. Overy, and an insightful afterword by Dr. David MacIsaac, a veteran of the Academy's Department of History and the original 1978 symposium. The current members of the department believe this edition can make an important and lasting contribution to the study of air power, especially as the 100th anniversary of manned flight approaches and the twenty-first century opens.

As editor of the updated volume, my overall intention was to keep as much of the original as possible. But the goals of the 1978 symposium and the resulting proceedings were limited. They were presented, in the words of the hosts, "to enhance both the teaching and public understanding of the record and potential of air power." Although this was successfully accomplished, the purpose of the new volume is more ambitious. Clearly, much has happened in the years since the symposium. Social, political, economic, and military events over the last two decades have fashioned a world much different than that of 1978. Air power has changed as well. Some would argue that air power has put us on the threshold of a military revolution no less dramatic than the one fashioned by the introduction of gunpowder on the battlefield. Whether this is true or false, it is clear that world leaders, military authorities, and the general public thinks of air power much differently now than at any other time in the history of airplanes. Events as recent as the Gulf War and Kosovo have demonstrated the influence and impact of air power in warfare in ways not apparent to even its most strenuous advocates of seventy years ago. In the simplest terms, this volume examines the promise and reality of military air power.

As a result, to be most useful, an updated symposium volume should mirror the best in contemporary air power scholarship. Quite simply, in the years since 1978 we have had more time to reflect on air power and its impact during the Cold War, in Korea, in Vietnam, and beyond. It is important to take advantage of this wisdom. Moreover, as a principal center for the teaching and study of air power history, the United States Air Force Academy can contribute powerfully to military professionals and policymakers by producing this volume. In sum, its publication helps validate an important Academy institutional goal. The new volume features an impressive group of air power shapers and scholars who will tell us not only where air power has been since its inception but also, by inference, may help provide a road map for the future. This updated book should enable a reader to draw relevant conclusions about the true revolution in warfare initiated by two Dayton, Ohio brothers on a blustery day in December 1903.

The first section of the updated volume lays the foundation for the book. Three superbly written essays survey the origins of aviation and speak to the growth and development of military air power. All three highlight the importance of World War I in the process. They make it clear, despite the United States' central role in the airplane's invention, that Europeans were subsequently responsible for much of its evolution. The essay on the French aerial contribution during World War I is new. General Lucien Robineau's insightful survey clearly shows why French military aviation and its associated industry led the world by war's end. The last essay, presented by one of the world's most famous air power authorities, effectively introduces a theme which runs either directly or implicitly throughout the volume. The late Eugene M. Emme speaks to the promises made by air power's earliest advocates. His comments regarding Giulio Douhet, Hugh Trenchard, and Billy Mitchell speak to the potential decisiveness of air warfare and serve as an excellent lead-in to the balance of the book.

World War II would became the first major proving ground for air power. That it failed to live up to the projections of its more strident advocates seems clear. Clear too, however, is the fact that modern warfare has forever been changed as a result of the use of airplanes in a variety of missions. Four essays in the volume deal principally with the impact of air power in World War II. They describe the complexities of air warfare and highlight the importance of doctrine, leadership, technology, strategy, and production. A principal point in all four is that although air power promised inexpensive and quick victory, the reality was too often attrition and lengthy combat. God was still on the side of the bigger battalions. As assessments of the air war experiences of the conflict's principal belligerents they are superb. Moreover, they form a central part of the volume's examination of the changing impact of air power over time.

The events of the war brought clear focus on the continuing interaction between air power technology, doctrine, and policy. These factors increased in importance during the years immediately after the war and during the early months of the emerging East-West confrontation. Three original essays are repeated in this section of the new volume. Along with other events, they highlight the intense interservice struggles over the control or use of strategic air power in the United States. They also demonstrate how military institutions adapt or fail to adapt to technological change. The last essay, by Robert Perry, provocatively introduces the air power quality versus quantity debate. This controversy, so much a feature of the last couple of decades, has only recently been muted by the apparent success of high-tech weapons in the Gulf and in Kosovo.

The final portion of this volume is brand new. The authors, each an acknowledged air power expert or practitioner, do an excellent job examining the impact of air power during conflicts since the end of World War II. In doing so they effectively address the volume's unifying theme. Evaluating air power's efficacy and impact from the Korean conflict through the Cold War and into Vietnam permits the reader to test the assertions of its more strenuous advocates. Thomas Keaney's examination of the Gulf War brings us almost to the present. Taken together, these surveys bring into focus the most important technological and doctrinal changes affecting modern air power. As a result, we are much closer to understanding whether or not it has fulfilled the promises of its prophets and forms the fundamental basis for an ongoing revolution in military affairs.

Mark K. Wells

Air Power and Warfare: A Historical Overview

Richard J. Overy

Air power history has come a long way in the two decades since the original symposium proceedings of this volume was published. Like aviation itself, air power history has become more sophisticated, more wide ranging, and more innovative. The exchange of information and ideas between academics, air power enthusiasts, and practicing airmen is now well advanced. As this present updated volume so clearly demonstrates, we all have something to learn from each other.

The history of air warfare has steadily moved away from a focus on air combat to explore the wider context in which air power was generated and deployed. Part of that context was dictated by political and social developments. Understanding, for example, the British decision to create a separate Royal Air Force (RAF) in 1917 has led historians to look at the political issues that pushed Prime Minister Lloyd George in that direction.[1] It was not a decision taken by the military chiefs on strategic grounds, but one taken by the political leadership to satisfy public demands for action following German bombing raids on London. The same might be said of the decision by the Roosevelt administration to expand the army air forces in 1940. The army had its own view of what air forces could and could not do. It is hard to believe that left to themselves they would have endorsed the strategy of large-scale long-range bombing. American commitment to air power during the war owed a great deal to political chance. If Franklin D. Roosevelt had decided to stand down in 1940, would a successor have embraced the air so enthusiastically?[2] Modern air power history has rightly emphasized the political element. Air warfare was not something predetermined by the character of the technology or the commitment of its champions, though both played a part. The evolution of air power was as dependent on the wider historical process as any other form of warfare. This was the case with the social reception of air power. Historians have developed the concept of "airmindedness" to describe the nature and extent of public interest in air power issues. The growth of a popular culture of air power in Germany in the 1920s or the Soviet Union in the 1930s helped fuel the development of large modern air forces.[3] Uri Bialer's work on the ambiguous attitude of the British elite to air power between the wars—seen both as menace and as protector—has paved the way to building a cultural history of air power.[4] Michael Sherry's work on militarism and public life in America has drawn substantially on this approach because air power has become such a central component of American military experience since 1941.[5]

The material aspects of air war history have not been neglected, but here the emphasis has moved away from the weapons themselves toward the economic and technical context in which the weapons are developed and produced. Few of the contributors to this volume ignore the material issues. Almost without exception, air strategy has been hostage to the ability of an industrial system to provide large quantities, or sufficient quantities, of high performance aircraft and the means to distribute and service them. Air power has relied critically on a fast-moving technological threshold, and thus the ability to harness scientific developments in operationally useful ways. The failure of Italian air power in World War II, despite remarkable advances in aircraft design and performance in the early 1930s, was largely due to the failure of materiel.[6] It should be observed here that the ratio between the size of the industrial base and the quantity of air equipment is a very imprecise one. Lucien Robineau, in his discussion of early French air power in this volume, makes the point that France achieved a remarkable record in air production despite a much smaller manufacturing base than either Britain or Germany. France happened to be a prewar pioneer of air technology and had a long tradition of highly skilled, high-quality machine production.[7] In World War II the contrast between German and Allied performance is hard to understand without recognizing that the output and quality of front-line aircraft in the Luftwaffe never matched what Germany was manifestly capable of producing.[8]

Social history has not been far behind. The social history of air power and air war may not look on the surface the most promising area of research, but new work in this field has opened up important perspectives on the whole subject. Mark Wells's own study of morale among the crews of RAF Bomber Command and 8th U.S. Army Air Force in World War II raises issues about the nature of combat and leadership which suggest that no study of air power can ignore the moral and human dimension.[9] There is much work to be done here. Recent studies of the German air officer corps, of blacks in the U.S. Air Force, and of the social and economic impact of bomb destruction in World War II suggest that a rich seam remains to the keen prospector.[10]

In the effort to locate air power firmly in the wider historical enterprise, the danger now exists that historians will lose sight of combat altogether. In the proceedings of a major air power conference held in Freiburg in 1988 on air warfare in World War II, only three entries out of thirty-four could be said to cover air combat in even the most general sense.[11] Air force doctrine was little better served, though there has been noticeably more research on doctrine than combat since 1978. There has, of course, been a good deal of new writing on both and it allows us to see the history of air power in a radically different way. For years the discussion of air power was dominated by the Anglo-Saxon preoccupation with independent bombing operations. Two things have changed this perspective. Since the 1970s, when B-52s bombed Cambodia, bombing has changed beyond recognition. Fast fighter-bombers with guided missiles bear scant resemblance to the fleets of heavy bombers carpet-bombing German cities or guerrilla hideouts. With the passage of time it is possible to see the age of the big bomber as a particular stage in air power development, and not as its only significant manifestation. What has always been called "strategic bombing" is now part of history.

The second development has been recognition of the sheer diversity of development in air power theory and practice. There was no orthodox path through either doctrine or combat from which air forces could be seen to deviate or to which they had to conform. Doctrinal choices were shaped by military experience, or by pressures of geography or politics. By recognizing that there are many paths it has proved possible, for example, to write a much more sympathetic history of French air power.[12] Indeed the definition of air power given by French Marshal Joseph Joffre in November 1914, cited in General Robineau's essay, could stand as a formal statement of the current air force mission. There are plenty of critics of strategic bombing who argue that the bombing strategy was a chimera, a distortion of the true principles of war, and who argue that a proper appreciation of tactical air warfare might have served the Western Allies better between 1941 and 1945.[13] French air historians are certainly at pains to point out that on balance French air doctrine drew more pertinent lessons from air combat in World War I than did the fledgling RAF. Recent studies of German air force doctrine have observed that in the first few years after 1919 the basic formula that proved so successful in 1939–41 was already more or less in place.[14] The history of air warfare since 1945 has confirmed the significance of tactical aviation over strategic operations. The Arab-Israeli wars showed tactical aviation at its most effective—as the Gulf War was to do later. By contrast most recent accounts of the Korean and Vietnam Wars, including the new essays in this volume, are highly critical of bombing strategy.

The shift away from strategic bombing as the apotheosis of air power, and the recognition of doctrinal diversity, have allowed historians to examine more critically many of the enduring myths about combat performance. There have been some surprising conclusions. One example may suffice here, and it comes, ironically, from the sphere of tactical air power. There has always been an assumption that fighter bombers used as battlefield weapons were capable of knocking out tanks and heavy ground installations, as they are today. The popular image of World War II of the Junkers Ju-87B dive bomber, the famous *Stuka* with its terrifying siren, hitting battlefront targets to clear a way for the advancing ground forces is one of the defining images of the conflict. It is now known that the moral impact of the *Stuka* outweighed its material affects. On the Western Front in France, troops panicked at the sight of diving aircraft and abandoned artillery when they appeared. In the *Barbarossa* campaign, recent Russian research has argued that Soviet tanks were destroyed mainly by artillery fire, which was remarkably accurate.[15] A direct air hit on a tank under battlefield conditions without guided weapons was the exception. This was certainly the case when the RAF began dive-bombing trials in 1939 and 1940, whose results were so disappointing that for the whole war RAF commanders argued that small targets were a waste of strategic effort.[17] In the Normandy invasion the Allies enjoyed overwhelming air superiority, and used waves of fighter-bombers with bombs, rockets, and cannon against German ground targets. While it appears that soft-skinned vehicles were highly susceptible to rocket or tracer fire, tanks could not be penetrated with current rocket technology, and a hit with a bomb was at best infrequent. German crews abandoned their tanks and ran for shelter because of what they thought aircraft could do, not from the reality.[18] The moral impact proved greater here, and in other battlefield situations. In this sense, Hugh Trenchard was closer to the truth about the moral impact of bomb attack than is usually thought.

The legacy of Trenchard still hangs heavily over the whole field of air power studies. Though the kind of bombing war he visualized is now part of history, the debate surrounding the nature and effects of bombing is as alive as ever. Three of the new contributions to this volume on the Cold War, on Vietnam, and on the Gulf War, all address the long shadow of the great bombing war of 1939–45. Although targets can now be hit with a force and accuracy impossible with conventional technology in 1945, Thomas Keaney makes the point that if the coalition air forces had been asked to take out a large and complex target system in 1991 they would still have experienced many of the problems of the bombing assault on Germany and one new one— the cost of maintaining a large enough fighter-bomber force to conduct sustained bombing operations. Most nations, except perhaps the United States, cannot afford to do it. An attrition war like Vietnam or World War II would not be possible even at modest loss rates. Neither pilots nor aircraft are readily reproducible as they were in the 1940s. One effect of this, as Mark Clodfelter points out, has been to draw air forces back to the idea of the "vital targets" first raised by Hal George and others at the Air Corps Tactical School in the 1930s.[19]

This begs the perennial question of what those vital targets are. This is an issue that has divided airmen and their critics since the whole question of what aircraft could do in war was first aired before 1914. The Clausewitzian answer, that aircraft should be concentrated against the enemy's main force, first in the air then against targets on or just behind the battlefield, has never entirely satisfied airmen, though it worked well enough on the Soviet front in World War II, and it was central to the early success of German air power in the same conflict. The most extreme alternative, that aircraft should be used with little discrimination to terrorize the enemy population into submission was regarded askance by most airmen, though it had powerful civilian champions, Winston Churchill included.[20] The changing technology after 1945, with atomic weapons mounted first on aircraft then in ballistic missiles, produced in effect just such a crude target system, since no one could pretend that nuclear weapons did not destroy entire population areas. Making cities or military centers the target hardly reduced the extent of the possible collateral damage, and both sides in the Cold War came to accept that the vital target might be the entire social and economic fabric of the enemy country.

This confrontation too is now history. The objective of "massive retaliation" sustained the air strategy adopted by Britain and the United States during World War II, that massive destruction of an enemy's social fabric and economy really could end a war by undermining the ability and will to carry on fighting. This was, of course, more the case with RAF Bomber Command than with the U.S. Army Air Forces. The British devoted more effort to industrial cities per se, though the recent claim that American bombing was so very much more precise than British exaggerates the extent to which American forces could do much more than area bomb precise targets.[21] The difference was that American airmen recognized much sooner than the

RAF that the enemy air force was the first vital target: destruction of that opened the way to other vital targets.

This was the secret of success in the bombing offensive. It was the strategic fighter, not the strategic bomber, that won the air war over Germany.[22] In the last year of war the bomber was much freer to pick out the "vital targets." The choice of oil, chemicals, transport, and iron and steel (among others), all capital-intensive industries incapable of improvised decentralization, proved the right ones. By 1945, German forces were unable to continue effective resistance because of the collapse of the foundation of the war economy. Existing bombing technology meant that this effect was secured with high casualties and a quite staggering degree of urban destruction, but it is difficult not to believe that if the 8th Air Force had been able to use laser-guided weapons and supersonic speeds the same targets would still have been hit. Bombing proved to be a heavy club rather than a rapier, but both are lethal.

If it finally worked in Germany, the question arises over why it failed in Vietnam and Korea. Conditions in the German war were unique. The search for a universal set of principles governing air warfare over the century blinded many airmen to the fact, forcefully expressed both by Tony Mason and Mark Clodfelter in their contributions to this volume, that no two conflicts are the same. This is not as obvious as it sounds, otherwise the American air forces would not have wasted millions of tons of bombs and the lives of hundreds of airmen trying to do to North Korea and North Vietnam what they had done to Germany, and even more devastatingly to Japan. The great strength of modern air power is its flexibility and versatility. It can fight on its own, or with other services. Its functions and doctrine must reflect that flexibility. The fact that air war has been fought over the century in many different ways is not just a reflection of deeply held doctrinal differences, though these did exist, even within the same air force, but derive from the fact that the circumstances of conflict have been very different. If the history of air power can teach us anything, it is the need for doctrinal and operational openness. In all examples where air power has been exploited within a narrow theoretical framework—in France in 1940 or the USSR in 1941—it has failed. This volume of essays succeeds in reminding us not only of the wider explanatory framework within which air power development must be understood, but of the necessary diversity of the air war experience both then and now.

Notes

1. See J. Sweetman, "The Smuts Report of 1917: Merely Political Window Dressing?" *Journal of Strategic Studies* 4 (1981), and M. Cooper, "A House Divided: Policy, Rivalry and Administration in Britain's Military Air Command 1914–1918," *Journal of Strategic Studies* 3 (1980).

2. On Roosevelt see Michael S. Sherry, *The Rise of American Air Power: The Creation of Armageddon* (New Haven, Conn.: Yale University Press, 1987), 77–89; on the interaction of politics and strategy see, for example, F. R. Kirkland, "The French Air Force in 1940: Was It Defeated by the Luftwaffe or by Politics?" *Air University Review* 36 (1985).

3. P. Fritzsche, *A Nation of Fliers: German Aviation and the Popular Imagination* (Cambridge, Mass.: Harvard University Press, 1992), and idem, "Machine Dreams: Airmindedness and the Reinvention of Germany," *American Historical Review* 3 (1993). On the USSR see, among

others, Von Hardesty, *Red Phoenix: The Rise of Soviet Air Power, 1941–1945* (Washington, D.C.: Smithsonian Institution Press, 1991).

4. U. Bialer, *The Shadow of the Bomber: The Fear of Air Attack and British Politics, 1932–1939* (London: Royal Historical Society, 1980); and R. J. Overy, "Air Power and the Origins of Deterrence Theory before 1939," *Journal of Strategic Studies* 15 (1992).

5. Michael S. Sherry, *In the Shadow of War: The United States since the 1930s* (New Haven, Conn.: Yale University Press, 1995). There is a stimulating account of postwar culture and the history of war in R. J. B. Bosworth, *Explaining Auschwitz and Hiroshima* (London: Routledge, 1993).

6. L. Ceva A. Curami, "Air Army and Aircraft industry in Italy 1936–1943" in Horst Boog, ed., *The Conduct of the Air War in the Second World War: An International Comparison* (Oxford: Berg, 1992), 85–107.

7. A recent example of air force history in which production and technology play a major part is John Morrow's excellent *The Great War in the Air: Military Aviation from 1909–1921* (Washington, D.C.: Smithsonian Institution Press, 1993). On French aircraft production in relation to German and British see pp. 344–47.

8. R. J. Overy, *The Air War 1939–1945* (New York: Stein & Day, 1981), chap. 7.

9. Mark K. Wells, *Courage and Air Warfare: The Allied Aircrew Experience in the Second World War* (London: Cass, 1995).

10. Horst Boog, *Die deutsche Luftwaffenführung 1935–1945* (Stuttgart: Deutsche Verlags-Anstalt, 1982); E. Beck, *Under the Bombs: The German Home Front, 1942–1945* (Lexington: University Press of Kentucky, 1986); 0. Groehler, *Bombenkrieg gegen Deutschland* (Berlin: Akademie-Verlag, 1990).

11. Boog, ed., *Conduct of Air War.*

12. For example, P. Le Goyet, "Evolution de la doctrine d'emploi de l'aviation française entre 1919 et 1939," *Revue d'histoire de la Deuxième Guerre Mondiale* 19 (1969); P. Vennesson, "Institution and Airpower: The Making of the French Air Force," in J. Gooch, ed., *Airpower: Theory and Practice* (London: Cass, 1995), 36–58.

13. For the latest in this line of argument see S. Garrett, *Ethics and Airpower in World War II* (New York: St. Martin's, 1993).

14. See Le Goyet, "Evolution," 22–23.

15. See, for example, articles by Jeffrey Grey and C. D. Coulthard-Clark in A. Stephens, ed., *The War in the Air 1914–1994* (Fairbairn, Australia: RAAF Power Studies Centre, 1994), 141–59, 163–76.

16. R. Steiger, *Armour Tactics in the Second World War: Panzer Army Campaigns of 1939–41 in German War Diaries,* trans. Martin Fry (Oxford: Berg, 1991).

17. On Luftwaffe weaknesses in 1940 see H. Faber, ed., *Luftwaffe: An Analysis by Former Luftwaffe Generals* (London: Sidgwick & Jackson, 1979), 201–6. On the RAF see R. J. Overy "Air Power, Armies and the War in the West 1940," Harmon Memorial Lectures in Military History, No. 32 (USAF Academy, Colo.: U.S. Air Force Academy, 1989), 9–11.

18. I. R. Gooderson, "Allied Close Air Support 1943–45" (Thesis, London University, 1994), 228–37.

19. C. Crane, *Bombs, Cities, and Civilians: American Airpower Strategy in World War II* (Lawrence: University Press of Kansas, 1993), 18–26.

20. See the essay by H. Boog in A. T. Harris, *Despatch on Operations* (London: Cass, 1995) on the attitude of German air officers to the use of air terror.

21. Crane, *Bombs,* passim. It should be recalled that only 43.6 percent of the bombs and mines dropped by RAF Bomber Command were aimed at industrial cities as such. Troops and enemy defenses took 12.4 percent, transport 14 percent, and oil 9.9 percent.

22. S. MacFarland, "The Evolution of the American Strategic Fighter in Europe, 1942–1944," *Journal of Strategic Studies* 10 (1987).

23. A. C. Mierzejewski, *The Collapse of the German War Economy, 1944–1945: Allied Air Power and the German National Railway* (Chapel Hill: University of North Carolina Press, 1988).

The British Dimension

R. A. Mason

On the broad theme of the impact of air power on twentieth-century warfare, three features in the British dimension have made a major contribution. They are, first, the presence in Britain even before 1914 of a clearly recognizable body of fundamental air power doctrine; second, the example of the first independent, unified air force; and, third, the formulation of concepts of tactical and strategic offensive air power.

The First Ideas

The first feature is the presence in Britain before 1914 of ideas about the application of air power. It is important to distinguish between speculative ideas, which these were, and systematic theories based on observed facts, which they could not yet be. The early British aeronautical enthusiasts, civilian and military, were a relatively small group. They exchanged ideas at meetings at the Aero Club, at the Aeronautical Society, at increasing numbers of flying exhibitions and, less frequently, at the Royal United Services Institution.

As early as 1893, Maj. J. D. Fullerton of the Royal Engineers had presented a paper at a meeting of military engineers in Chicago, in which he prophesied that the impact of aeronautics foreshadowed "as great a revolution in the art of war as the discovery of gunpowder," that future wars may well start with a great air battle, that "the arrival of the aerial fleet over the enemy capital will probably conclude the campaign," and that "command of the air" would be an essential prerequisite for all land and air warfare.[1]

Although Major Fullerton does not seem to have included such prophetic statements in his addresses to British audiences during the next twenty years, his engineering friend, F. W. Lanchester, expressed a similar view in the *Aeronautical Journal:* "Under the conditions of the near future, the command of the air must become at least as essential to the safety of the empire as will be our continued supremacy of the high seas."[2]

In 1909, *Flight* became the official journal of the Aero Club, "Devoted to the Interests, Practice and Progress of Aerial Locomotion and Transport."[3] On 27 February, *Flight* published the first international survey of military aviation, by Maj. George O. Squier of the Signal Corps of the United States Army.[4] In 1911, the magazine *Aeroplane* under the dynamic editorship of Charles Grey began its stern monitoring of developments in British military aviation. These journals, together with occasional contributions in the *Journal of the Royal United Services Institution,* were the primary breeding ground for British ideas about air power.

On 15 May 1909, the editorial in *Flight* was titled "Britain and the Command of the Air" and expressed concern at the nation's vulnerability to hostile aircraft even at their current stage of development, quite apart from the advent of "all-weather aircraft."[5] In May 1911, Capt. C. J. Burke wrote the first article on air power to be published by the *Journal of the Royal United Services Institution,* initially concentrating on the airplane as a reconnaissance vehicle, but then reaching the same conclusion as his civilian friends: "May not the command of the air be as important to us in the future as the command of the sea is at the present moment?"[6] Yet this idea was not the prerogative of English theorists.

At the same time as he wrote that article, Captain Burke had reviewed a book for the *Aeronautical Journal* by the French General H. Frey, who had posed the question: "May not the command of the air be of such importance that the power who loses it may be forced to sue for peace?"[7] But Captain Burke then concluded his review with a very different idea: "No one can question the need for the fourth arm at the present minute, and if aviation continues to advance at its present rate, a new service will be a necessity."[8]

During this period, the *Aeronautical Journal* surveyed the French and German, but not the Italian, aviation press. So there is no evidence that the British coterie heard about Giulio Douhet's first thoughts in 1910 on *Problems of Air Navigation,* which included his proposals for a separate service.

Then, in 1913, Colonel Fullerton addressed the Royal United Services Institution on the theme of aeronautical progress. After examining the concept of the command of the air, he concluded that "A separate organization, with its own commander-in-chief, is essential for success and it is to be hoped that this will be realized before it is too late to take action in the matter."[9]

Not surprisingly, his view prompted stern disagreement from the major general who was chairing the meeting and from several members present. In the following year, Lanchester, now a member of the Aeronautical Society and a regular contributor to the other journals, began his well-known series of articles in the magazine *Engineering*. He foresaw the employment of aircraft well behind the conventional battlelines, both on land and at sea, and stressed the further implication of offensive air power: "It is safe to say that if during a battle it is found practicable to conduct air-raids and air attacks systematically over a considerable belt of territory in the rear of an enemy's lines, this belt will require to be defended."[10]

Therefore, by 1914, five fundamental ideas of air power had been formulated within the United Kingdom. First, air power could contribute enormously to land and naval operations; second, command of the air was as essential to Britain as was command of the sea; third, to achieve command of the air an independent service needed to be established; fourth, air power could reach out far beyond the lines of battle and strike at targets in the enemy's homeland; and, fifth, by such offensive action the enemy could be forced to divert essential resources to his own air defense.

Moreover, one can trace the personal connection between those who formulated the ideas and those who in World War I attempted to put them into practice. For example, Captain Burke, then a pilot in the Air Battalion of the Royal Engineers, became the commander of No. 2 Squadron, Royal Flying Corps, in 1914, and later commander of No. 2 Wing in France, working for Col. Frederick Sykes and alongside Lt. Col. Hugh Trenchard. But, as so often in the history of warfare, from the formulation of ideas to their successful application in battle was to be a long and arduous passage. The visionaries simply did not wield sufficient military, economic, or political influence in 1914 to procure the equipment necessary to implement their ideas.

In 1912, a technical subcommittee of the Committee of Imperial Defense had emphasized that the contribution of both military and naval aircraft in future wars would primarily be reconnaissance.[11] No attention was paid to the likely needs of home defense or to offensive action by aircraft. Consequently, the Royal Aircraft Factory at Farnborough was directed to produce slow and stable aircraft with an engine maximum of 100 h.p., ideal for reconnaissance in peacetime but wholly unsuited to hostile battlefield conditions. Nor was enthusiasm stimulated by the alleged attitude of Gen. Douglas Haig:

> I hope none of you gentlemen is so foolish as to think that aeroplanes will be able to be usefully employed for reconnaissance purposes in war. There is only one way for a commander to get information by reconnaissance, and that is by the use of cavalry.[12]

Fortunately for the future development of British air power, naval aviation enjoyed the vigorous support of Winston Churchill as First Lord of the Admiralty from 1911 to 1915. From the outset, the Naval Wing of the Royal Flying Corps regarded aircraft as an extension of the offensive and defensive power of a fleet: for attacks on naval units at sea, dockyards and other shore installations, and for the protection of British units afloat and ashore. Navigational instruments and bombsights were developed, and, because of the envisaged range and payload required, more powerful engines and airframes were commissioned from a variety of civilian companies. But much depended on the support of Churchill and the enthusiasm of relatively junior officers; between them stood many admirals whose interests were far more traditional and who were suspicious of what a Captain Neumann of the German airship battalion in 1908 had called "excessively optimistic expectations, fantastic conclusions and impossible schemes."[13]

Somehow a climate and an organization had to be created which would permit the implementation of air power ideas despite the presence of an unsympathetic military and naval hierarchy.

The Third Service

The Royal Flying Corps (RFC) was constituted in 1912 with a Naval Wing, Military Wing, and a Central Flying School to train both army and naval pilots. An Air Committee was established to coordinate the contribution of the two parent services, but within a very short time the wings began to develop more in isolation than in harmony. The separation of the Royal Naval Air Service (RNAS) was officially recognized on 1 July 1914. So, on the outbreak of war there were two British air forces: one dispatched to France intended to provide long-range reconnaissance for the army; the other located in Britain and in Belgium with a very new responsibility for the air defense of the United Kingdom but with imprecise ideas about its potential contribution to naval operations.

Many cogent and coherent explanations have been offered for the momentous decision by the British government in 1917 to create a unified Royal Air Force (RAF).[14] The RAF historian has stressed the wasteful competition for resources, the duplication of effort in some cases, and the mutual neglect in others, all of which almost inevitably accompanied the presence of two autonomous agencies frantically seeking to meet the ever-expanding demands of commanders for more aircraft, more crews, more technicians, and more supporting equipment.[15]

A series of boards and committees were established in an attempt to resolve these problems, but they were not invested with executive authority and were dependent on the goodwill of the individual service and civilian members. However, a further board, under Lord Cowdray, was established in April 1917 with greatly increased powers. It could organize and maintain a supply of aircraft; it could appoint and draft its own staff; and it did have its own building at the Hotel Cecil on the Thames Embankment.[16] RFC and RNAS staffs worked side by side under their respective directors, and it was seen at the time as a natural step toward an independent Air Ministry.[17]

But the Cowdray Board remained absolutely dependent on the War Office and the Admiralty for such things as nontechnical stores, armaments, and airfields. Its advice was given by soldiers and sailors back to their own services, and it had no power to allocate men and airplanes and certainly none to provide for home defense or independent operations. And, sadly, as in earlier days, cooperation and provision were constantly bedeviled by the personal rivalries and jealousies of senior commanders, politicians, newspaper publishers, and industrialists—so much so that it became cynically known as the "Hotel Bolo," after a well-known enemy agent who had done a great deal of harm to the Allied cause. In short, the Cowdray Board could only provide the equipment; it could not say how or where it should be employed. Internal evolution alone could not produce that kind of authority, and without it the potential of air power remained stultified.

Another view of the creation of a unified RAF holds that the British government's decision in 1917 was taken in panic in the wake of the German bombing of that summer and autumn.[18] Certainly, airship and later four-engine Gotha raids had disproportionately affected British morale, perhaps only locally in initial impact, but spread nationwide by the press and thrust into politics by Members of Parliament who had been vociferous critics of aerial policies for a considerable period.[19] One airship raid on Hull in 1915, for example, caused widespread panic and prompted the local MP to write to Mr. Arthur Balfour:

> Citizens of all classes are in a state of great alarm; the night after the raid a further warning was given and tens of thousands of people trooped out of the city. The screams of the women were distressing to hear. Could you let us have half a *dozen* aeroplanes?[20]

Then, on 2 April 1916, Scottish morale was severely impaired when a Scotch Whisky bonded store was destroyed near Edinburgh.

It was, of course, the Gotha airplane attacks in 1917 which produced the heaviest casualties, the greatest panic, and the strongest criticism of the aerial defenses of the United Kingdom. But they also provoked the greatest indignation and the most vociferous demands for reprisals. It was certainly in response to widespread public dissatisfaction that Gen. Jan Smuts was directed on 11 July 1917 to examine "the defence arrangements for home defence against air raids and the existing general organization for the study and higher direction of aerial operations."[21] His recommendations on UK defense were presented to the government eight days later and immediately implemented.

His second report, presented to the government on 17 August, has been called the "Magna Carta of British Airpower." Smuts traced the previous attempts at coordinating army and naval air services and stressed the inability of the existing Air Board to embark on a policy of its own. He then continued:

> The time is however rapidly approaching when that subordination of the Air Board and the Air Service can no longer be justified. Essentially, the position of an Air Service is quite different from that of the Artillery Arm . . . [it] can be used as an independent means of war operations. Nobody who witnessed the attack on London on 7 July could have any doubt on that point. Unlike artillery, an air fleet can conduct extensive operations far from, and independently of, both Army and Navy. As far as at present can be foreseen, there is absolutely no limit to the scale of its future independent war use. And the day may not be far off when aerial operations with their devastation of enemy lands and destruction of industrial and populace centers on a vast scale may become the principle operations of war, to which the older forms of military and naval operations may become secondary and subordinate.[22]

Smuts had been advised that aircraft production in the next twelve months would far surpass the joint requirements of the Army and Navy, and, therefore, an air staff to plan and direct independent operations would soon be necessary. Moreover, he warned, "The enemy is no doubt making vast plans to deal with us in London if we do not succeed in beating him in the air and carrying the war into the heart of his country."[23]

To realize the full potential of air power, he recommended the creation of a separate air ministry, air staff, and air service.[24] He then concluded his report with a sentence almost identical to one used by Lanchester ten years before:

> It is important for the winning of the war that we should not only secure air predominance, but secure it on a very large scale; and having secured it in this war we should make every effort and sacrifice to maintain it for the future. Air supremacy may in the long run become as important a factor in the defense of the Empire as sea supremacy.[25]

I have no evidence that Smuts ever thought about air power before 1917. We know that he was misled into expecting a large surplus of aircraft; we know that he was convinced by exaggerated estimates of German intentions; and we know that some of the enthusiasts he consulted looked far beyond the immediate capability of strategic bombing. His closest confidante was Gen. G. F. R. Henderson, who, in turn, had moved steadily closer to the views of men like Lanchester. The latter, in the 1916 reprint of his articles from the journal *Engineering,* argued strongly that the aeronautical arm was a national affair, because not only would it tax national resources to the uttermost but "because it is the arm which will have to be ever ready, ever mobilized, both in time of peace and war: it is the arm which in the warfare of the future may act with decisive effect within a few hours of the outbreak of hostilities."[26]

Therefore, in 1917, air power was freed from the constraints of army and navy priorities partly by the force of unique circumstances, partly by mistaken interpretations, partly even by the self-seeking of opportunists, but above all because in Britain ideas were maturing into theories and visions into forecasts. Air power was given the chance to become a living, self-developing organism endowed with a voice, a brain, and a limb.

There are several other features of the British experience before World War II worth noting. Among them are the use of air power to control under-developed areas, or the struggle over naval aviation which had such a debilitating effect on the evolution of British naval air power, or the constant fight to persuade a democratic government in peacetime that its defensive insurance policy should keep pace with the growth of the external threat.

The influence of imperial responsibilities on the RAF between the wars still awaits comprehensive analysis. Sufficient here to say that from 1919 until the beginning of the armament program in the 1930s, more than half the RAF's squadrons were based overseas and of those remaining in the "metropolitan" area, approximately one-third were allocated to naval duties, one-third to army cooperation, and one-third to home defense, which itself was a misleading title.

The internecine struggles between the Royal Navy and the RAF for control of naval aviation have, on the other hand, been well documented in several official histories and biographies.[27] The Air Council feared, with some justification, that the creation of a Fleet Air Arm would be the first step toward the disintegration of an independent RAF. They were, therefore, opposed in principle to the establishment of strong naval air institutions because of fears that their own authority would be undermined. Resources dedicated to naval aviation varied, but were always a small percentage of the whole, reflecting the overall allocation of effort summarized above. Meanwhile, no naval staff was responsible for long-term studies of naval aviation, and no naval air lobby existed either to fight the political, technical, and economic pressures which tended to restrict progress or to challenge "the Admiralty's habit of associating the battleship's well-being with their own."[28] Consequently, it is hardly an over-simplification to state that British naval aviation between the wars fell between the Scylla of a sorely pressed Air Council and the Charybdis of an Admiralty denied the power of air-minded admirals.

Offensive Air Power

Trenchard, while working for Gen. Douglas Haig as General Officer Commanding (GOC) Royal Flying Corps in France, consistently demanded offensive tactical action by his air crews. His attitude is best summarized in a memorandum which he addressed to General Haig in September 1916:

> An aeroplane is an offensive and not a defensive weapon. Owing to the unlimited space in the air, the difficulty one machine has in seeing another, the accidents of wind and cloud, it is impossible for aeroplanes, however skillful and vigilant their pilots, however numerous their formations, to prevent hostile aircraft from crossing the line if they have the initiative and determination to do so. . . . British aviation has been guided by a policy of relentless and incessant offensive. Our machines have continually attacked the enemy on his side of the line, bombed his aerodromes, and carried out attacks on places of importance far behind the lines. It would seem probable that this has had the effect so far on the enemy of compelling him to keep back or to detail portions of his forces in the air for defensive purposes . . . the sound policy, then, which should guide all warfare in the air would seem to be this: to exploit the moral effect of the aeroplane on the enemy, but not let him exploit it on ourselves. Now this can only be done by attacking and by continuing to attack.[29]

Note, however, that this was not an argument for air operations independent of land fighting, but simply a proposal for the best way of giving air support to it, by forcing the enemy air on the defensive and keeping it there. Trenchard, in 1916, was strongly opposed to the use of air power independently of other military operations.

Inevitably, losses of men and materiel seemed heavy, but as Lanchester commented, "The defence of modern arms is indirect: tersely the enemy is prevented from killing you by your killing him first."[30] During the German spring offensive of 1917, for example, when the enemy enjoyed temporary air predominance, the RFC lost fewer than eight men a day, against a daily average for the British Army as a whole of ten thousand killed or missing.[31]

The impact of the RFC on the enemy ground troops in return for these losses was carefully recorded at headquarters.[32] Except when obscured by fog, gas, or very low clouds, modifications to German defenses were photographed daily, prompting regular German concern. German troops were reluctant to dig trenches by day and frequently assumed that very low-flying air attack was directed against their own dugouts. The association of spotter aircraft and highly accurate artillery bombardment was particularly resented when accompanied by the belief that German air crew were reluctant to give battle. "The RFC pilots," on the other hand, "seem to seek air combat whether it is necessary or not."[33] Nevertheless, Trenchard's RFC was discovering an inherent paradox of offensive air power. Attacks on enemy targets, either tactical or strategic, will undoubtedly force him to divert more resources to air defense. But the more successful the policy in forcing the enemy on to the defensive, the more difficult and costly it becomes to inflict proportional damage on the original targets.

Meanwhile, the RNAS had conducted long-range bombing operations intermittently since the early days of the war, first against airship sheds and then, in 1916, against industrial targets in Alsace, Lorraine, and the Rhineland. The strategic activities of No. 3 Wing RNAS from October 1916 to April 1917 were curtailed by bad weather and by constant pressure from the RFC for assistance. Nevertheless, contemporary reports of German attempts to organize air defenses against these raids clearly indicate that they also were forcing the diversion of resources away from offense to defense, although with little effect. RNAS staff realized that their bombers' immunity would inevitably be threatened as the German air defenses improved, and, therefore, even before the end of 1916 they were planning to develop long-range escort fighters and modified fighter bombers.[34]

The RNAS anticipated what the Independent Force and RAF were later to discover: a second paradox of offensive air power is that concentration of force is required for maximum offensive effect, but concentration of force in that age could only be achieved by large numbers of aircraft and repeated attacks, which in turn provided the opportunity for the defending air force to concentrate its own fighter squadrons to maximum defensive effect.

Hard thinking about air power employment was not confined to tactics. Under the direction of Lieutenant Commander Tiverton, experiments were carried out with bombsights, ballistic trajectories, and a variety of long-range navigation aids.[35] The activities of the RNAS staff in this period clearly illustrate that, contrary to the views of the official history of World War II, the problems of long-range aerial navigation were fully appreciated and were being addressed.[36] Moreover, special attention was paid to practical training and target acquisition because experience had shown that it was quite easy for five squadrons to set out to bomb a particular target and for only one of those five ever to reach the objective, while the other four in the honest belief that they had done so, had bombed four different villages bearing little, if any, resemblance to the one they desired to attack.[37]

Later that year, Tiverton presented a second paper to the Air Board which appears to contain the first analysis of strategic targets based on the scientific principles usually associated with the operational analysis of World War II. He identified chemical plants as the key industrial targets because of the dependence of the German war industry on them and their vulnerability to air attack. Further, he studied individual factories to identify the departments whose destruction would have the greatest effect and studied their areas to assess the bomb loads required to achieve the necessary amount of destruction. Not surprisingly, he concluded that success could only be achieved by concentrated and repeated attacks.[38] These ideas were incorporated in the policies of the infant RAF and relayed to Trenchard, now commander in chief of the Independent Force in France, in May 1918.

There was, however, one problem. The Air Board believed that in 1918 Britain would have a strategic bombing force of 2,000 aircraft, of which 1,000 would be serviceable at any one time, each carrying nine bombs or approximately 1,000 lbs. In fact, Trenchard never had more than 100 aircraft under his command from June to November 1918. Rarely were twelve aircraft serviceable in a squadron, and combined operations by more than one squadron were seldom undertaken because of difficulties of inter-unit coordination and lack of preparation and training time.[39]

Trenchard identified his problems in his first report to the Secretary of State for Air on 2 July 1918:

> I took over the tactical command of this Force on the sixth of June, and the plan on which I decided to work was to attack a large number of objectives in Germany so as to force the enemy to disperse if possible his defensive forces at various points, and then to concentrate for two or three days and nights on the same objective. This plan, however, was unable to be carried out in its entirety. Wind, together with the necessity for training new squadrons and new pilots . . . [and] the few squadrons at my disposal, prevented the plan being entirely carried out.[40]

Nor did matters improve. From June to November the entire force was grounded completely by weather for almost 50 percent of the time, quite apart from the numerous occasions when the aircraft sought secondary targets because of weather over the primary targets.

At the third session of the Inter Allied Aviation Committee held at Versailles on 21 and 22 July 1918, the French delegate asked what weight of projectiles each of the Allied aviation services could drop in twenty-four hours between July and December 1918.[41] Trenchard prepared a set of notes for Sykes, who was the British delegate at this session, stating that he could give such figures but that they would not mean very much. He attached a table which included the theoretical weight in tons which the aircraft of the Independent Force could drop daily for each of the specified months. The nearest the Force ever got to the estimate was 3.5 percent.[42] No one at headquarters was surprised because the table had carried the following note: "These figures are purely theoretical and can in no way expect to be borne out by fact."[43]

It has been frequently pointed out that Trenchard was less than enthusiastic about his role,[44] but contemporary evidence clearly illustrates that he discharged his duties with a clear eye to the practical difficulties which faced him: in this case, numbers, serviceability, weather, distance, the need for French goodwill, and frequent Allied requests for shorter range support. When GOC of the Royal Flying Corps, he maintained a meticulous collection of intelligence reports on the effects of the Force's raids under the headings of "British Official Report," "German Official Report," "Materiel Results," and "Moral Effects."[45] The fundamentals of his strategy were spelled out on the first page: "Though materiel damage is as yet slight when compared with moral effect, it is certain that the destruction of 'moral' [morale] will start before the destruction of factories and, consequently, loss of production will precede materiel damage."[46]

It is, therefore, not surprising to find that when Trenchard was directed by the military representatives of the Supreme War Council at Versailles in the autumn of 1918 to produce a "methodical plan" for the proposed allied strategic bombing force, it began as follows:

> There are two factors—moral effect and materiel effect—the object being to obtain the maximum of each. The best means to this end is to attack the industrial centres where you:

a. Do military and vital damage by striking at the centres of supply of war material.

b. Achieve the maximum of effect on the moral by striking at the most sensitive part of the whole of the German population—namely, the working class.[47]

Here was a definition of the Third Dimension of Warfare. Some roots lay partly in the ideas formulated before 1914, others in the use of air power in indirect deeper support of the land battle, others in the angry demand for reprisals for German attacks on Britain, others in the need to find employment for the expected thousands of additional aircraft. Other roots perhaps lay in the technological fact of life of 1918 that morale seemed much easier to damage than materiel, or even in the inexorable implication of democratic warfare that all who contribute to a war effort, military and civilian alike, may be said to be justifiable military targets.

Wherever emphasis is placed, it is clear that even by 1919 the British had contributed conceptual ideas, organizational example, and offensive operational experience which were to have a strong influence on the evolution of air power later in our century. These are the three major permanent features of the "British Dimension."

Notes

1. Quoted by Alfred F. Hurley in the appendix, "Additional Insights," to his *Billy Mitchell: Crusader for Air Power* (Bloomington: Indiana University Press, 1975), 142.
2. F. W. Lanchester, "Aerial Flight," *Aerodynamics*, vol. 1 (London: Constable, 1908), vi.
3. *Flight*, 2 Jan. 1909, 1.
4. George O. Squier, "The Present Status of Military Aeronautics," ibid., 27 Feb. 1909, 121ff. This paper was previously presented in December 1908 to a New York meeting of the American Society of Mechanical Engineers.
5. *Flight*, 15 May 1909, 272.
6. Capt. C. J. Burke, "Aeroplanes of Today and Their Use in War," *Journal of the Royal United Services Institution* 55 (May 1911): 624–29.
7. Capt. C. J. Burke, review of *L'Aviation aux armées et aux colonies* in *Aeronautical Journal* 58 (April 1911): 98.
8. Ibid., 99.
9. *Journal of the Royal United Services Institution* 57 (1913): 333.
10. F. W. Lanchester, "Engineering," 27 Nov. 1914, later included in *Aircraft in Warfare, The Dawn of the Fourth Arm* (London: Constable, 1916).
11. An excellent comprehensive analysis of the findings and long-term influence of the technical subcommittee is in Neville Jones, *The Origins of Strategic Bombing* (London: William Kimber, 1973), 36ff.
12. This comment is attributed to General Haig in July 1914 by Sir Frederick Sykes in his autobiography *From Many Angles* (London: Harrap, 1942), 105. While there is no reason to doubt that Haig held such sentiments, it should be remembered that Sykes's testimony was that of a man embittered by his personal rivalries with both Haig and Trenchard before, during, and after World War I.
13. Captain Neumann, "The Possibility of Making Use of Balloons and Motor Airships in the Navy," *Marine-Rundschau*, July 1908, trans. in *Journal of the Royal United Services Institution* 52: 1502.
14. For example, see James M. Spaight, *The Beginnings of Organized Airpower* (London: Longmans, 1927). Among analyses of the events leading to the creation of the RAF, this study

by a British civil servant may not have been given quite the attention it deserves. Despite a classical eulogy of independent air power in the opening chapter, largely included in Eugene Emme's *The Impact of Air Power* (Princeton, N.J.: Van Nostrand, 1959), the greater part of the book offers a well-balanced, detailed, and comprehensively documented account of the press, parliamentary, and other pressures which had a powerful cumulative influence on the British government's decision in 1917.

15. Summarized in *Royal Air Force Air Publication 125,* 1936, 290–95.
16. Spaight, 92ff.
17. *Hansard,* House of Lords, 21 Dec. 1916, vol. 23, col. 1070.
18. A. D. Divine, *The Broken Wing* (London: Hutchinson, 1966), 105.
19. The contribution of Pemberton-Billing, Joynson-Hicks and other members of both Houses of Parliament to the incessant debates on the application of British air power between 1914 and 1918 is comprehensively described in B. D. Powers, *Strategy without Slide Rule* (London: Croom Helm, 1976).
20. Quoted in Squadron Leader C. J. Mackay MC DFC, "The German Air Raids in England," RAF Staff College, Andover, 1924.
21. Appendix II to Cabinet Minutes WC233, 24 Aug. 1917.
22. Ibid.
23. Ibid.
24. Ibid.
25. See note 2 above.
26. F. W. Lanchester, *Aircraft in Warfare, The Dawn of the Fourth Arm* (London: Constable, 1916), 202.
27. H. Montgomery-Hyde, *British Air Policy Between the Wars 1918–1939* (London: Heinemann, 1976), is the most comprehensive and best documented account of the RAF viewpoint of the implications of the RN-RAF struggles. On the naval side, the official RN histories by Capt. Stephen Roskill received a most valuable supplement in 1979 with the publication by MacDonald and Janes of London of *The Bomb and the Battleship* by Geoffrey Till, Senior Lecturer at the Royal Naval College, Greenwich.
28. Till, chap. 15.
29. HQ RFC Memo, 22 Sept. 1916; quoted in full in Higham, *Military Intellectuals in Britain 1918–39* (New Brunswick, N.J.: Rutgers, 1966), 253–56.
30. Lanchester, *Aircraft in Warfare,* 40.
31. Quoted in Divine, 141.
32. "What the Germans say about the RFC": typescript collection of German letters, prisoner of war reports, and Army orders retained in the Trenchard files at the RAF Staff College Bracknell (hereafter cited as Bracknell Papers).
33. Ibid., several similar comments.
34. Jones, 120ff.
35. Ibid., 142.
36. Sir Charles Webster and Noble Frankland, *Strategic Air Offensive Against Germany 1939–45,* vol. 1 (London: HMSO, 1916), 48.
37. Report by Lieutenant Commander The Lord Tiverton, submitted to the Air Board on 3 September 1917, quoted in Jones, 146.
38. Second Tiverton report, 2 Nov. 1917, quoted in Jones, 154ff.
39. "Experiences of Bombing with the Independent Force," lecture by Wing Commander J. E. A. Baldwin DSO OBE to the Royal Air Force Staff College, 1922; printed in AP 956, December 1923, Bracknell Papers.
40. Trenchard to Weir, IFG/79, 2 July 1918, Bracknell Papers.
41. Annexure to *Proces Verbal,* third session, International Allied Aviation Committee, p. 2, Bracknell Papers.

42. Ibid., Table A, together with monthly classified reports by Trenchard to Weir, June–November 1918, Bracknell Papers.

43. Ibid.

44. See, for example, Montgomery-Hyde, 43–44.

45. "Confidential Results of Air Raids on Germany 1 Jan–30 Sep 1918," DAI no. 5, Bracknell Papers.

46. Ibid., 1.

47. Undated typescript duplicate originating from IF HQ in Autumn 1918 retained by Trenchard in his own IF file, Bracknell Papers.

The Continental Experience

Edward Homze

The Pre–World War I Era

The Wright brothers' initial exploits with heavier-than-air flight in 1903 were an incentive for the major military powers of Europe to discuss the application of the airplane in modern warfare. Prior to the Wrights' experiments, the Italian and Prussian armies had concentrated on balloon and airship development.[1] Both armies believed until 1911 that the airship was superior to the airplane in range, load, and speed. The Russians too, believed in the airship's superiority, and in 1906 their War Ministry rejected an offer to purchase one of the Wright's airplanes. The French, however, exhibited extensive interest in heavier-than-air flight. Continued successes by French fliers in 1907 spurred the Prussians to set up an aviation technical section of the general staff. By August 1908, the Prussian General Staff questioned the army's policy of rejecting the airplane in favor of the airship and recommended that the Army actively support the development of the flying machine by constructing its own airplane.[2]

From 1908 to 1911 a combination of factors focused on the airplane's development. Technological improvements, the founding of a number of aircraft companies, the formation of public lobbying groups such as the German Air Fleet League and the Imperial All-Russia Aero Club (both started in 1908), and the lively interest of prominent individuals such as Prince Heinrich of Prussia and Grand Duke Alexander of Russia gave impetus to the airplane among military circles.[3] It became patently clear that the airplane had potential military value for reconnaissance and communication. By 1911, the major European continental armies had contracted for the purchase of airplanes and were actively training fliers.

The last three years of peace before World War I were characterized by the gradual assimilation of air machines into the organizational structure and doctrines of the European military establishments. Although appropriations for aviation were modest compared to total armament expenditures, the pace of aeronautical rearmament had quickened. When hostilities commenced in August 1914, each of the major European powers had a few hundred aircraft fit for active service, and the crucial base for the aviation industry had been laid in each nation.[4]

Organizationally, aviation had been incorporated into the military structure on the basis of its intended use. Herein lay the greatest weakness of prewar military aviation. Given the generally accepted expectations of a short war and the limited role aircraft would play, the consensus among military authorities was that the aircraft would primarily be used for reconnaissance—an improved cavalry. Secondary duties such as artillery spotting and message carrying were also recognized.

The further possibility of the airplane's use as an attack weapon was considered, for there had been some experimentation with bombing before the war. The Italians had dropped some 2-kg grenades during the Libyan War in 1911, and the Germans had ordered some of their planes equipped with 5- and 10-kg bombs in mid-1913; but there was no systematic study of the problems involved and surprisingly little theorizing on the subject. Presumably, the load-bearing factor for the flimsy airplanes of the era precluded further speculation. The airship, with its greater lifting capacity, appeared to be a much better possibility as a bomber. The appalling picture of cities destroyed by airships had long been a part of the European literary tradition of "voices prophesying war." More profitable speculation was centered on the arming of aircraft with light-weight machine guns for defense or attack. The development of the machine gun and the growing reliability and safety of aircraft meant that two of the three necessary ingredients for effective aerial combat—the armament and the gun platform—had been worked out before the war; the third ingredient—tactics—would soon follow.[5]

The War

World War I was a major turning point in military aviation on the continent, perhaps the turning point. In four bloody years of combat aviation evolved from an oddity to a military necessity. Very much in the manner that the military had absorbed and mastered the technology of the railroad for war purposes in the nineteenth century, a process that took decades, the military absorbed and mastered aviation technology in the incredibly short period of four years. In the areas of doctrine, tactics, organization, and weapons development, World War I established patterns for the future.

Some idea of the quantum jump in aviation during the war can be gleaned from a few statistics. When the war began, the French had 150 to 200 pilots and 24 squadrons; when it ended, they had 320 squadrons and 4,398 aircraft. French industry produced 52,000 airframes and 92,000 motors. Germany produced approximately 44,000 airframes and 48,000 engines. Italy, which had 106 aircraft and 5 dirigibles when it went to war in May 1915, ended the conflict with 1,778 aircraft and 22 dirigibles. The Italian air industry produced approximately 12,000 aircraft and 20,000 engines. Even the Russians, with a much smaller industrial base than the other major European countries, produced nearly 5,600 aircraft during the war.[6]

Development of supporting materials grew at the same astonishing rate. Before the war, for example, not a single European military service had an aerial bomb, bomb rack, bombsight, or release mechanism on aircraft; but four years later, their air services not only had the bombs, but the bombers to carry them considerable distances, the technical proficiency and equipment to aim and release bombs, and the organizational structure and military doctrines to plan mass bombing attacks.

It is clear that the European powers surmounted the thousands of problems associated with the expansion, organization, and development of aviation during the war. The intermeshing of the development of tactics, weapons, doctrine, and organization was so pronounced that it is difficult to analyze any strand separately. However, an analysis of the organization of the military will serve as a framework.

The most obvious changes in organization were the evolution of a more compatible structure to fit the new service and the enormous increase in size of all units. At first, the usual army cavalry organization was applied as befitted a reconnaissance arm. Flying units were equally distributed to ground commanders. No thought was given to concentrating forces or differentiating additional functions. Gradually the limitations of the traditional organizational structures were recognized and changed. A more flexible pattern was developed to fit the rapidly increasing functions and potential of the air services. This was a difficult task since the new flying services were often outranked by the senior services, suffered from a lack of appreciation at higher headquarters, and were hindered by a healthy dose of the fliers' own impudence.[7]

The Germans and the French were the trendsetters in redistributing and concentrating their aircraft into fighter, reconnaissance, and bombing units. The Germans had also placed all of their air arm under a separate organization in October 1916, but this was more administrative than operational since most units were still under the control of local army commanders except for some fighter and bomber units directly under the General Headquarters. The French persisted in keeping their units tied to local commanders, but the Italians showed a great deal of ingenuity by retaining most of their long-range reconnaissance and bomber units under the command of a central Army Headquarters.

The Italians were also quick to see the advantages of massing air power over key ground operations, but they were surpassed by the French in the latter stages of the war. The French, who had concentrated on short-range tactical bombing under the control of subordinate military formations, had not achieved much success until they massed forty squadrons into the First Air Division over Soissons in 1918.[8] Although the Europeans had learned by the end of the war how to mass air power, to organize it with more flexibility, and to handle larger aggregates of men and materials, they still had not resolved how to command and use air power most effectively. What they had learned from tactics did not solve their problems.

World War I opened up the entire spectrum of tactics in air power. From fumbling improvisation to controlled experimentation, to standardization and, in a few cases, to deft mastery, the Europeans learned their trade of making war in the air. In fighter tactics alone, the long, arduous melees of 1915 gave way to the brief, deadly, orchestrated dogfights of 1918. In bombing, the sporadic ventures over enemy lines evolved into the raids of 1918 involving hundreds of aircraft. Not only did the numbers increase, but the types and functions of bombing raids changed as the Europeans used bombers for close support, interdiction, and strategic bombing.

Although the Germans and Italians achieved some success in strategic bombing by forcing their opponents to divert sorely needed resources to home defense, the psychological effect was probably more important than the material damages done—a point noted by the critics of air power. The war was to end before any convincing proof of the offensive power of aircraft was gathered; this reinforced the skeptical attitude of senior officers of the army and navy about the ability of the air arm to carry out independent missions. The results of tactical bombing were also not too impressive, but there was no denying the effectiveness of close-support aircraft. At the Somme and Verdun, air support was used extensively, but it was not until 1917 when the Germans introduced specially built ground attack fighters that the assault fighters came into their own. The other European powers quickly followed, and the French in particular appreciated the possibilities of close support missions.[9]

Fighter tactics showed a slow but steady progression throughout the war; and, probably because of the wartime propaganda and the inherent drama of the heroic dogfight, they were the most publicized part of the war. The struggle for aerial supremacy was a bitter one and one which seesawed back and forth as each side brought out new aircraft and tactics. The struggle was intently followed by the military as well as the general public, for in a grim and impersonal war, this was a flamboyant and personalized form of combat that captured the imagination of all. Like mythical heroes of the past, these twentieth-century "knights of the air" flung themselves into dangerous jousts ending in victory or flaming defeat. These dashing, gallant aviators were ready-made heroes for the publicity men. Enormous amounts of time, money, and energy were devoted to fighter development and tactics, and the results were forthcoming.

Speed, sturdiness, reliability, and maneuverability of fighters were rapidly improved, but the biggest steps were the introduction of the synchronized machine gun and the standardization of its production. A number of patents for synchronized machine guns had been taken out in Europe before World War I, but none was in operation when the war started.[10] The crude Garros deflector system was tried with some success in March 1915, but the first workable system was designed by Anthony Fokker for Germany a few months later. In the hands of such pilots as Oswald Boelcke, Max Immelmann, and Ernst Udet, the tractor type fighter revolutionized fighter tactics. The Allies matched the Germans a year later when they too introduced a hydraulic synchronized gear system to the fixed-gun format.

The standardization of production which allowed for identical performance of aircraft in units promoted formation flying, the second most important change in tactics. By later 1915 the Germans had achieved this, followed by the French, British, and Italians in the next year.[11] Although there would be many more changes in aircraft and tactics in the last two years of the war, the broad lines of tactics for the fighters had been worked out by 1917.

Fighter pilots may have grabbed most of the headlines, but the flyers who took the same risks and produced even more results were the reconnaissance airmen. From the first Battle of the Marne until the last shot was fired, these airmen gathered vital information for their commanders. The mounting of cameras on aircraft and the growing sophistication of the photo interpreter opened up new vistas for ground commanders planning their operations. At sea, the military potential of the airplane in extending the visibility of surface ships was soon appreciated, but it would not be until World War II that the full potential of over-water aerial reconnaissance would be utilized. Strategic photo-reconnaissance was also conducted by European armies with notable success. The techniques employed during World War I may have been primitive, but they marked the beginning of what may have been the most important contribution of the aircraft to modern warfare.[12]

The Russians may have gained more from their Civil War than World War I. The Imperial Air Service was markedly inferior to the other major air services. Equipped largely with foreign models, it was chronically short of supplies and poorly handled during the war.[13] Their organization, patterned after the French, was closely tied to ground support tasks, but, ironically, their chief success was in strategic bombing. Igor Sikorski's famous four-engine bomber, Il'ya Muromets, established a precedent of directly supporting frontal operations with large bombers.[14] By the time the Bolshevik Revolution broke out, the Air Service was in wretched condition. The Civil War that followed was dominated by land engagements; but the fluid battlelines, great distances, and lack of good ground and sea communications greatly influenced Russian interest in maintaining air communications and in concentrating their forces. While leaders of the Red air fleet appreciated the need for air power, they found it impossible to secure. On the whole, the Russians viewed ground support operations as the most valuable form of air power, but they did not lose sight of the need for centralizing some of their air power for tactical or strategic goals. The experience gained in the Civil War, while modified by ideas inherited from the Imperial Air Service and the Germans, remained dominant in the formation of the Red Air Force.[15]

In summary, most of the weapons and ideas used in World War II had their origin in World War I, except for devices such as radar and the atomic bomb, which depended on technological progress in other scientific fields. Furthermore, as historian John W. R. Taylor pointed out, ". . . it is clear that almost every basic tactical and strategic application for air power had been tried out, at least experimentally, by the end of the 1914–18 War. Advances since then have been concentrated mainly on refining the weapons in terms of both aircraft and equipment."[16]

The Postwar Years

After World War I, the military airmen concentrated on three major activities: first, an analysis of the war; second, a justification of air power in the defense structure, preferably as an independent and equal service branch; and third, a general consideration of stalemate in warfare. The shocking disparity between the ends sought and the enormous price in blood and material paid for the meager results obtained during World War I had not only sharply reduced the independence as well as the prestige of the military in the eyes of the general public and the politicians, but also had illustrated the narrowmindedness of modern strategic thinking.[17]

It is easy to see the seductive charm that Giulio Douhet's theories had for airmen grappling with these concerns, since he seemed to supply the answers to their problems. His strategy appeared revolutionary, bold and in tune with the new industrial age and the experiences of World War I, when in fact it was none of these. Douhet had borrowed heavily from contemporary prewar sea strategy; his observations about modern industrialism were inaccurate; and his emphasis on strategic bombing was derived from the weakest example of air power during World War I.[18]

A word of caution about Douhet. Many, if not most, of the key officers and officials in the air services of Europe during the interwar period had neither heard of him nor read his works, but, like all great theorists, Douhet had synthesized and articulated a body of thought that had occurred in whole or in part to many others.

The evidence supporting Douhet's major assumptions—the capability and destructive power of the heavy bomber, the impotence of air defense, and the fragility of a modern industrial society in the face of heavy bombing—was thin and inconclusive. Like most prophets, Douhet was long on prognostications and short on facts, but his theories had a sweeping boldness and grandeur that his critics could not match. Airmen on the continent especially found him useful in arguing for an independent air force and supplying a conceptual framework for the next war, but they ran afoul of many well-entrenched vested interests, bureaucratic inertia, and lots of evidence drawn from World War I.

The prevailing pattern for the European air forces from 1919 to 1936 was to fight out with the other military services and the governmental bureaucracies the issues of an independent air force and use of strategic bombing, only to have the first idea accepted and the second rejected. In Italy in 1923, in Sweden in 1926, in France in 1928, and in Germany in 1933, independent air forces or air ministries were set up, clear recognition of the new role of air power, but the drive for a strategic bombing force failed everywhere.[19]

The reasons why the Europeans ended with basically tactical air forces by the mid-1930s vary from country to country, but in general they have this much in common: they were land powers; they were in agreement on the direction and pace of aircraft technology; and, militarily, they were traditional and conservative. Germany and Russia were the classical cases of continental land powers, but even the French and the Italians, both of whom had had long associations with the sea, were by then conditioned to think in terms of continental land power. In short, although they could appreciate the potential of strategic air power, their immediate interest and security appeared to depend on fielding mass armies to fight decisive battles along their frontiers.

The prospects of developing a true strategic bomber—a multiengine, long-range aircraft with a big bomb load—also seemed dim. Before such an aircraft could be built, many tough technical problems had to be solved. New, powerful engines had to be developed, as well as better fuels, more accurate bombing systems, and improved long-range navigational and radio equipment. Such an aircraft would have had to meet the standards of reliability and serviceability essential for their use in operational units, and, above all, every one of these factors had to be combined into a successful aeronautical design. Despite the fact that the Italians and Russians in the early 1930s and the Germans in the late 1930s conducted some amazing long-range flights, only the Russians were willing to gamble a disproportionate amount of their scarce materials and factory capability to build a large fleet of heavy bombers.[20] The results for the Russians were not encouraging, since the speed, ceiling, and range of the Tupolev TB-3 proved markedly inferior to the warplanes coming off the drawing boards by the mid-1930s.

The French tried another tack. Instead of building a strategic bomber, much of their effort was concentrated on developing a multipurpose "battleplane" which may have owed something to Douhet for its origins but nothing in its final form. The French effort produced only small, slow, and heavily armed aircraft designed to support ground operations.[21] The Italians opted for medium bombers, as did the Germans. Medium bombers were more appropriate for the European-scale war they were planning. The Germans, however, did design one other solution to the problem—the dive bomber. The dive bomber was a bridge, an interim solution, to cover the technological deficiencies that had arisen in medium and heavy bomber development. A cheap, quick way to achieve maximum bombing punch with a minimum use of resources, the dive bomber was a calculated-risk aircraft designed to serve until either a superior heavy aircraft or a new generation of medium bombers could be developed.[22]

In sum, there was a consensus among the European military staffs that the aviation technology of the 1930s, especially in terms of engine development, was not mature enough to deliver a strategic bomber that could become the capstone of a complete air force. Later, pressures of rapid rearmament, shortages of fuel and raw materials, and limited production facilities would strongly militate against a decision to build a fleet of heavy bombers.

In addition to technological objections, the Europeans had many doctrinal reasons for preferring tactical over strategical air power. Older, ground-oriented officers were in the key command and staff positions; and, although they were comfortable in accepting aviation as an auxiliary arm, they resented the "young Turks" who wanted aviation as an independent arm with a strategic bombing role. In France, the army's doctrine of defense dominated and eventually perverted air theory and practice. Although the increasing importance of aerial bombardment was recognized, the emphasis remained on tactical bombing. Even the establishment of an air ministry in 1928 and the independent *Armeé de l'Air* in 1933 did not appreciably alter air power. Still, France was at least able to maintain its aerial superiority until the depression, when a combination of financial difficulties, war-weariness, political polarization, and the influence of André Maginot and Philippe Petain accentuated a defensive strategy at precisely the time Germany started rearming and a major breakthrough in aerial technology had occurred.[23] By the time the French woke up and started to rearm for offensive air operations again, they found themselves behind the power curve in production, equipment, and strategy. By the spring of 1938, too much valuable time had been lost. French aircraft production had dropped to less than a hundred monthly, French spending for aviation had plummeted to 19 percent of the total defense expenditures compared to 54 percent in Great Britain, and French appreciation of air power had sunk so low that the chief of the general staff, General Maurice Gamelin, could comment, "... the role of aviation is apt to be exaggerated, and [that] after the early days of war the wastage will be such that it will more and more be confined to acting as an accessory to the army."[24]

The Italians did not fare much better than the French, despite the brief presence of Douhet in the ministry and the unbounded enthusiasm of Mussolini for a revolutionary, Fascist air force. Under the energetic Italo Balbo, the Italians built a formidable air force by the late 1920s and early 1930s, but it still was a conventional force. The rhetoric of the Fascist regime might have been Douhetan, but the aircraft were not.[25]

The Russian experience was different in that the Red Air Force never became an independent service, even though a strategic bomber force was built in the 1930s. The Red Army kept the air force tied closely to the ground forces, and the doctrines of tactical bombing and close support prevailed. The strategic bombing advocates did make some serious inroads in these doctrines. In 1926, theoretician A. N. Lapchinskii pressed for strategic bombing under the guise of independent air operations. A lively debate ensued, with some Russians arguing the necessity of an independent air force. They never took the extreme position of some of their Western counterparts that the next war could be won solely with massive long-range bombardment, but they were effective to the extent that their government began building a fleet of heavy bombers. By the May Day parade of 1933, 50 TB-3s were seen over Moscow and a year later 250 appeared.[26] By 1936, the Soviets had reorganized their heavy bombers into a strategic force—the first of its kind in the 1930s. The chain of events then swiftly reversed this trend. The purges by Stalin, the Russian experiences in Spain and in Asia, and the rapid technological advances in tactical aviation by the Western European powers led the Soviets to conclude that their strategic bombing doctrine was erroneous. They changed direction and entered World War II with the most tactically oriented air force in Europe.

Perhaps the experience of the Luftwaffe is the most revealing. Originally organized as an independent force and headed by the second most powerful man in the country, the Luftwaffe was tendered the kind of preferential treatment reserved for a favorite son. The political leadership favored a strategic force and, in fact, cleverly cultivated the image of such a force at home and abroad. Hitler soon became the most adroit manipulator of the "bombing scare" technique as he bullied his neighbors with the threat of wholesale destruction from the air. But the professionals of the Luftwaffe, many of whom were drawn from the army, had come to the same conclusions as those of the other European professional staffs—that a strategic bombing force was desirable but not feasible and that a tactical force was a sounder and easier choice. Their overwhelming emphasis on tactical rather than strategic bombing continued throughout the history of the Luftwaffe.[27]

In summary, the most striking aspect of the thinking of the Europeans from 1919 to 1936 on the nature of the next air war was the disparity between the popular and the professional estimates of that conflict. The Douhetans won the debate with the general public, but lost it with the professional military. There was a widespread popular fear of the destruction of civilization through mass bombing and use of gas. Yet those charged with the responsibility of planning for the next war rejected this view. Aside from their brief flirtations with the Douhetan theory, the professionals held to a more balanced, rational, and less exaggerated view of air warfare, a view based on a war they had fought in, studied, and hoped would be like the last one. Neither the popular nor professional view of the character of the future air war was correct, as the real wars of the 1930s would show.

Warfare from Ethiopia through Poland

Starting with the Ethiopian war and through the first phase of World War II, the European states fought a series of wars on three continents that ultimately either overthrew or drastically modified their principal ideas about aerial warfare. The Italian experience in Ethiopia proved little. They fought a colonial-style war against a feudal regime that could not protect itself against air power. The lessons learned from the experience were scant, save for the imperative need for air transport in a country largely devoid of modern land communications.[28] Spain, however, offered an entirely different perspective. For nearly three years the major European air forces tested their equipment, tactics, and personnel there in a combat situation similar to what they might expect in the immediate future. There was a surprising degree of uniformity in how they perceived the results of their experiences in Spain. Strategic bombing was downgraded and tactical bombing emphasized, perhaps unduly. High-level bombing was found to be ineffective, while low-level, close support bombing was judged effective and an absolute necessity for successful ground operations. The Spanish Civil War experience convinced the Russians, Germans, Italians, and French of the need to integrate the work of their air forces with that of their ground units. Naval aviation, long-range bombing, and even air defense were now slighted. The concept of strategic bombing was decisively downgraded in favor of the more orthodox view of combined military operations.[29]

In France, the concept of the multipurpose battleplane was brought into question, and the experience in Spain contributed to French indecision in development. More prototypes were ordered, adding to the confused welter already existing in the development program while production languished. Although the French digested the lesson of tactical bombing behind the battlelines, they were much slower in understanding the need for close ground support. Indeed, the results of the Spanish Civil War seemed to the French to validate the lowly estate of air power in their defensive strategy.[30]

The impact of the German experience in Spain on their tactics, development, and theory were profound. The Luftwaffe and the German Army had been organized along conventional lines, but both were groping for a new style of warfare. The formation of the new Panzer divisions in late 1935 coincided with the "dive-bombing craze" in the Luftwaffe, and the experience in Spain seemed to confirm the views of the "young Turks" in the ground forces and in the air arm that the wave of the future lay in a combination of armor, infantry, and air power. More than anything else, the Spanish war helped to weld the Luftwaffe to a tactical concept of operations geared to direct air support.[31] The successes of the Condor Legion precluded the evolution of a more independent, strategic type of air force, with the impact seen most clearly in the German production and development programs, where they swung into mass production of tactical aircraft. Spain convinced the top leadership in Germany that, like the proverbial gunfighter, they had achieved the technological "drop" on their neighbors, and they intended to use their advantage.

In Asia and in Spain, the Soviet Union had engaged in wars that tested its equipment and tactics. In China and later in Mongolia in 1939, the Soviets found themselves involved in some of the largest air battles fought since World War I. Hundreds of airplanes were used, and losses were heavy on both sides; but the Russians found they were equal to the Japanese except in the cases of the newer single-engine fighters. Again, as in Spain, the accent was on tactical air power, and the need of the Soviets for new fighters and close support aircraft was glaring.[32] The winter war in Finland confirmed these results and spurred the Russians to reorganize and reequip their air force. By the time the Nazis struck at the Soviet Union, the Soviets were caught in the middle of their modernization cycle, but in terms of doctrine the Soviets were firm. They were committed to a doctrine of air power that stressed the integration of air power with their ground forces, while rejecting a reliance on air power as a single strategy to win a war.[33]

The first year of World War II seemed to confirm the theories of the European air forces. The brilliant successes of the Luftwaffe in Poland, Norway, the Lowlands, and France were attributed to an aggressive, offensive-minded tactical air force. Curiously, the same air force had value as a strategic deterrent. From Munich to the Battle of Britain, the key to the Nazis' successes was their ability to pursue three objectives simultaneously. First, they were able to deter conventional military operations by issuing warnings of strategic instability. Second, they deterred all-out air attacks by promises of restraint and threats of retaliation. Third, they were able to isolate combat zones for their tactical air attacks.[34] As long as the Luftwaffe seemed to possess a strong deterrent value, it was far more formidable than in actual practice. Once it had to prove that value in the Battle of Britain, the inadequacies of the Luftwaffe were fully exposed.

In looking back at the conventional experience with military aviation from Kitty Hawk to the Battle of Britain, two observations stand out most clearly in my mind. The first is the astonishing pace of the technological development of aviation and the ingenuity and perseverance of the Europeans in using it to make war. I am reminded of a few lines George Bernard Shaw wrote in *Man and Superman* in the year of Kitty Hawk: "In the arts of peace Man is a bungler. I have seen his cotton factories and ... machinery ... they are toys compared to the Maxim gun and submarine torpedo boat. There is nothing in Man's industrial machinery but his greed and his sloth: his heart is in his weapons."

The second observation is the seemingly inexplicable failure of the Europeans to understand and properly conceptualize air power. Granted, by the yardstick of human experience two generations of time represent too brief an interval to expect much conceptualization. Still, the disparity between the ingenious military utilization of air power and the ineffective theorizing about it is striking. One can only conclude that Rebecca West was correct in 1937 when she wrote, "Before a war military science seems a real science, like astronomy, but after a war it seems more like astrology."[35]

Notes

1. Robert Saundby, *Air Bombardment: The Story of Its Development* (New York: Harper & Brothers, 1961), 6–7.

2. John Howard Morrow, Jr., *Building German Airpower, 1909–1914* (Knoxville: University of Tennessee Press, 1976), 17.

3. Ibid., 23–25, 45; for early interest in Russia see Robert A. Kilmarx, *A History of Soviet Air Power* (New York: Praeger, 1962), 4–5.

4. There seems to be little agreement on the total number of aircraft available in August 1914. Using German records, Morrow, 87, claims that the French had 300 first-line aircraft out of 600, the Prussians 295–320 fit out of 450, the English 160 and the Russians 400 with about half fit. William Green and John Fricker, *The Air Forces of the World* (New York: Hanover House, 1958), 98, 112, 244, list Germany with 281 aircraft and 9 Zeppelins, the French with 160 aircraft and 15 dirigibles, and the Russians with 244 aircraft. Arch Whitehouse, *The Military Airplane: Its History and Development* (New York: Doubleday, 1972), 3–9, cites another set of figures.

5. Whitehouse, 3–8, 22–23, 31–38.

6. Production figures for France are from Robert Krauskopf, "French Air Power Policy 1919–1939" (Ph.D. diss., Georgetown University, 1965), 1–2; for Germany, Werner Schwipps, *Kleine Geschichte der deutschen Luftfahrt* (Berlin: Haude & Spenersche, 1968), 69–71, 76–77, and J. A. Gilles, *Flugmotoren 1910 bis 1918* (Frankfurt: E. S. Mittler, 1971), 123–24; for Italy, "Italian Air Force: An Official History," *Aerospace Historian* 20 (Winter/December 1973): 178–79, Piero Vergnano, *Origini Dell'Aviazione in Italia 1783–1918* (Genova: Edizioni Intyprint, 1964), 91; and for Russia, Alexander Boyd, *The Soviet Air Force Since 1919* (New York: Stein & Day, 1977), 6, and Kilmarx, 15.

7. Robin Higham, *Air Power: A Concise History* (New York: St. Martin's, 1972), 29–30; Basil Collier, *A History of Air Power* (London: Weidenfeld & Nicolson, 1974), 78.

8. Charles Messenger, *The Art of Blitzkrieg* (London: Ian Allan, 1976), 29.

9. Bill Gunston, *Fighters, 1914–1945* (London: Phoebus, 1978), 30–31.

10. Ibid., 15–16; Whitehouse, 63–67.

11. Higham, 35–36.

12. Whitehouse, 72–79.

13. Kilmarx, 13–14.

14. Ibid., 24; Boyd, 4–5.

15. Kilmarx, 50.

16. John W. R. Taylor, *Combat Aircraft of the World from 1909 to the Present* (New York: Putnam's, 1969), 35.

17. Bernard Brodie, *Strategy in the Missile Age* (Princeton, N.J.: Princeton University Press, 1959), 7.

18. For Douhet's theories see: Giulio Douhet, *The Command of the Air*, trans. Dino Ferrari

(New York: Coward-McCann, 1942) and Frank J. Cappelluti, "The Life and Thought of Giulio Douhet," (Ph.D. diss., Rutgers University, 1967). A few of the many writers who discuss Douhet's influence are: Brodie, 74–83; Edward Warner, "Douhet, Mitchell, Seversky: Theories of Air Warfare" in Edward Mead Earle, ed., *Makers of Modern Strategy* (New York: Atheneum, 1967); Higham, 67–72; Ken Booth, "The Evolution of Strategic Thinking," in John Baylis, Ken Booth, John Garnett, and Phil Williams, *Contemporary Strategy: Theories and Policies* (London: Croom Helm, 1975), 30–32; B. H. Liddell Hart, *History of the Second World War* (New York: Putnam's, 1970), 590–93; Hanson W. Baldwin, *Battles Lost and Won* (New York: Discus Books, 1968), 79–81; Hilton P. Goss, *Civilian Morale under Aerial Bombardment, 1914–1939* (Maxwell Air Force Base, Ala.: Air University, 1948), 70–72.

19. On Poland, see Michael A. Peszke, "The Operational Doctrine of the Polish Air Force in World War II: A Thirty-year Perspective," *Aerospace Historian* 23 (Fall/September 1976); and for Sweden, Klaus-Richard Bohme, "Swedish Air Defense Doctrine, 1918–1936," *Aerospace Historian* 24 (Summer/June 1977).

20. Taylor, 613–15.

21. Krauskopf, 89–92, 101–2; Robert J. Young, "The Strategic Dream: French Air Doctrine in the Inter-War Period, 1919–1939," *Journal of Contemporary History* 9 (October 1974), 65–66.

22. Edward L. Homze, *Arming the Luftwaffe: The Reich Air Ministry and the German Aircraft Industry, 1919–39* (Lincoln: University of Nebraska Press, 1976), 165–67.

23. Krauskopf, 95–99; Young, 63–67; Russell H. Stolfi, "Reality and Myth: French and German Preparations for War, 1933–1940" (Ph.D. diss., Stanford University, 1966), 49–50.

24. As quoted in Anthony Adamthwaite, *France and the Coming of the Second World War* (London: Cass, 1977), 162.

25. George H. Quester, *Deterrence before Hiroshima: The Airpower Background of Modern Strategy* (New York: Wiley, 1966), 75; "Italian Air Force," 181–83.

26. John Erickson, *The Soviet High Command: A Military-Political History, 1918–1941* (London: St. Martin's, 1962), 382–83; Taylor, 614.

27. Reichminister der Luftfahrt und Oberbefehlshaber der Luftwaffe, Generalstab, "Luftkriegführung," L. Dv. Nr. 16, 1936, Lw 106/12, in Dokumentenzentrale des Militärgeschichtlichen Forschungsamtes, Freiburg; also reproduced in Karl-Heinz Völker, *Dokumente und Dokumentarfotos zur Geschichte der deutschen Luftwaffe*, Beitrage zur Militär- und Kriegsgeschichte (Stuttgart: Deutsche Verlags-Anstalt, 1968), 9:466–86; "Die Entwicklung der deutschen Luftstrategic," von Rohden document (4376-447) in the Bundesarchiv/Freiburg.

28. Higham, 81; Green and Fricker, 171; George Werner Feuchter, *Der Luftkrieg* (Frankfurt: Atheneum, 1962), 143.

29. Kilmarx, 146–47; Kenneth R. Whiting, "Soviet Aviation and Airpower under Stalin, 1928–1941," in Robin Higham and Jacob W. Kipp, eds., *Soviet Aviation and Air Power: A Historical View* (London and Boulder, Colo.: Brassey's & Westview, 1977), 52–54; Robert Jackson, *The Red Falcons: The Soviet Air Force in Action, 1919–1969* (London: Clifton Books, 1970), 55–56; Boyd, 69.

30. Krauskopf, 371; Young, 67; M. Astorkia, "Les leçons aériennes de la guerre d'Espagne," *Revue Historique des Armees* 4 (1977).

31. General der Flieger a.D. Paul Deichmann, *German Air Force Operations in Support of the Army* (USAF Historical Study No. 163, June 1962), 55.

32. Whiting, 57–59; Jackson, 61–62; Boyd, 86–87.

33. Richard E. Stockwell, *Soviet Air Power* (New York: Pageant, 1956), 14; Raymond Garthoff, "Soviet Attitudes towards Modern Air Power," *Military Affairs* 19 (Summer 1955): 77–78.

34. Quester, 101.

35. As quoted in B. H. Liddell Hart, *Europe in Arms* (London: Cassell, 1937), 199.

French Military Aeronautics before and during the Great War

Lucien Robineau
Translated by Richard Lemp, Department of English, U.S. Air Force Academy

In the aeronautical landscape of the Great War, France occupies a distinct place, both in the war and in aeronautics. With 1,400,000 soldiers killed and more than a million wounded out of a population of barely 39 million, France holds the unenviable record of the highest casualty rate of any country engaged in the war. For more than fifty months, the bitterest fighting levied by the Allies against the Kaiser's soldiers took place on her territory and it was there that nearly all of the war's devastation took place. Separated forty-four years earlier from the two provinces annexed to the German Empire—Alsace and a good part of Lorraine—this same land, from the first weeks of the conflict, had seen ten of its departments invaded and occupied until the Armistice. Most of its energy, mining, textile, and iron and steel resources were occupied for the duration of hostilities, to the point that the French found it necessary to bomb their own industrial installations in an attempt to deny the enemy their use.

This Great War was by consensus the first technological war in history. Among the newest and most technical components was the airplane. France had taken a leading role in the conquest of the air. From early on, sometimes with more enthusiasm than judgment, the French military had taken an interest in flying machines and their possible application in the war. The airplane had had its prophets and precursors. Between 1914 and 1918, the French Army was to discover principles, such as the offensive character of the airplane or the virtues of air superiority, and modes of action that it would apply to the maximum technical capacities of the moment. In the matter of logistics, it would establish an organization that was not always the best. It would, however, also engage an efficient system for the recruitment and training of all flying personnel, such that, at the end of hostilities, the extended branch was the size of a separate service. In very little time, France would develop a powerful aeronautics industry that, by the end of the war, would make its military aviation first in the world. None of this came without difficulty, not only because of the mistakes that are inevitable when it is a question of innovation or showing the way, but also because opposing vested interests and political intrigues sometimes worked against efficiency in the procurement of materiel.

France, Pioneer in the Military Conquest of the Air

Without repeating the entire history of the conquest of the air, it might still be useful to recall some of the distinguished achievements attributed to Frenchmen. It was in France that the first hot air and hydrogen balloons took flight, virtually at the same time. Not as well known, however, is that the first passenger in such a vehicle, Giroud de Villette, returning from his first ascent on 17 October 1783, observed: "From the instant, I was convinced that this machine would be very useful in an army to discover the enemy position, its maneuvers, its line of march, its movements, and by these signals, communicate to ground troops allied with the machine."

Thus began military ballooning. Two companies of balloons served the armies of the Republic from 1793 to 1799, at the sieges of Maubeuge and Charleroi, at the Battle of Fleurus and pursuing the enemy across the border. But these units were mothballed by the Directory and deactivated by Napoleon Bonaparte in 1799, possibly due to their difficulty in following the rhythm of a war of movement. From Lieutenant Meusnier's time—1785—came the thought about making balloons steerable. But there had to be an adequate engine for such an idea to be truly attainable. Such maneuverability became a reality in the closed-circuit flight of Capts. Charles Renard and Jean Krebs on 8 August 1884. The flight demonstrated, by returning to the point of origin after a course of appreciable length, that one could henceforth master a trajectory, overcome a reasonable wind, and choose the landing site.

Beyond dispute is the primacy, merit, and glory due the Wright Brothers for having accomplished the all-time first of powered, controlled, and sustained flight. Let us likewise not neglect to mention the name of French engineer Clément Ader for at least two reasons. First, at the time when everyone, everywhere, was trying to escape earth's gravity, he discovered how to establish the relationships that were to connect mass, wing surface, and the motor force of a heavier-than-air airframe in order to permit such a machine to take off from level ground, demonstrated by his experiment of 9 October 1890. The second reason has undoubtedly a greater importance for the military. Clément Ader had proclaimed his faith in the future of airplanes by signing a contract with the minister of war on 3 February 1892, a contract impossible for him to honor, because it contained several clauses that remained beyond attainment for practically another twenty years. Among other things, the contract specified that any aircraft would be able "to carry aloft, in addition to the pilot, a passenger or his equivalent weight in munitions; climb to several hundred meters; fly for six hours at a speed of at least fifteen meters per second, either 54 kilometers per hour or 33.6 miles per hour; follow a predetermined course and pass over a fixed point." Such a contract at least proved that the French government doubted nothing! Even more interesting is the *Memorandum on Military Aviation* which Ader concurrently sent to the minister. It included the following chapters: the founding of two schools: aviation (pilots) and avionics (engineers); the creation of an arsenal for the construction of airplanes; the creation of an aviation army; and aerial tactics and strategy.

The chapter on aerial tactics and strategy was developed in a book published in 1908, entitled *Military Aviation*. Its visionary character remains impressive. It called for specialization of aircraft according to mission and air superiority preceding all other action: "the major preoccupation of the tactician should be the opposing air force that he must seek out, attack, and combat unto its complete defeat... aircraft shall first be directed towards the enemy air force and, once neutralized, then turned on ground forces." The book also pointed out the necessity to set up an anti-aircraft defense (vertical artillery), deployment of torpedo planes (bombers) for strategic and tactical objectives, including the concept of dissuasion by terror: "Torpedo planes will become genuine terrors... their formidable power and the fear of seeing them appear will inspire prudent reflection in statesmen and diplomats, dispensers of war and peace."

Let us note that the impossible pact imposed on Ader in 1892 was proposed in exactly the same terms to the Wright Brothers in 1906 when the French military thought to assure itself of their competition and acquire their airplanes. The American inventors declined such a ridiculous proposal, not having, in their own words, the slightest intention of breaking their necks to please the French minister of war. A consortium of bankers led by Lazare Weiller persuaded them in the spring of 1908 to market their flying machines in France, where aviation was becoming more and more popular. Wilbur Wright came to Le Mans, then to Pau, where climatic conditions were more favorable. Having broken in France all existing records and having seen his airplane improve—a more powerful engine, wheels, and an airspeed indicator—he founded the world's first flight school in Pau at the very beginning of 1909, soon followed by other builders.

During this time, performance continued to progress. Cross-country flights, initially flown by Wilbur Wright, Henry Farman, Louis Blériot, and Hubert Latham in 1908, became current practice. This aspect of flying began to attract the military, especially after Louis Blériot had crossed the English Channel in July 1909. The French Army ordered twelve airplanes (four of them Wright aircraft) in September 1909, and sent ten officers for the Aero Club of France pilot's license at the four schools instituted by the builders: Antoinette, Blériot, Farman, and Wright. The Ministry of War's interest was about to be rapidly confirmed. In the September 1910 maneuvers, airplanes were judged more efficient than dirigibles and a report mentioned that "aviators went beyond what was asked of them, beyond hopes that could have been imagined." The director of engineering, General Roques, who shared supervision of aeronautics with the director of artillery, declared with enthusiasm, "Airplanes are as indispensable to armies as cannon and rifles. It is a truth one must freely confess, at the risk of having to suffer it by force."

Close on the heels of these events, the military ordered forty additional airplanes, twenty Blériot and twenty Farman, and potential competitors were invited to build aircraft for the future needs of the military, as those needs were then apparent. The 1910 experiment predicted only two possible uses of aircraft: observation for artillery and intelligence-gathering for command. That is why the competition aircraft were defined as armored three-seaters configured for a pilot, an observer, and a third man who could function either as a machine-gunner, a radio operator, a photographer, or a mechanic. Naturally, neither radio nor photography was sophisticated enough at that moment for aeronautical application and the engine was already taxed to lift two men—not to mention armor and the fuel for the required three hundred kilometers; the competition's results were unconvincing. But, of course, hope is not necessary for enterprise, nor success for perseverance.

Thus perseverance continued. In October 1910, ballooning and aviation were regrouped under a Permanent Inspector of Aeronautics. At year's end, two military flying schools had opened and fifty-eight pilots—six from the Navy—had graduated. In 1911, a military pilot's certification had been created, whose check-rides were much closer to the probable missions of air units. Following the year's exercises, these units were organized into flights, provided common materiel, and the wings of the airplanes bore the round tricolor.

1912 saw the first reasonably organized attempt to specify the military role of the flying resources that were already defined, if not abundant (five flights). General Roques, Permanent Inspector of Aeronautics, authored the following note in which he first proposed, quite naturally, to supply armies, army corps, cavalry divisions, fortresses, and ports with flights of airplanes intended to keep them informed and provide liaison. He also proposed—to no one's surprise—flights to adjust artillery fire in both siege and campaign conditions. But, above all, referring to Ader, he suggested—extraordinary for the times when experience was nil—forming ten flights of eight aircraft each,

> that would constitute the generalissimo's air force and would be composed of combat aircraft of the strongest caliber, capable of carrying 300 kilos up to 300 kilometers . . . these independent flights, while completely capable of strategic exploration, would be tasked with going afar to destroy enemy aircraft and to proceed to useful destruction by launch of projectiles.

This precursor—who envisioned a plan to procure 832 airplanes when he had but a handful of units—laid the cornerstone of a building that the war would soon make necessary. Industry, the high command, and enthusiastic researchers followed the movement. Industry had produced 1,350 aircraft and 1,400 engines in 1911, 1,425 aircraft and 2,220 engines in 1912. General Joseph Joffre, army chief of staff and future commander in chief, had directed the exercises of 1912 (sixty airplanes) and 1913. In these two years, the Army had twice ordered 400 aircraft. Michelin concentrated its efforts on aerial bombardment, bombs, and targeting systems. At Blériot and Morane-Saulnier, pilots were attempting to install on board armaments and promote firing along the flight axis, including firing through the propeller arc. In August 1914, the flying schools had produced 657 pilots, of whom about 350 were in the flights. These 24 flights had an inventory of 141 aircraft of the 1,250 that the Army had procured before Germany declared war on France on 3 August 1914.

Aeronautics in the Crucible of Operations

Aeronautics began its operational life armed with certain assets, but also faced cumbersome handicaps due to some faltering steps related to organization. In the beginning of 1914, ballooning and aviation were separated, thus dissolving what had initially seemed a natural unity of action. Also replaced was the Permanent Inspection of Aeronautics, the tool as well as image of the unity of action, by a "Twelfth Directorate" in the Ministry of War after air forces had been placed under the direct authority of army corps commanders or commanders of fortified locations. Such wavering reflected the rivalry existing between the Artillery and the Corps of Engineers over the use of aircraft and predicted the disputes over doctrine that were to long taint the history of French aviation and impede its efficiency.

With respect to French military aviation, critical historians have often observed:

At the end of the First World War, everything had been invented in the matter of air warfare including organization, principles of employment, doctrine, etc. . . . and that everything was forgotten, from the Armistice on, resulting in the disastrous events at the beginning of the Second World War.

This is in fact not the case. It is true that, for more than four years, the high command, the chiefs of staff, officers, and aviators had the time to think about the use of airplanes. Those responsible had not erred in noting the results of the actions they had ordered, what mistakes they had made, and, consequently, what corrective measures they could envision to improve the role of air power. No one is infallible, however, and advice was never unanimous. On the contrary, every time it was a question of the application of air power in war, debate raged as much over employment itself as over the structures most favorably disposed for it. Ideas definitively applied were so because their proponent had enough influence on the decision maker at the appropriate level. Sometimes these ideas were indeed right, but at other times they were not, or the timing just wasn't right. In the matter of military art, there is no eternal truth. As illusory as it was to predict strategic bombing of enemy cities and industrial centers—by definition deep inside its territory—at a time when the operating radius of aircraft barely allowed penetration over enemy lines, it would have been just as intellectually faulty not to have considered the legitimacy of such a mode of action when technological advancement would enable it. Inversely, command authorities would have been guilty of not fostering such progress, given that political considerations are often the first step and the only condition of a fruitful technical or industrial research effort.

In reality, everything that could be was applied to the development of aeronautical technology throughout the Great War, a technology that evolved significantly between 1914 and 1918. Combat experience proved the validity of new tactics and an extended strategy. It was a mistake consequently to accept as gospel truth results of 1918 that were clearly tentative, as it was to admit once and for all the virtual infallibility of, "the prestigious leaders who had led our armies to victory." Distribution of air forces into a General Reserve, responsible solely to a commander in chief, and into "organic elements" assigned to major army units, was a solution adapted to the 1918 situation and no longer related to the art of air warfare in 1939–40.

Dirigibles, Captive Balloons, and Airplanes

On 3 August 1914, French military aviation consisted of dirigibles, captive balloons, and airplanes, all in rather limited number: dirigibles, because of their sensitivity to the wind; balloons, because they were considered ill-suited to a war of movement; airplanes, because of their young age and the rapid evolution of their characteristics that resulted from their newness. In the beginning, all these aircraft had the common task of observation, in the larger sense. This role was weak for the first group, great for the second, and immense for the third.

Only ten dirigibles made flights in support of ground troops, and only until the end of 1916. Their vulnerability to ground fire (including friendly fire) was evident very early and most of their missions took place at night. The first missions were long-range reconnaissance, then, as the stabilized fronts made such intelligence less urgent, missions were solely bombing sorties. Dirigibles engaged only at the discretion of the commander in chief. The first reconnaissance mission took place on 8 August 1914, and the first four 155mm shells were dropped on a German railway station on the 10th. The dirigibles' objectives were bivouacs, communications lines, and, above all, railway depots. Dirigible missions were long—up to eight hours—and, in the end, they carried 1,300 kilos of bombs. Some were equipped with an upper platform from which a machine-gunner, linked to the aircraft commander by telephone, provided defense against enemy aircraft. Six of the initial ten aircraft had been destroyed from a variety of causes when the dirigibles were transferred to the Navy in February 1917, when airplanes began to emerge more efficient as bombers. All total, dirigibles had flown only sixty-three sorties. Disappointing in their above-ground role, they were to be of greater service to the Navy, which used them in some forty missions at sea in the fight against submarines and in minesweeping operations. For the same kind of missions, only closer, the Navy put hydroplanes to use. At the Armistice, there were 1,264 crewed by 11,000 men.

The role of the captive balloon was to reveal itself of foremost importance, as much for adjusting artillery fire as for the discovery of enemy batteries and their destruction by counterbattery assault. They were also vital for general battlefield surveillance, thus enabling early detection of enemy offensive preparation. However, when combat actually began, there were only four companies of captive balloons assigned to the four fortifications. The high command had withdrawn all the others in 1911, having judged that balloons would be useless in a war of movement. Moreover, the available aerostats were the spherical kind that did not handle well in the wind and were very uncomfortable for the observer. It took the combined efforts of the *Grand Quartier Général* (High Headquarters, the *GQG*) and some very talented men (José Barès, M. Saconney, Albert Caquot) to quickly correct the situation. Ten companies—30 balloons—were at the front from October 1914, 10 others at the beginning of 1915, 36 at the end of the year, and 75 (250 balloons) shortly thereafter. The quality of the material was initially improved by simple imitation of the German *Drachen* (Dragon) that eventually gave way to the *saucisse* (sausage). Captain Albert Caquot then perfected, in several successive versions, a model of his own invention adopted by all Allied armies—and even by the Germans themselves!

Balloons ascended to a distance of five to ten kilometers from the lines and to a relatively high altitude (on the order of 700 to 1,000 meters). The balloons had a wire communications link on one side to winch operators standing by in case emergency maneuvers were necessary due to attack by enemy aircraft or artillery, and on the other, a link to the artillery batteries they were responsible to inform. From September 1915, special army and army corps balloons were directly linked to the command posts of these major units. The first "wireless" installations appeared during the Battle of the Somme in the summer of 1916 and a year later balloonists were monitoring the transmissions of aircraft patrolling their sectors. In order to assure a continuous and uniform surveillance, observers spent entire days in their baskets, equipped with photographic gear. From 1916 on, they also carried parachutes.

It is not necessary to underscore the utility of aerostats; that utility was never more evident than when they were absent. For example, during the German attack on Verdun in February 1916, when air superiority belonged to the enemy, French artillery was essentially blind for days. On the other hand, these balloons were able to play a very active role in the battle of movement that characterized the final period of the war, thanks to an organization that gave them the capacity to shadow the displacement of ground divisions. The balloon corps had a strength of 10,000 men in 1918 (2,000 officers and observers). One thousand one hundred balloons had been built, including more than 200 for the Allies.

In this trilogy, airplanes occupied the largest and most spectacular place by virtue of their numbers, their increasing capabilities, the variety of their roles, the evolution of their modes of engagement, and the changes their intervention brought to general strategy. Their numbers grew quickly and considerably: the 24 flights of 141 aircraft on 3 August 1914 had already grown to 31 flights with 183 airplanes on 1 September. There were 756 aircraft and 661 pilots in service to the armies of the North and Northeast on 15 August 1915; 1,103 aircraft in the spring of 1916; 1,854 a year later; 2,602 on 1 November 1917 (820 pursuit aircraft, 1,274 observation planes, and 508 bombers); 3,300 in the spring of 1918; and 3,608 in November (1,423 pursuit aircraft, 1,639 observation planes, and 514 bombers). At the time of the Armistice, for the 288 flights active on all fronts, French aviation counted a total of 11,836 aircraft, including those of the flight schools (3,400) and those belonging to the reserves (3,800). Such numbers should be treated with caution, however, for quantity alone would be incapable of establishing air power and one always gets the raw materials industry is capable of providing. Until midway through 1917, the majority of reconnaissance planes were obsolete, even those fresh out of the plant. (The Farman F40, with rear propulsion, was very vulnerable.) The same thing was true of the bombers that, for a long time, only flew night sorties because of their slowness and limited ceiling. Even the fighter situation was troublesome: obsolete and destroyed aircraft had to be replaced and the Nieuport XVII—that equipped most of the flights—was henceforth outclassed by the Albatross D.II and D.III. The Spad VII, on the scene at the end of 1916, would—when its Hispano-Suiza engine was finally operational—contribute to remedying this situation along with the Spad XIII. With the "Baby Nieuport" (7,200), these two aircraft were produced in the largest numbers—3,500 and 7,300, respectively. Air superiority see-sawed from the Germans to the Allies as a function of the technical or tactical improvements each side applied. Among other examples were the propeller-synchronized firing of the Fokker; the rockets of Le Prieur; the appearance of a model temporarily superior to those of the enemy; flight in formation and so on. Not until the second half of 1917, when airplanes of a completely new generation became operational, was it possible to decrease the vulnerability of reconnaissance and bombardment aviation, increase range as well as transport capabilities, and to reengage daylight bombing, all of which permitted introducing aircraft into the battle zone more efficiently and with new tactics. The two-seat Salmson 2 A2, Breguet XIV A2 (reconnaissance), and Breguet XIV B2 (bomber) aircraft were, with the three-seat Caudron R XI combat aircraft charged with protecting them, instruments of a renewed employment doctrine, thanks to which air superiority remained with the Allies until the Armistice.

The role of the airplane in the overall battle, limited to observation at the beginning of hostilities, had diversified since 1914, thanks to a "forward" organization that was generally efficient enough to adapt to the landscape of ground operations. After aerial observation had provided the commander in chief intelligence on German movements of 2 September 1914, and thus enabled him to win the first Battle of the Marne, General Joffre, who had indicated his strong interest in aviation before the war, was at the same time convinced of its utility and of the need to organize it better. He created a position on his personal staff of a "Chief of Aeronautical Forces at the High Headquarters of the Armies" (*Grand Quartier Général des Armées*), whose charter was detailed by an Instruction of 13 December 1914:

> Intervenes in the employment of air power for the armies as an advisor to the commanding general. Consolidates requests for the assignment of personnel and materiel among the several units of Aeronautics. . . . Regulates the employment of bomber flights according to the instructions of the commanding general in chief.

In fact, with the Germans having dropped bombs on a city in the east the day war was declared, the French saw themselves obliged to counterattack by bombing the dirigible hangar in Metz-Frescaty a few days later. French bomber aviation was born on 27 September 1914 with the formation of the first group of three flights of Voisin LA 5, the GB 1 aircraft. At the same time, it had become customary to exchange a few cartridges of carbine or revolver fire when encountering an enemy aircraft in flight, if for no other reason than to prevent intelligence gathering. The first success came on 5 October 1914, when a crew of the observation flight V 24 aboard a Voisin equipped with a machine gun shot down an Aviatik. A repeat performance ushered in a new breed of aviators, namely, fighter pilots.

Thus, three months after the outbreak of hostilities, the principal action areas of aviation were open. General Joffre signed the memo on 10 November 1914 that defined those areas:

> Aviation is not only, as one might have already suspected, an instrument of reconnaissance. It has become, if not indispensable, at least extremely useful in adjusting artillery fire. In addition, it has shown by dropping powerfully explosive projectiles that it has been likewise capable of acting as an offensive arm, either for long-

range missions or in conjunction with other troops. Finally, it has the task of pursuing and destroying enemy aircraft.

This note, evidently drafted by the chief of the Aeronautical Service at the *Grand Quartier Général,* was counterpoint to the report that the same Major Barès had addressed only a short time before to the commander in chief:

> aviation is an arm [not a service] a distinctly offensive arm, either in the pursuit of enemy aircraft or in the destruction of troops, encampments, or fortifications by means of projectiles. It may be assigned distinct missions [independent] at more or less great distances or attack in conjunction with other troops

Already formulated with a certain vigor, the role thus defined for aerial forces was to become more precise with experience and material progress under the impetus and thanks to the personality of the successive chiefs of the Aeronautical Service of the *GQG* who naturally articulated the aviation policy of their commanders in chief: Major (later Lieutenant Colonel) Barès with Joffre (September 1914–February 1917); Major du Peuty with Robert Nivelle (February–August, 1917); Colonel (later General) Duval with Philippe Pétain (August 1917–November 1918).

The refinement of this role was to lead to the progressive elaboration of employment doctrine for air units, not without bitter controversies, but, finally, with the effectiveness sanctioned by success in combat. The formulation process of this doctrine bears such importance that it is appropriate to devote a separate section to it, as is also the case for the schools that, faced with rapidly growing needs, had to provide well-trained personnel. The organization of the "rear"—whose fluctuations were the source of serious difficulties—will also receive special attention.

Employment Doctrine

It had been understood very early that the airplane was cut out for an offensive role against aircraft and against the ground. The necessity of air superiority, at least local and timely, had imposed itself at Verdun as a situation ascribing the conditions for the outcome of all other action, in the air or on the ground. However, a number of months elapsed before bombing missions were directed against the destruction of enemy aircraft on their own fields (with the exception of the very first attack on a dirigible hangar on 14 August 1914, so decided simply because it was a military objective within the range of airplanes) and it was by the action of fighter aircraft in aerial combat that this result was first pursued. As for the use of a large number of aircraft in observation and reconnaissance missions for the benefit of artillery or infantry movement, this was the very basis of the recognized utility of aviation.

Early on, it appeared logical to assign to ground units flights responsible for these missions, flights regularly known as "army corps aviation" that were always employed under the direction—sometimes distant—of the commanders of army aeronautical units. By contrast, leadership of fighter and bomber units was the object of constant dispute. The Army air commanders hoped to use these units for close protection of their observation aircraft, local air cover for their troops, and, on their orders, direct intervention in the battle. The chief of aeronautics of the *GQG* generally found it preferable to keep them under his personal control in order to achieve concentrations or movements on the strategic level, that of the commander in chief. Such disputes not only did not end during the entire war, they continued after the Armistice. Solutions adopted (or imposed) as a function of technical progress and the conduct of operations proved to be wise, in view of their effectiveness.

Bombardment

It is certainly more natural to think about strategic air actions when separated overseas from the theater of operations than when the battleground is on one's own territory against an adversary who is from the same continent. This explains why the French and British views on the subject were different. Nonetheless, the French conceived and executed strategic actions against Germany very early (perhaps too early).

The few attacks that did take place against cities occurred against Karlsruhe on 15 June 1915, then against Trier and Sarrbrücken in reprisal for German attacks against London and Paris. Major Barès, opposed to any renewal of such attacks on moral grounds, but also because of their limited range, said, "That would add the odious to the useless." It was evidently desirable to attempt to weaken the enemy's armament industry, hence the initiative to engage a platform capable of reaching the Ruhr (the "SN" bomber) while a certain number of raids were effected. First, before the formation of the first bombardment group, it was the isolated actions of crews launching projectiles on their own initiative against targets of their choosing from a list compiled by the *GQG*. Then, after May 1915, it was a concerted effort led by the massed aircraft of the group (18) on the same objective and during a restricted time frame, but executed successively and individually by crews (for example, bombing the BASF chemical plants at Ludwigshafen on 27 May). Finally, however, counterattack from German fighters demonstrated the vulnerability of bombers that were too slow and ill-defended. This was true despite relatively massive actions concentrated in both space and time and even when bombers flying in formation supposedly protected themselves by side-shooting defensive firepower. Accordingly, at the end of 1915 bombing missions on the other side of the lines took place only at night. In fact, it was quickly understood that the reduced operating radius, low load capacity, and impossibility of providing escort to protect these aircraft made the strategic venture fairly untenable for a long time, at least if such a notion has the connotation of deep penetration against vital resources of the enemy country. At the same time, there were systematic attacks during the war against factories, mines, and the network of deposits in the iron basin of Alsace and Lorraine, as well as in Luxembourg, regions from which the German arms industry drew more than 75 percent of its iron-ore resources. The actual tonnage of munitions dropped never achieved any significant impact. It was, by the way, the same for the British, who, during the entire war, only dropped 800 tons of bombs on Germany, and was likewise true for the Germans who dropped about a hundred tons of bombs on England and less than thirty on Paris.

The relative ineffectiveness of this mode of action rather quickly led French command authorities to restrict the use of bombardment groups (four in the summer of 1915, five in February 1916, twelve in 1918) to indirect battle support, targeting military objectives immediately behind the front lines: encampments, depots, train stations, railroads and trains, landing zones, later airfields, attacked for the first time in July 1916 and continuously thereafter. Little by little, it became progressively evident that bombardment aviation should be employed in mass, always against the same kinds of targets, always in support of ground combat, with an increasing concentration on enemy airfields. The realization of this idea became possible with the arrival, in growing numbers in 1917, of robust, powerful modern aircraft that were finally well adapted to these interdiction missions, to which were added direct battlefield intervention missions, sometimes in an independent manner. This concept of bomber employment assumed a situation of air superiority that they competed to obtain and that fighter aviation secured and maintained, thanks to a renewed employment doctrine that also depended on mass action. Colonel Duval, *aide-major général* to the commander in chief and chief of aeronautics of the *GQG* in 1917 and 1918, was instrumental in combining these two at last kindred specialties of combat aviation, first by creating wings, then groups, and finally an air division, whose manning and results were impressive.

Fighters

Fighter aviation, born almost by accident in October 1914, was to gain momentum in 1915 and earn its spurs at Verdun and over the Somme during all of 1916. Before Verdun, it had no tactical doctrine, although from January 1916 Major Barès, chief of aeronautics of the *GQG*, had begun to define one: "It is necessary to accustom aviators to fly and to fight in groups. . . . All future operations of Escadrille N.3 will take place with three aircraft. Furthermore, the commander of Escadrille N.3 will train his pilots to engage in larger and larger groups, up to and including the entire flight."

At Verdun, the Germans preceded their initial offensive with a large concentration of flights (270 aircraft) that, in securing control of the sky, prevented French balloon and airplane activity, thus denying French artillery any intelligence and, consequently, any effectiveness. The counterattack had consisted in rapidly assembling a comparable concentration of 15 flights under a single commander and deploy the planes in multiple patrols of five planes each to counter, by constantly occupying the sector, the enemy tactic of the *Jagdstaffeln* missions of six airplanes in two stages. Such tactics, coupled with the technical surprise of engaging *Le Prieur* missiles (the first air-to-air rockets that proved effective against the *Drachen*), immediately proved their validity. From this came profound changes in the employment of fighter aviation, henceforth comprised of "combat groups," likewise charged, on direct order from the *GQG*, to machine-gun ground troops.

It was a complete change, soon confirmed at the Battle of the Somme. Up to that point, single combat was the rule. In fact, some select pilots were authorized to continue the practice under the title of *rondes de chasse*, "fighter rounds," and a kind of mythology had given birth to the special category of "War Ace."

The whole time the two opposing armies were mired in endless trench warfare, the frequent offensives launched by one or the other usually succeeded in gaining or losing but a few yards painfully exchanged for the lives of tens of thousands of men. The results of such unspeakable butchery could not be published in the popular press: morale in the rear could not have withstood it and censorship reigned. It therefore seemed necessary to show tangible results and real successes. Hence the exalting of the exploits and victories of air combatants whose numbers—less than 20,000—paled in comparison with the millions of soldiers mobilized. But this minority waged a neater war and whose incontrovertible conclusions could keep hope alive for a happy resolution of the conflict. The results thus made public distinguished an elite, cited by name in the communiqué announcing the fifth enemy plane shot down. On these elite men, universally glorified as heroes like medieval knights, fame conferred the title of "Ace." Thus passed to posterity the names of George Guynemer, Rene Fonck, Charles Nungesser, and many others, in France and among all the warring nations. In France, the phenomenon of the "Ace" undeniably contributed, by its mythological character, to perpetuating a regrettable tendency toward individualism, in spite of opportune changes adopted by the command from the beginning of 1916.

From 1917 forward, with the implementation of a mass doctrine for air power, under the impetus of Colonel (soon General) Duval, fighter groups were widely integrated with bombardment groups in order to constitute a combat aviation that proved its worth in the course of the battles of 1918.

The Air Division

One hesitates to recall that a massive employment of air power presupposes the existence of a massive number of airplanes. Such a situation existed from the autumn of 1917 forward. General Duval (whose name remains closely associated with this new strategy) stated:

> The tactics of the commanders of Army Aeronautics limited aviation to a very modest role: put into the air observation planes necessary for ground operations and disperse ahead of these aircraft a defensive screen of small patrols of fighter aircraft cruising a various altitudes.... It was not possible to engage 3,000 airplanes by following the same principles as for 1,500 or 2,000. From the moment that I admitted mass as an instrument of combat, I had an obligation to organize the mass.

His guiding principle, with the purpose of assuring mass and concentration, was to unify, permanently and under the same command, groups that had previously been put together only temporarily, in order to give them the training and cohesion necessary for an organized action. His goals: guarantee freedom of movement for observation aircraft on the front lines; seek out enemy aircraft and destroy them; and intervene in mass, systematically and continuously, over the rear areas and in the battle itself. Realization of these goals consisted in successively creating combat wings: February 1918, uniting three squadrons of four flights of fifteen airplanes (180 planes and 2,000 men); bombardment wings: February 1918, with three squadrons of three flights of fifteen aircraft (135 planes and 1,700 men); groups: March 1918, combining a combat wing with a bombardment wing (315 planes in each group); and the air division: on 14 May 1918, marshaled from groups, consisting of two fighter wings and two daylight bombardment wings (270 bombers and 360 fighters) to which were added two night bombardment groups, all total, nearly 1,000 aircraft.

The air division was to function at the army group level. Its organization permitted its deployment by groups and, with Duval having admitted the experimental nature of this configuration, with a certain flexibility. From the formation of the first group in March 1918, aircraft were launched against ground troops and against airfields in mass formations of 80 to 200 airplanes; every Allied offensive was supported, from the summer on, by a concentrated air power of 300 to 1,000 planes, to culminate, in September, under the command of Brig. Gen. Billy Mitchell, in a force of 1,500 airplanes (900 French, 450 American, 160 British) at the capitulation of Saint-Mihiel.

The effectiveness of this new air power strategy was never better attested than by the appreciation of Gen. Ernst Von Hoeppner, commander of the combined German air forces (*Kogenluft*): "The enemy attacked simultaneously on virtually the entire front. We could no longer, as we could in 1917, concentrate all our forces on a given point to there secure mastery of the air. Everywhere, we were in a state of numerical inferiority."

It was one of the essential elements of victory. It was made possible, however, only by the availability and sufficient number of qualified personnel.

The Schools

During the entire war, twelve schools were to produce 17,000 pilots and 2,000 observers. An impressive number of mechanics and other specialists (machine-gunners, photographers, radio operators, meteorologists) would also graduate to make up the 90,000 men who represented French aviation at the end of the war.

Up until 1916, a pilot left for the front as soon as he had his certificate in hand; his future depended a great deal on his innate skill and on luck. From 1917 on, a technical school furnished future aviators—before putting them at the controls of an airplane—all the information about flying that could be taught in ground school. The program was as follows: (1) technical school and selection; (2) military pilot certification from one of the schools, eventually civilian, controlled by the Inspector of Aviation Schools; (3) attendance at a combat aircraft lead-in school, according to the student's given specialty; (4) attendance at a qualification school; (5) attendance at one of two specialty applications schools (aerial gunnery or bombing); and (6) assignment to the Training Division Group, which was the pilot reserve unit prior to assignment to a flight.

The entire instructional program required five to six months, instead of the forty days that had been the rule before. The flying schools had to respond—as is always the case in war—to two needs: satisfy the increasing number of flying billets as units expanded and replace losses. In four years, the number of units had risen from 24 to nearly 300 flights, as the number of aircraft per flight had gone from 6 in 1914 to 15 in 1917, and losses had always been significant, not only because of the enemy, but also because of an elevated accident rate as well. Records indicate that of the 5,500 pilots and observers put out of action during the war, 35 percent of the losses were due to accidents—including 300 at flight schools. Combat losses were also much greater in the last months of hostilities: 2,327 pilots killed, wounded, or missing between May and October 1918 (practically 400 per month), as many as the total staff of March 1917. The performance of these organizations could be measured in the number of pilots they produced: 134 in 1914, 1,848 in 1915, 2,698 in 1916, 5,609 in 1917, and 6,900 in 1918. French flying schools were, at the end of the war, an impressive factory making use of more than 3,000 airplanes, a workforce of nearly 20,000 people, and logging, in 1918, 360,000 hours of flight.

Mechanics (thanks to whom the less-than-reliable engines of the era keep working), were also tasked with making repairs to the damage frequently inflicted on the airframe, repairing the canvas and repainting it. They had been recruited from the civilian auto industry and, in special schools—in minimum time—received the additional training needed to master the aeronautical technology of the day.

The Rear Organization

Already discovered in this era was the importance of what would later be called "logistics." The frontline conducted military operations and communicated needs the rear was expected to meet. Naturally, these were changing needs, given that they involved adapting to combat techniques whose principles were discovered experimentally. The same was true for requirements evolving from enemy initiatives. These needs, above all, concerned materiel (airplanes, munitions, fuel) in quality as well as in quantity. They obviously relied on the aeronautics and armament industries such as they were and such as they were to become. The interest of industrialists, who wanted to make their production profitable once it was underway, might oppose too rapid an evolution of the performance demanded of aircraft or engines, or, even more, procuring materiel from abroad when such a move was judged beneficial to fill the gaps at home. The influence of certain professional lobbies—as in intervening in the political sphere—sometimes interfered with the function of organizational structures designed to satisfy purely military needs, notably by provoking untimely upheavals in the established order.

The organization established to furnish the aeronautical arm the means of making war changed not less than five times in four years. Even more, in its last form the organization would have two heads, with the first dismissed after six months for political reasons, in spite of his ample capability. As mentioned before, the results had been—in spite of everything—rather remarkable, at least in the amount of materiel produced. But it was only from 1917 on that the quality of the aircraft was at a height equivalent to their number, and, in 1918, at peak production, an airplane left the plant every fifteen minutes, an engine every ten minutes, day and night. The airplanes (Spad XIII, Spad XII-cannon, Breguet XIV, Salmson, Farman F50, Caudron RXI) were modern, their engines powerful (200 to 300 horsepower), their armament effective, and their equipment (radio, cameras) compatible.

The organization was alternately controlled by a political organization—the *Undersecretariat for Aeronautics*—or dominated by a military directorate—the *Directorate of Aeronautics,* 12th Directorate of the Ministry of War, the *General Directorate of Aeronautics,* depending on whether the Minister of War was a politician (Alexandre Millerand, Paul Painlevé, Georges Clemenceau) or a general (Joseph Gallieni, Roques, Louis Lyautey). That also made six ministers in four years.

The military solution had several drawbacks: the director (except for General Hirschauer, from October 1914 to September 1916) had no aeronautical qualification and therefore could not assert any real authority over the captains of industry, he generally (for lack of a mediator) opposed the frontline chief of the Aeronautical Service (who was committed to urgency and could have his own ideas about the definition of materiel). The civilian solution at least had the advantage of representing a single authority simultaneously governing the Aeronautical Service of the *GQG* and the services responsible for coordinating industrial activity. Organizational harmony, however, depended on the personality of the undersecretary of state and the relationship he could establish with the frontline chief of aeronautics, a relationship that was not always without clash. His position would have been improved by ministerial rank, notably in relation to the minister of armaments to whom he was occasionally subordinate.

Circumstances were fortunate that the last two undersecretaries of state (Daniel Vincent and Jacques-Louis Dumesnil) would both have previous military experience at the front, in aviation, and dealing with concrete problems encountered by units, that they were young and dynamic, that one succeeded the other without animosity, and that they had each discussed confidential and follow-up reports with the chief of aeronautics of the *GQG,* General Duval, who likewise had the intelligence to prosecute the situation globally and give priority to general interest.

Under the authority of the undersecretary of state, through the ministry's director of aeronautics, an Aeronautical Manufacturing Service had the responsibility for overseeing all production consigned to private industry (3,000 factories of all sizes), coordinating industrial activity, and standardizing production. A Technical Section was charged with studies on various categories of materiel: airframes, engines, armaments, munitions, equipment, mobile materiel, and the like, for their development and quality control.

In the final analysis, the results obtained by this organization were remarkable and they are well known. All total, 52,150 airplanes and more than 92,000 engines had left the factory during the war, of which more than half came out during the course of 1918 alone. A significant number—264—of different prototypes had been studied, leading to 38 kinds of operational aircraft put into service. This would seem excessive only if forgetting that the British tested 309 prototypes for 73 kinds of aircraft in service or that the Germans had 609 test models for the 72 kinds of aircraft eventually assigned to operational units.

As today's historian comes to appreciate the accomplishments of French military aeronautics throughout the course of World War I, including its industrial and administrative aspects, he should keep in mind the precise circumstances that characterized these first years of the twentieth century now finishing in space.

One must especially remember that the airplane was in its early infancy in 1914, that aviators were rare and generally considered by the public to be unthinking daredevils, that piloting the fragile machines of the era remained an exploit and was, in any case, a tricky sport whenever there was a little wind. In such a way, one could better measure the progress achieved by an industry that took off in its totality in the *Blériot* of the Channel and the *Farman* of Issy-les-Moulineaux, and in four years, was succeeding with airplanes of mostly metallic framework and whose general design was long to remain valid. One will also measure the accomplishments of this industry, built from the most modest of workshop beginnings and developed in a few months on the order of a technological revolution, all the while noting that, with men in the workforce largely mobilized, it employed thirty percent women at a time when they were little removed from their work as housewives.

The military command, although perhaps deserving reproach for its conservatism, had finally, with the highs and lows already mentioned, assimilated aviation rather well as a combat arm. General Pétain, commander in chief of the French Armies of the North and Northeast, had written in July 1917: "Aviation has taken on a vital importance, it has become one of the indispensable factors of success. One must be master of the air."

Wishing "to give to this arm the maximum power attainable in the time we have available before the end of the war," he also declared that he was waiting for, "planes, tanks, and Americans" before planning to launch the future offensives. Truthfully, it was a question of compensating, by new modes of action, the numerical inferiority of Allied forces facing a German army strengthened by divisions freed from the cessation of hostilities on the Eastern Front. In fact, direct intervention in the battle—in a practically independent way—by Duval's air groups during the German offensives of Spring 1918, as they were in support of the Allied offensives in the closing months of the war, had let the entire range of the offensive capacities of air forces employed in mass under a centralized command be a matter for all to see.

It was a revolution in the art of war and the introduction of a distinct air operation was changing strategy. It is true that the different modes of this application of air power have given rise to discussion, as they did even at the time. Some only wanted to see in a unit that was as important as the air division a simple pool of forces or a general reserve comparable to the general artillery reserve where army commanders would have been able to draw resources earmarked for the support of their own operation. Such was not the vision of the creator of the air division, for whom this great unit, in order to preserve the effectiveness resulting from the two principles of mass and concentration, should remain under the control of a commander in chief, or, as a minimum, the commander of an army group. He admitted, however, the exploratory nature of his venture, and remained open to suggestions for improvement. Thus the notion of a separate air force, represented by the air division, coexisted with maintaining isolated combat groups at the armies' disposal and sent into action by their air chiefs.

This general disposition of air forces was undoubtedly adapted to the circumstances of the end of the war: a predominance of separate units, an offensive situation, the carefully considered choice of a concept favoring direct action at the front and excluding long-term strategic action against the enemy's economy and morale. From General Duval's experience, it should not have been considered unalterable.

Bibliographic Note

José Barès, Jean Castex, and Louis Laspalles. *Général Barès, créateur et inspirateur de l'Aviation.* Paris: Nouvelles Editions Latines, 1994.

Claude Breguet and Emmanuel Breguet. *La reconnaissance aérienne et la bataille de la Marne (30 August–3 September, 1914).* Paris: RHA no. 1, 1987.

Charles Christienne, Patrick Facon, Marcellin Hodeir, and Pierre Lissarrague. *Histoire de l'aviation militaire française.* Paris: Lavauzelle, 1988.

Christian Delporte, *Les pertes humaines dans l'aviation française (1914–1918).* Paris: RHA no. 3, 1988.

Jean-Pierre Dournel. *L'image de l'aviateur française (1914–1918), une étude du milieu des aviateurs d'après la revue La Guerre Aérienne.* Paris: SHAA, RHA, 1974–75.

Général Duval. "Pourquoi fut créée la Division aérienne." *Revue de l'armee de l'Air* (1935).

Capitaine Pierre Eienne. "Réserve générale et Division aérienne." *Revue de l'armee de l'Air* (1935).

Patrick Facon. *Arme ou armée, aviation réservée ou aviation organique? L' aéronautique militaire à l'épreuve de la Premiere Guerre mondiale.* Paris: RHA no. 4, 1994.

———. *Le comité inaterallié de l'aviation, ou le problème du bombardement stratégique de l'Allemagne en 1918.* Paris: RHA no. 3, 1990.

———. *Un exemple de l'adaptiation de l'arme aérienne aux conflicts contemporains: la division aérienne.* Paris: SHAA, RAE, 1984–85.

———. *Aperçus sur la doctrine d'emploi de l'aéronautique militaire française (1914–1918).* Paris: SHAA, RAE, 1984–85.

———. "L'aviation française au Chemin des Dames." *Revue Aviation Magazine* (1987).

———. "La bataille aérienne de Verdun." *Revue Aviation Magazine* (1986).

Marcel Jeanjean. *La vie en escadrille pendant la Guerre de 14–18.* Paris: RHA no. Hors série, 1969.

Alain Morizon. *L'aviation française en 1916.* Paris: RHA no. 2, 1965.

———. *L'aviation française en 1917.* Paris: RHA no. 4, 1967.

François Pernot. *1914–1917, ou l'aviation militaire à l'épreuve de la Grande Guerre.* Paris: RHA no. 3, 1993.

Simone Pesquiès-Courbier. *L'aéronautique militaire française 1914–1918*. 2 vols. Paris: Revue Icare, 1978.

———. *La politique de bombardement des usines sidérurgiques en Lorraine et au Luxembourg pendant la Première Guerre mondiale*. Paris: RHA no. 4, 1984.

———. *1914: le général Joffre et l'aviation*. Paris: RHA no. 1, 1984.

Général Lucien Robineau. *Ader, prophète en son pays?* Paris: SHAA, 1990.

———. "Guynemer: Et provoquera les plus nobles èmulations." *Revue Icare* (1987).

———. *Histoire des ècoles de pilotage (1912–1962)*. Rochfort: VIème Congrès du Commandement des Écoles de l'armée de l'Air, 1987.

———. *Lazare Carnot et les compaignes d'aérostiers de la République*. Paris: RHA no. 2, 1989.

———. *De 1919 à 1939, la pensée militaire française à l'école de la Grande Guerre*. Istanbul: Colloque de la Commission internationale d'Histoire militaire, 1993.

———. *La Première Guerre mondiale dans les airs*. Boulogne-Billancourt: Université interâges, 1988.

———. "Le colonel Charles Renard (1847–1905)." *Revue Air Actualités* (1987).

Général Voison. *La doctrine de l'aviation française de combat au cours de la guerre (1915–1918)*. Paris: Berger-Levrault, 1932.

The American Dimension

Eugene M. Emme

The Continental powers and England gained considerable wartime experience in the exercise of air power before the United States belatedly entered World War I and then had to create an air force "virtually from whole cloth." The United States had only just acquired some measures of world influence at the turn of the century. It had inherited new responsibilities in the Philippines and Cuba from Spain, and soon completed the Panama Canal. As historian Alfred Thayer Mahan had argued, and the Naval War College understood, the "New Navy" was still the "first line of defense." The "dreadnaught," or battleship, was now the capital ship. The submarine was just coming out of its experimental stage. Coastal artillery still remained in place as the key to a second line of defense. The army had begun a major reorganization featuring a general staff and a war college which studied the classic principles of war derived from the great battles of the past. The validity of that experience and, ultimately, all American defense arrangements would be called into question by the arrival of the airplane and the powered balloon called the airship.[1]

The Beginning: 1903–1917

Military aviation received its decisive impetus in December 1903 from Orville and Wilbur Wright. In world military history, balloons such as the American vehicle flown in action at San Juan, Cuba, in 1898, had made a mark, but never as lasting as that to be made by the successor vehicles of the Wright invention.[2]

While perfecting their flying machine at Dayton, Ohio, in 1905, the Wrights twice offered their airplane, or exclusive use of their pending patents, to the U.S. Army. They explained that they considered their Flyer practical for scouting and communications, but that any potential commercial use must await further development beyond their present resources. The Army's Board of Ordnance and Fortifications brusquely turned down each offer of the Wrights because it still smarted from the bad press over the failure of the aerodrome built by Dr. Samuel P. Langley with an unpublicized federal grant of $50,000. Rebuffed by their government, the Wrights stopped flying, fearful that further exposure would invite the theft of their hard-earned innovations.[3]

A New York congressman and some Ohioans stirred the interest of Theodore Roosevelt and his administration in both the Wright Flyer and its implications. A miniscule Aeronautical Division of the Signal Corps, organized on 1 August 1907, became the government's instrument for staying in touch with aeronautical advances.

The Signal Corps soon was advertising for "a practical means of dirigible aerial navigation" and set up a balloon facility at Fort Omaha, Nebraska. By December 1907, the Signal Corps was seeking bids for one flying machine. On 8 February 1908, the Wrights agreed to deliver one Flyer within 200 days.

In July 1908, Lt. Benjamin Foulois, who always had one eye cocked to the future, submitted his thesis to the Signal Corps School at Fort Leavenworth on the "tactical and strategical value of balloons and aerodynamic flying machines." In a future war, Foulois wrote, an air battle would influence "the strategic movement of hostile forces before they have actually gained contact."[4] As Foulois explained later, he had mainly elaborated upon the doctrine in the Infantry Manual by inserting aviation whenever tactical employment of the cavalry or artillery had been called for. The first fruit of the thesis was his assignment to aeronautical duty.[5]

In August, Wilbur Wright began his spectacular flights in France, from which he would go on to make over a hundred more in Europe, all demonstrating the superior flight control of the Flyer over any European aircraft. But the first demonstration at Fort Myer, Virginia, also in August, was that of the impressive nonrigid airship of Thomas Scott Baldwin of California. His flight engineer was Glenn Curtiss, a builder of motorcycles and of the airship's 20 hp engine. The Army bought the airship, designated it "Signal Corps I," and later shipped it to Fort Omaha, where it was used to check out a few airship pilots and to provide demonstrations at Fort Leavenworth and at air shows and state fairs. In 1912 it was sold for scrap.[6] Also in 1908, the German Army accepted its first large rigid Zeppelin, and Kaiser Wilhelm was to declare Count von Zeppelin the "greatest German of the century."[7] Later, in Nazi days, a German history of flight was to publish a picture of the Wright brothers' "German grandfather."[8]

On 9 September 1908 Orville Wright made two spectacular flights at Fort Myer, one of fifty-seven minutes, the other of over an hour, to be followed by other demonstrations before Washington officials and thousands of onlookers. The most qualified army officer was Lt. Thomas Selfridge, who had worked with Alexander Graham Bell and associates and had already flown an aircraft. When he flew with Orville, the Flyer crashed from a height of seventy-five feet, killing Selfridge. Orville Wright was seriously injured, thus postponing completion of the army acceptance flights until 1909. It was little wonder, given the fragile state of the flying art and its barely organized sponsorship in the Signal Corps, that Congress rejected a budget request of the secretary of war in 1908 for $500,000 for army aeronautics.

Already increasingly evident in the public arena were the nonmilitary speculations about the potentialities of military power to be served by an air weapon not tethered to surface forces. In 1908, H. G. Wells published *The War in the Air,* which, even before Louis Blériot's hop across the English Channel, depicted "The Battle of the North Atlantic" and "How War Came to New York" in Italian-style airships. In Wells's account, the United States was attacked because:

> It was known that America possessed a flying machine of considerable practical value, developed out of the Wright model; but it was not supposed that the Wash-

ington War Office [sic] had made any wholesale attempts to create an aerial navy. It was necessary to strike before they could do so.[9]

Orville Wright completed the trials of the Flyer at Fort Myer in July 1909 in the presence of President Howard Taft and the secretaries of the War and Navy Departments. The rebuilt Wright Flyer exceeded all Signal Corps specifications, remaining aloft seventy-two minutes and averaging 42 mph with a passenger. The U.S. Army soon had the first and only military airplane in the world, a short-lived and singular technological lead never again enjoyed by the United States until the appearance of the B-17 and the later atomic bomb. As agreed in the contract with the army, Wilbur Wright proceeded to check out Lts. Frank Lahm and Frederic Humphreys at College Park, Maryland. Although both officers were then transferred back to their respective nonflying line organizations, others, including Henry H. Arnold, were later given instruction at locations such as Dayton, Ohio; College Park, Maryland; and Augusta, Georgia.

After the International Air Meet in Rheims, France, in 1909, the flying machines of dedicated mechanics and sportsmen generated a flying boom around the world. In September of that year, Orville Wright flew a record altitude flight of 1,600 feet above Berlin, Germany, while, on the same day, Wilbur flew around the Statue of Liberty in New York. Soon many airplanes were built in the United States, and, even if they did not fly very well, they replaced balloons at county fairs.[10]

The first congressional appropriation of $125,000 in 1911 for army aviation ended the Wright Flyer era. Five new aircraft were ordered, and a permanent flying school was soon established at North Island, San Diego. A few experiments, beyond those involving only higher and further flights, had long-term significance but were not immediately pursued by the army. The low-recoil machine gun developed by Col. Isaac Lewis and fired from an aircraft by Capt. Charles DeF. Chandler, chief of the Aeronautical Division, was to become a standard air weapon in Europe. A bomb sight was tested, bombs dropped, and airborne photography and radio tests made. Twelve of the first forty-eight officers assigned to army aviation were killed in accidents. Pusher aircraft of the Wrights and Curtiss were dropped in favor of tractor aircraft (with the propeller in front of rather than behind the crew) such as the Curtiss JN-1 or "Jenny." This tractor evolved after 1914 into the basic trainer used throughout the war to come, and later was used as a bomber by marine aviators and as a plaything by hundreds of civilians.

On 18 July 1914, Congress authorized the Aviation Section of the Signal Corps, with a strength of 60 officers and 260 enlisted men. After the outbreak of war in Europe in August, the First Aero Squadron was created on 1 September 1914, under Captain Foulois. In his "History of Rockwell Field," Maj. Henry H. Arnold later wrote about the First Aero Squadron:

> This was the first operating unit of any kind ever organized. . . . The question now arose for the first time as to whether a flying officer of limited administrative experience or non-flying officers of considerable administrative experience in the Army should not be placed in command of such a squadron. It is also to be noted,

however, that this question had not been satisfactorily solved even several years later [1916].[11]

The First Aero Squadron was always provisional until 1917, in the sense that it did not have a full complement of planes, even when three other squadrons were organized on paper. Also apparent was the need for greater understanding of the potential of combat aircraft, beyond the obvious reconnaissance mission. Lieutenant Thomas D. Milling summarized very well the relationship between doctrine and equipment at that time when he observed, "Our doctrine has been consistent since 1913 within the limits of our equipment."[12]

On the day before the founding of the First Aero Squadron, a British Royal Flying Corps reconnaissance plane spotted the armies of German Gen. Alexander von Kluck's "inward wheel" heading southeast to Paris. This intelligence, soon confirmed, led to a series of battles called the "Miracle of the Marne."[13] The halt of the German offensive and the eventual "race to the sea" created the trenches of the Western Front. Within a few months, the Army Signal Corps issued its first specification for a reconnaissance two-place biplane with a speed of 70 mph. Twelve bids were received, but the lack of a reliable engine thwarted procurement of the desired airplane.[14]

At first, few Americans, including military leaders in Washington, seriously believed that the war in Europe would involve the United States. With the sinking of the *Lusitania* in May 1915, unrestricted German U-boat operations greatly increased military concern despite President Woodrow Wilson's strict neutrality posture.[15] Hindering American understanding of the war, especially its evolving air operations, was the inability of neutral observers to penetrate the cloak of secrecy laid down by both sides.

The U.S. Navy knew about Dr. Langley and the Wright brothers, Navy Lt. George Reed having almost flown in the tests at Fort Myer. Later he flew with Army Lt. Frank Lahm at College Park. To handle the queries about aviation being directed at the navy secretary's office, Capt. W. I. Chambers was made its air coordinator. Along with Glenn Curtiss, the developer of the first practical seaplane, Chambers brought the airplane into the navy, which had expressed no interest in aviation until convinced it could help the fleet.

Chambers arranged the first ship-to-shore flight by Eugene Ely, a pilot employed by Curtiss in one of his firm's pushers, from a plank platform on the cruiser USS *Birmingham* at Hampton Roads on 10 November 1910. Two weeks later, Curtiss informed the secretary of the navy that he would provide free flight training for an officer at his winter camp on North Island, San Diego. Lieutenant T. G. Ellyson was detailed and became the first naval aviator. In early January 1911, Ely landed on a platform on the USS *Pennsylvania* at anchor in San Francisco Bay and soon took off again. Later in the month, Curtiss made the first successful hydro-airplane flight with his "Silver Fish" off North Island, Ellyson assisting in the preparations. In February, Curtiss taxied his seaplane out to the *Pennsylvania,* was hoisted aboard and then returned to the water to taxi back to North Island. It was a persuasive demonstration, and in March, Congress appropriated $25,000 for naval aviation. The Wright Company now offered to train one navy pilot, contingent upon the purchase of one airplane for $5,000. Lieutenant John Rodgers was sent to Dayton, to become Naval Aviator No. 2. By 8 May 1911, the U.S. Navy had purchased three airplanes: the Curtiss "Triad," a "hydra-terraairplane" to whose float Curtiss had added wheels for both land and water landings, a Curtiss pusher, and a Wright Flyer.[16]

Captain Chambers was directed to set up an experimental station at Annapolis, where Lt. John Towers and others were training. There and on exercises, experiments went on in the application of the airplane to navy needs. Off Cuba, Towers confirmed that submerged submarines could be seen from the air; other experiments went on with radio telegraphy, photography, and water-based operations to include the testing of catapults. At the Washington Navy Yard, Naval Constructor Holden C. Richardson worked on hull designs for seaplanes, a wind tunnel for aircraft design, and flight testing. From 1912 on, year-round flight training and operations were located at the first Naval Aeronautics Station, Pensacola, Florida.

In 1912, Marine Lt. A. A. Cunningham began flight training at the Burgess and Curtiss factory at Marblehead, Massachusetts, and became Naval Aviator No. 5. Later called the "Father of Marine Corps Aviation," Cunningham and his associates soon were engaged in exercises in Cuba with an Advance Base Brigade.[17]

One of the earliest steps taken by the United States after the "guns of August" began firing in Europe, was to create in March 1915 the first federal agency responsible for coordinating and stimulating aeronautical research. A rider to the Naval Appropriations Act of 1915 created the National Advisory Committee for Aeronautics (NACA) which could never live down its navy birthright in the eyes of some army airmen. Modelled after a British body founded in 1910, the NACA was to examine and make recommendations "on the problems of flight, with a view to their practical solution," a general and unwarlike charter in keeping with the neutral position of the United States. The twelve-men membership of NACA included the chiefs of the Army Signal Corps and its Aviation Section, the director of Naval Aviation and its Constructor, the chiefs of the Weather Bureau and the Bureau of Standards, and professors interested in aerodynamics as it grew out of fluid mechanics.[18]

At the first NACA meeting in the office of the secretary of war, chaired by Brig. Gen. George Scriven, chief signal officer, on 23 April 1915, the membership considered his previously submitted position that the problem "most requiring attention involved military aviation and national defense." "Nothing," he said, "will so readily bring order from chaos as the carefully considered decisions [sic] of this Advisory Committee."[19] But the NACA was to become concerned with the technical problems of civilian as well as military aviation. Its recommendations bound none of its members, and, in unmilitary fashion, it elected its own chairman.

A month after NACA's first meeting, the first German Zeppelin attacks on London highlighted a capability unavailable in the United States. The committee, for its part, modestly surveyed research capabilities nationwide, gathered what basic knowledge it could in Europe, and began to issue its widely used bibliographies. In 1916, the NACA undertook some policy initiatives by inviting aircraft engine manufacturers to discuss the problems of attaining more powerful and more reliable aircraft engines, by recommending a government air mail service, and by seeking the creation of a laboratory at an army-navy aircraft proving field, which became Langley Field at Hampton, Virginia, in 1917.[20] The dozen or so employees of NACA at Langley, however, did no research in its wind tunnels until after the war.

Another scientific initiative, this time by the National Academy of Sciences in 1916, prompted President Wilson to establish a National Research Council (NRC) to engage scientists on defense problems, particularly submarine detection. Once the United States entered the war, some scientists put on uniforms. Among them was Maj. Robert Millikan of the California Institute of Technology, the head of the Signal Corps Science Research Division.[21]

In November 1915, Maj. William Mitchell, then a general staff officer, apparently prepared a survey of national defense needs in aviation. He claimed that aviation would be particularly useful as "a second line of defense," by acting as a backstop to the navy when attached to harbor and coastal defenses, by carrying on reconnaissance and spotting for artillery, and by destroying attacking aircraft and submarines. Army aviation should be increased, said Mitchell, to 46 officers, 243 enlisted men, and 23 aircraft of various capabilities. By 1916, and at his own expense, Mitchell began taking flying lessons during his off-duty time.[22]

In early 1916, congressional support for aerial rearmament greatly accelerated when the First Aero Squadron quickly wore itself out in supporting Gen. John J. Pershing's Punitive Expedition against Pancho Villa. An emergency appropriation for the Aviation Section of $500,000 was followed by the enormous sum of $13 million, a figure nine times the total of all funds which had been received by army aviation to date. (Incidentally, Captain Foulois must have had a typing pool larger than the Aviation Section, since so many original and carbons of his report on the demise of the First Aero Squadron are scattered through the files in the U.S. National Archives.)

In April 1917, when the United States entered the war in Europe, its army aviation had 131 officers (mostly pilots), 1,087 enlisted men, but no aircraft capable of combat. Naval aviation had forty-five float seaplanes, six flying boats, three land planes, and one blimp, none ready to operate with the fleet. Almost ten years had passed since the army accepted its Wright Flyer, but American air power was almost nonexistent, with a handicraft industry, no organized planning or research and development, and very little knowledge of aviation progress in Europe.

World War I: 1917–1918

American air power in the Great War was scarcely born when it was demobilized. A nightmare for that air power ensued from the utterly rash promises by industrial, military, and political leaders in 1917 that thousands of American planes would gain perpetual air superiority, darken the skies over Berlin, and end the war. The first American-built but British-designed DH-4s reached France in May 1918, unready for operations and often damaged in transit. The American model had a reputation as a "flaming coffin" until the gas tank between the pilot and observer was repositioned after the war. To the hundreds of airmen arriving for flight training in France, it appeared that their presence was designed more to raise that country's morale rather than to get on with the air war. When promised first-line European aircraft were not delivered, the American airmen who eventually qualified in Allied flight schools had to take whatever aircraft were offered them. Those American airmen who got to the front did a great job with what they had. James Lea Cate's assessment seems sound: "Had the war dragged on into 1919, the boasts might have been made good."[23]

German Field Marshal Paul von Hindenburg and Gen. John J. Pershing credited the Allied victory to the waves of fresh American infantrymen whose assaults cracked the Western Front. But it was also true that those ground forces were protected by air actions that denied superiority to the Germans over the battlefront. In the rear areas, by the fall of 1918, British bomber crews attempted to strike at the center of cities in the Reich. The resulting panic caused the German government to ask for an immediate halt to the bombing raids as part of its armistice proposals.[24]

Perhaps it was indeed inopportune, as Raymond Fredette has observed, that General Pershing just missed witnessing the first bombing raid by German Gotha bombers on London on 13 June 1917.[25] A vivid demonstration of air power's potential might have been most persuasive. The Gotha bomber raids on England, for example, helped to spur the creation of the independent Royal Air Force (RAF) in the midst of the decisive phase of the war on the Western Front.[26]

Professor-General Bill Holley has treated very well the incredible history of the American aircraft production program in his *Ideas and Weapons*. The haste and waste in the program offered lessons that were well learned in time for the World War II buildup. Hampering the World War I American air effort as well were the requirements to mobilize and train tens of thousands of raw recruits after the United States had entered the war, not to mention the problems of organizing and staffing the higher direction of the air effort.[27]

A few highlights from the American experience in the Great War may be suggestive. The splendid biography of Billy Mitchell by Colonel-Professor Al Hurley provides a clear understanding of Mitchell's early air power role. Mitchell got himself to Spain and then to Paris four days after the American declaration of war. Fluent in French, he wangled his way to that nation's share of the front, absorbing briefings on air employment and taking lessons in flying its latest aircraft. Mitchell seemed more influenced by his three-day visit early in May with British airmen, principally Maj. Gen. Hugh Trenchard.[28] Trenchard impressed on Mitchell the concepts of "forward action" and the "relentless offensive."[29] For various reasons, Mitchell's reports to Washington about all this had little impact.

Mitchell also had contributed to the preparation of French Premier Alexandre Ribot's request for American resources that became the bottom line for the take-off of the ambitious U.S. aircraft construction program. Ribot asked for an American "flying corps" of 4,500 planes, 5,000 pilots, and 50,000 mechanics to be sent to France in 1918. American acceptance of this goal led to sending the Bolling Mission to Europe to determine what kinds of airplanes should be built. Its prompt recommendations included ideas on the strategic bombing of enemy industries. Top priority was given to the production in the United States of the British DH-4 reconnaissance bomber and the American all-purpose Liberty engine, with the second priority, pursuit planes, to be purchased in France and England. In the meantime, Pershing made Lieutenant Colonel Mitchell the Aviation Officer of the American Expeditionary Forces (AEF). Soon, however, Mitchell was subordinated to Brig. Gen. Benjamin Foulois. Eventually, the leadership and talent Mitchell showed as chief of the Army Air Service, First Brigade, on the American sector of the front won him fame. General Trenchard's early impression of Mitchell was noteworthy: "If only he [Mitchell] can break the habit of trying to convert opponents by killing them, he'll go far."[30]

Those few American squadrons which reached France by late 1917 served with French and British units after they had been organized and trained. General Pershing refused to flesh out depleted and tired Allied air units with Americans. After April 1918, a few American squadrons began to operate in support of their own forces. While news from the trenches was drab and bloody, the individualism of air combat made heroes of Eddie Rickenbacker, Frank Luke, and others. Contrary to Hollywood's later dramatization, however, aerial combat involved a lot more than glamorous dawn patrols and was fully subject to the vagaries of the weather and the fragility of the flying machines themselves.

Pershing made Brig. Gen. Mason Patrick, a West Pointer, his chief of Air Service, AEF, in May 1918. Eventually Patrick assigned Mitchell the leadership of all American air units with the First Army. The struggle for the St. Mihiel salient offered the best example of air power's potential on a battlefield. Mitchell's plan to gain air superiority required 1,500 planes, only 609 of them piloted by Americans, the rest being drawn from Trenchard's Independent Force, along with a French air division, and other Allied squadrons. Only a third of the force directly supported the First Army; the rest, in two brigades, struck at the flanks of the salient and at the German Air Force facilities in the rear of the salient. Pershing praised the action's success, and all airmen saw it as a model for the effective concentration of air forces.

In the remaining Allied offensives, Mitchell usually had only American squadrons at his disposal and used them mainly in close support and counterair roles. German air opposition persisted to the end. Meanwhile, Trenchard's Independent Force bombed German targets in an effort that gained momentum from September onward. The Armistice aborted planning for a much larger bombing campaign by the Inter-Allied Independent Air Force under Trenchard, who would have been responsible to Marshal Ferdinand Foch, the supreme commander.

How is one to evaluate the limited American effort? Statistics are one measure. The U.S. Army Air Service in France constituted 10 percent of the Allied air forces, dropped 139 tons of bombs, and reached as far as 160 miles behind the German lines.

Some 237 American airmen were killed in battle; no figures for greater operational, training, and other losses are available. There were 58,000 army airmen in France, 20,000 in training in England, and some in Italy. A total of 10,000 army aviators completed flight training, but one must also note that 27,000 officers and men of the Air Service had been assigned to obtaining the spruce used in the manufacture of aircraft.

Over 3,000 DH-4s and 7,800 training planes had been produced, a total of some 11,000 aircraft against the 27,000 planned. Of the 1,005 aircraft in American air units at the front, only 325 were American made. There was no lack of doctrine, leadership, or courage for the employment of American air power in France, only the absence of the equipment and the manpower at the right time and place.[31]

For its part, U.S. Navy aviation concentrated on the development of the HS series of flying boats. The Royal Naval Air Service had used some of those flying boats, two-engine long-range Curtiss "Large America" flying boats, to score a unique success by shooting down, at sea, two German naval dirigibles in May and June 1917. American naval aviation operated out of twenty-seven bases in Ireland, England, France, and Italy. On anti–U-boat patrol, the navy reported attacks on twenty-five U-boats, sinking or damaging a dozen.[32] Operating with the Northern Bombing Group in France, the mission of the naval airmen was expanded to bomb German submarine and dirigible installations with DH-4s. Round-the-clock bombing was being discussed when the Armistice intervened. In the Italian theater of war, navy, marine, and some army pilots flew Caproni bombers in Allied air units against Austrian targets.

The U.S. Navy's air force had grown to a total of 6,716 officers and 30,693 men in Navy units, and 282 officers and 30,000 men in Marine Corps units. Of these, 18,000 had been sent abroad.[33]

Despite the employment of air power and its rapid development during the war, for many observers, air power had yet to prove itself in warfare as a military and naval instrument. As America demobilized her military forces after the Armistice, the contrast between reality and vision would set the tone for its military aviation during the next twenty years.

Nascent Air Power: 1919–1937

With the conclusion of "the war to end wars," "the long armistice" began.[34] From its position of isolation, the United States tried to secure peace through the Washington Naval Treaties of 1921–22 and the Kellogg-Briand Peace Treaty six years later. In preparation for a presidential election during a deepening depression in 1932, the Hoover administration, supported by Army Chief of Staff Douglas MacArthur, considered a ban on all submarines and aircraft carriers for submittal to the World Disarmament Conference in Geneva.[35] In an increasingly nationalistic world, fantasies about disarmament abounded while Congress investigated "merchants of death" and the impact of the airplane on national defense. Congress tried to perpetuate peace by passing the Neutrality Acts prohibiting the sale of armaments to any belligerent while Adolf Hitler tore up the Treaty of Versailles by announcing the existence of the Luftwaffe and universal military service in Nazi Germany.

In the United States, the postwar demobilization was chaotic. American airmen returned from France to help answer congressional inquiries about the failure of the billion-dollar aircraft construction program. Of the 200,000 men in the wartime Army Air Service, only 10,000 officers and enlisted men remained on duty by June 1920. A year later, the aircraft inventory was 1,100 DH-4s, 1,500 Jenny trainers, 179 SE-5 pursuits, and 12 Martin MB-2 bombers. There were fewer than 900 active army pilots and observers. Sixty-nine of these were killed in 330 flying accidents in 1921 alone. Ninety percent of the aircraft industry was bankrupt. The Army Reorganization Act of 1920 was a crushing blow to Army Air Service expectations. Although authorized strength was set at 1,516 officers, 2,500 flying cadets, and 16,000 enlisted men out of a total postwar army of 280,000, there was no money to recruit to these levels or to purchase many new airplanes. The airmen would have to make do with Liberty engines until late in the 1920s.[36]

Most frustrating to army airmen who were usually junior in rank and rarely West Pointers, was the prevailing dim view of the future of military aviation held by those who managed the purse strings and the promotion lists. Frustration soon turned into a struggle not only with the general staff and the secretary of war, but also at the summit on Capitol Hill, where the fate of the postwar services was being deliberated. The fundamental issues then, as now, inevitably involved the White House.

The first postwar congressional dialogue on aviation centered on whether all federal aviation activities should be centralized in a cabinet-level Department of Aeronautics. Foulois and Mitchell at least agreed on this possible step. But it proved impossible to achieve unified command of "air power" whether it operated over land or over the sea, despite the precedent in the creation of the RAF in England. Billy Mitchell soon directly challenged the navy's long-standing claim to be "the first line of defense," by asserting that his bombers could sink any battleship. The celebrated and highly publicized sinking of the unsinkable German battleship *Ostfriesland* seemed to justify Mitchell's claim, but his oral "bombs" led President Warren Harding to note that Mitchell gave the admirals "apoplexy"; and later President Calvin Coolidge was provoked into calling him a "God-damned disturbing liar."[37]

After 1923, a single department of defense with an independent air force became the central issue in the American air power story. Mitchell, after cooling-off trips to Europe in 1922 and to the Far East in the first half of 1924, launched even more inflammatory attacks upon the navy's admirals and the army's general staff. Having succeeded in alienating every responsible person with authority to help him, he was transferred into "exile" at San Antonio. The crash of the navy dirigible, the *Shenandoah,* gave Mitchell the occasion he wanted to assure his court martial. He accused the Navy and War Departments of "incompetency, criminal negligence, and almost treasonable administration of aviation." To undercut the airman's charges, President Coolidge created the Morrow Board, which met, heard all of the familiar witnesses, and reported out before Mitchell's trial. The framers of the Army Air Corps Act of 1926 would attempt to remove some complaints on flight pay and promotions and gave the Air Corps a spokesman by authorizing an assistant secretary of war for air. In the meantime, Mitchell resigned and continued to express his views.[38]

Billy Mitchell's legacy was permanently ingrained in the Army Air Corps. Not the least of his marks was made by the corpus of his many papers on the role of airplanes in national defense. Defining "air power" as "the ability to do something in the air," Mitchell's central idea was that air forces rendered armies and navies obsolete because they could achieve a decision in war by directly attacking "vital centers" of an enemy nation. After he resigned, he continued to spread the gospel, wherever possible, that "the airplane is the arbiter of our nation's destiny."[39] Colonel Hurley's biography and Dr. Frank Futrell's monumental work on the history of air force thought permit me little opportunity to say more.[40]

The beliefs of Mitchell, however, made it absolutely unnecessary for him to quote Hugh Trenchard or, if he knew them, the theories of Giulio Douhet, later collected in *The Command of the Air*. Mitchell met and talked with Douhet, although where and when he did remains not fully clear. It would be interesting to know if there is more on the Mitchell-Douhet connection. We have more evidence about the views of Douhet's associate, Count Gianni Caproni, which were communicated to Americans in 1917.[41] At any rate, General "Hap" Arnold's later judgment on Billy Mitchell seems fair enough: despite his political failings, no one should ever forget that Mitchell was ahead of his time in his ideas on the employment of air power.[42]

Demobilization proved equally as disruptive to naval aviation. During the war, that aviation had been loosely organized and was mothered by various bureaus of the navy. Most of its pilots were reservists, and few remained on active duty after demobilization. The navy had only 319 active naval aviators in June 1920, with 3,296 inactive, including reservists.[43]

Chief of Naval Operations Adm. William Benson vehemently opposed the proposal in 1919 to combine army and navy aviation. Billy Mitchell lashed out directly at Benson, decrying his shabby outlook on the "ugly duckling" of the navy. More air-minded admirals on the navy's General Board advised the secretary of the navy in June 1919 that "a naval air service must be established, capable of accompanying and operating with the fleet in all waters of the globe."[44] The president of the Naval War College, Adm. William S. Sims, gave Congressional friends studies which argued that a superior fleet of aircraft carriers, similar to those developed by the Royal Navy, "would sweep the enemy fleet clean of its airplanes, and proceed to bomb the battleships, and torpedo them with torpedo planes. It is all a question of whether the airplane carrier, equipped with eighty planes, is not the capital ship of the future."[45]

Naval aviation got another boost when the NACA recommended that the War, Navy, Post Office, and Commerce Departments have separate bureaus of aeronautics, to be coordinated by a top-level board of civilians. This NACA proposal, which smacked of retaining for NACA a postwar policy role in all aviation, was rejected by the Joint Army and Navy Board. The secretaries of war and navy successfully refused any connections with the NACA by creating an Aeronautical Board to consider policy questions regarding the roles and missions of aviation in both services.[46] In February 1920, Admiral Benson agreed to give bureau status to naval aviation. The new status was not to be public knowledge, however, until 10 August 1921, after the sinking of *Ostfriesland* by Mitchell's bombers in July.[47]

Admiral William Moffett, chief of the Navy Bureau of Aeronautics until he was killed in the crash of the dirigible *Akron* in 1933, was a different personality from Mitchell. Moffett, an academy graduate, worked within the system to put aviation into the corpus of the navy. With the help of airmen executives who were all Annapolis products after 1922, the navy learned to operate aircraft carriers, which evolved from surface auxiliaries into capital ships in a task force. With the commissioning of the first make-shift aircraft carrier, *Langley*, the Bureau of Aeronautics (BuAer) was underway. Moffett, who now bore the brunt of Billy Mitchell's attacks on the navy, came to regard him, he said, as a man "of unsound mind and suffering delusions of grandeur." One wonders what Moffett and Maj. Gen. Mason Patrick, again chief of the Army Air Service, said to one another about Mitchell at meetings of the NACA.[48]

The navy steadily advanced its sea-air capabilities after an NC-4, one of the three flying boats built during the war, completed in May 1919 the first trans-Atlantic flight from Newfoundland to Plymouth, England, by way of the Azores and Lisbon.[49] The Naval Appropriation Act for Fiscal Year (FY) 1920 had already funded conversion of a collier into the *Langley*, which used aircraft landing hooks and deck cables developed by the British and catapult launchings. Also authorized in 1920 was the procurement of two merchant ships as seaplane tenders, the construction of one rigid dirigible (later the *Shenandoah*), and the purchase abroad of another (the ill-fated British R-38). A third dirigible, the *Los Angeles*, was acquired from Germany as part of the reparations settlement. By 1923, flights from the *Langley* had begun, and the fleet exercises in Panama used patrol squadrons. The next year, while he was flying mail to the Canal Zone, Army Lt. Odas Moon, bombed the *Langley* with ripe tomatoes and delayed a fleet exercise for a day.[50]

BuAer's greatest achievement in the 1920s was the development of the aircraft carrier as a part of the navy's capital ship construction program. Two battle cruiser hulls, permitted under the Washington Naval Treaties of 1921–22, became the aircraft carriers *Saratoga* and *Lexington*, commissioned in 1928. The completely new carrier *Ranger* appeared in 1934, followed by *Enterprise* and *Yorktown* in 1936, and the promise of another carrier, *Wasp*.[51]

All major navy ships had catapult scout planes, and flying boats were not neglected. The "flying aircraft carriers," the *Akron* and *Macon*, however, were expensive disasters. C. G. Grey, editor of the English journal, *The Aeroplane*, once quipped: "The airships breed like elephants and aeroplanes like rabbits."[52] The navy could afford no more of the airships, which most Army airmen always considered unworthy combat vehicles.

Since the Washington Naval Treaties forbade the United States to build a major naval installation in the Philippines, Pearl Harbor became the major port for the Pacific Sea Frontier. Fleet exercises after 1931 included the *Saratoga* and *Lexington*, although the cost of fuel was a major constraint. In the 1932 exercises, aircraft from the "Sara" and the "Lex" successfully "bombed" Pearl Harbor. Supplemented by the three new carriers, the carrier force was generally divided by 1936 between the Atlantic and the Pacific Sea Frontiers. Only the *Langley*, converted to a seaplane tender, was ever on station in Asiatic waters.

After the Japanese Army, supported by carrier aircraft, invaded China in 1937, the U.S. Navy's efforts to equip its sea-based air power with up-to-date aircraft, to lay keels for more aircraft carriers, and to train manpower became most urgent. In the fleet exercises of 1939, naval aviation was deemed to be "fast reaching a high state of readiness." Still, in 1940, the Japanese carrier force had grown to ten.[53]

When the MacArthur-Pratt agreement of 1931 affirmed that the Army Air Corps was to be responsible for the land-based air defense of the United States and its possessions, Marine Corps aviators were required to become qualified on aircraft carriers. By 1934, sixty of the hundred-odd regular Marine Corps aviators had served on aircraft carriers and had gained experience on more up-to-date aircraft. In 1933, however, the creation of the Fleet Marine Force to seize shore bases for naval operations basically altered the mission of Marine Corps aviation to one of close air support for amphibious operations. By 1939, the number of active Marine Corps aviators had grown to 245, plus reservists, but those numbers would soon double again and again.[54]

Meanwhile, the Army Air Corps got its first veteran flyer and nonacademy graduate as its chief in the person of Maj. Gen. James E. Fechet in 1927. In the post-Mitchell period, every army airman was still a rebel, but he maintained a low profile. One of Fechet's aides, later Gen. Ira Eaker, has said that Fechet approved "more special projects to keep the air effort in the headlines than any of his predecessors." The Pan-American Goodwill Flight and the in-flight refueling endurance flight of the *Question Mark* occurred during Fechet's regime.[55]

Fechet's successor, Benjamin Foulois, sounded a theme that rapidly would become more than a theory, telling an Army War College class that air power was "the strength of a nation in its ability to strike offensively in the air. . . . The real effective air defense will consist of our ability to attack and destroy the hostile aviation on the ground before it takes to the air."[56] Four years before, when defending fighters had failed to intercept a bomber attack during maneuvers in Ohio, Maj. Walter Frank concluded that "a well planned air force attack is going to be successful most of the time." In the classrooms at the Air Corps Tactical School, Lt. Kenneth Walker was credited by his students with originating the theorem that: "A well organized, well planned, and well flown attack will constitute an offensive that cannot be stopped."[57] A similar emphasis on the power of the air offensive prevailed in Europe, and Douhet was not its sole author. The most famous acknowledgment of the power of the air offensive came from Prime Minister Stanley Baldwin of England who, enroute to the World Disarmament Conference in Geneva in 1932, stated: "The bomber will always get through."[58]

Without a doubt, President Franklin D. Roosevelt's assignment to the Army Air Corps of the task of flying the air mail proved a turning point in 1934. General Foulois took up the assignment with a "can do" attitude. Beginning in the depths of a severe winter across the nation in late February, nine army airmen were killed within three weeks while flying the mail. The press deemed the Army Air Corps to be incompetent or ill-equipped. Roosevelt's Postmaster General eventually renegotiated air mail contracts with the same airlines, which had merely changed their names. But there were at least two important consequences for army aviation. First, a War Department board under Newton Baker reviewed once again the status of aviation in the army. It restated that the Air Corps should remain in the army and recommended that the War Department buy aircraft directly from industry through bid contracts or design competitions. Secondly, the White House set up a Federal Aviation Commission under Clark Howell, a publisher, to consider once more the idea of a separate air force.

Before the Howell Commission reported out, the army chose a solution recommended earlier by two of its own boards and gave the Air Corps a mission not tethered to other army forces by establishing a provisional General Headquarters Air Force (GHQAF). Brigadier General Frank Andrews become head of the GHQAF on 1 March 1935. The GHQAF was headquartered at Langley Field, Virginia, with other wings at Barksdale Field, Louisiana, and March Field, California, to support the "tactical mission" of coastal defense.[59] In the meantime, Lt. Col. Henry H. Arnold had led a flight of ten Martin B-10 bombers to Alaska, returning to Seattle nonstop over water on a 8,290-mile round trip. The B-10, the first prototype of the modern bomber, had closed cockpits, retractable landing gear, and a speed faster than that of contemporary fighters.

Sustaining the continued struggle of the Army Air Corps to develop and procure heavy bomber forces was the dynamism of the revolution in flight technology in the early 1930s. It suffices to stress here the appearance of the Boeing 229, designated the XB-17, which flew nonstop from Seattle to Wright Field in August 1935—2,100 miles at 232 mph with four modest-sized engines in flush wing-mounts. To army airmen from Generals Oscar Westover and Frank Andrews on down, the XB-17 "was a vision of the promised land." Earlier, in May 1934, Air Corps arguments with the army general staff had prevailed and had secured the mission of "the destruction by bombs of distant land and naval targets." The Boeing Aircraft Company had then begun "Project A," a more advanced bomber with a range of 5,000 miles, and a speed of 200 mph with a 2000-lb bomb load. The resultant X-15, contracted for in June 1935, flew in 1937. It was underpowered but contributed to the ultimate B-17 and the B-29. In October 1935, the War Department contracted for the XB-19, a forerunner of the wartime B-29 and postwar B-36.[60]

In testimony before the Howell Commission, most of the senior Air Corps representatives had supported the idea of giving the GHQAF a fair trial before seeking a further reorganization. But some of the heady thoughts on the primacy of air power in modern war as taught by the majority of the instructors at the Air Corps Tactical School were freely expressed by Maj. Donald Wilson, Capts. Harold George and Robert Olds, Lts. Kenneth Walker and Laurence Kuter, and others. To George, air power was "the immediate ability of a nation to engage effectively in air warfare."[61] To Walker, "An Air Force is an arm which, without the necessity of defeating the armed forces of the enemy, can strike directly and destroy those industrial and communications facilities, without which no nation can wage modern war."[62]

It is generally conceded that between 1933 and 1937 the army and the navy had not been ungenerous in funding their respective air arms within the fiscal constraints imposed by the state of the national economy and inevitably slim budgets. The Army Air Corps justified its infant heavy bombers in terms of coastal defense rather than by trying to sell a concept of a strategic air offensive against some specific enemy. From a budget of $6 million for FY 1936, the Army Air Corps only received $3.5 million for aircraft procurement for FY 1938, the year the German Luftwaffe rose like a phoenix to dominate the diplomatic balance of power in Europe. A numbers game also may have come into vogue in the selection of aircraft in this period. In 1936, the Air Corps was directed to order more airplanes for the dollar, or more two-engined Douglas B-18s rather than fewer B-17s.[63] The interregnum came to an end for the Army Air Corps in September 1938, however, thanks to ex-Corporal Adolf Hitler. He remembered trench warfare.

Take-Off: 1938–1941

Erosion of the "long armistice" had been underway in Europe ever since Nazi Germany falsely claimed in 1935 that its new Luftwaffe already had "air parity" with the RAF. Nazi Germany's aggrandizement was transparent in 1936 with the reoccupation of the demilitarized Rhineland and the commitment of the Condor Legion to the Spanish Civil War. In the United States, the interregnum persisted despite President Roosevelt's attempt to alert public opinion concerning the stark portents for the future of peace reported by his ambassadors and attaches in London, Paris, Berlin, and Tokyo. In September 1937, Roosevelt tested the public's readiness for a policy change by calling for an active "quarantine of aggressors," a thought that was not well received nationwide. FDR was branded a "war monger" by the isolationists; Congress responded by extending the Neutrality Act. But a series of international crises were to prompt small changes in American military policy, if only to prepare adequate defenses for the continental United States.

In December 1937, Japanese bombers intentionally sank the USS *Panay* and machinegunned its lifeboats in Chinese waters. The United States only protested, and the undeclared war by Japan continued. However, American rearmament began when Roosevelt soon called for augmenting American defenses because of the threats "to world peace and security." U.S. Navy aviation got the first boost with the passage of the Naval Expansion Act in May 1938, which marked a significant step toward a "two-ocean navy" and which provided for a 3,000-plane program to move carrier aircraft beyond the biplane era. From this legislation came the navy's first modern production fighter, the Brewster Buffalo, the precursors of the Grumman F4F and TBF, and the Douglas SBD, the dive bomber which would win the Battle of Midway four years later.[64]

For the Army Air Corps, the thrust to expand its initial force of B-17s continued to fare poorly under a president who had once been assistant secretary of the navy. At Langley Field, the GHQAF under General Andrews was developing heavy bomber operations with thirteen B-17s. In May 1938, three of these bombers departed Langley Field and intercepted the Italian liner *Rex,* 725 miles out of the port of New York. It was a good navigation job by Lt. Curtis LeMay.[65] The next day, pictures of a B-17 at mast-height alongside the *Rex* appeared on the front pages of the *New York Times* and other East Coast newspapers. The navy blew a fuse. The primary mission of the B-17 in national strategy was scrubbed when the word was passed down from on high that Army Air Corps planes were limited to operational flights not to exceed 100 miles from shore. Secretary of War Harry Woodring laid it on the line when he directed that no production B-17s be procured in FY 1940. Deputy Chief of Army Staff Gen. Stanley Embick stated simply: "Our national policy contemplates preparation for defense, not offense . . . Defense of sea areas other than within the Continental Zone, is a function of the Navy." It was little wonder that General Andrews told the National Aeronautical Association convention in St. Louis that the United States was no better than a fifth or sixth rate air power in the world.[66]

The so-called Munich Crisis of September 1938 was the turning point in American air power policy before World War II. England and France appeased Hitler because of their fantastic belief that, in the event of war, the German Air Force was more powerful than the other Western air forces combined, plus that of the Soviet Union.[67] Everyone forgot that air operations from German bases could barely reach England, but the collective action required of both England and France proved impossible. It was a triple tragedy: Hitler's bloodless victory gave him the Czech "Little Maginot Line"; it put him in command of the German Army General Staff which had been prepared to depose him; and, it encouraged more adventures by Hitler, who believed that those "worms of Munich will not fight."

It is difficult to recreate the climate of Munich. Douhet virtually became a household word in France and England. There had not been enough gas masks, air-raid shelters, or hospital beds in Paris or London during the crisis. "Peace in our time" was the scrap of paper which British Prime Minister Neville Chamberlain brought back from the Munich conference, for which the mobilization of the English fleet and the activation of the Maginot Line had been to no avail. There were only four armed Spitfires, and the relative weakness of the Royal Air Force and the French Air Force presented grim alternatives.[68] Hitler was correct. Neither Britain nor France really fought until they were attacked, which was also true of the Soviet Union and the United States. But that is another subject.

Two days before the Munich outcome, President Roosevelt dispatched Harry Hopkins, the director of the Works Progress Administration, on his first secret fact-finding mission to survey the capacity of the American aircraft industry. Hopkins, with the deputy chief of the Army Air Corps, General Arnold, reported that American production was almost 2,600 planes of all types per year.[69] After Munich, Roosevelt confided to Hopkins that he was "sure then that we were going to get into the war and he believed that air power would win it."[70]

On 14 November 1938, Roosevelt outlined the "Magna Carta" of American air power, as Arnold termed it, at a top-level White House conference. The president wanted at least an army air arm of 20,000 planes and an annual production of 24,000 aircraft. Only this would influence Hitler. Congress, Roosevelt opined, might only approve 10,000 aircraft, of which 3,370 would be combat effective types and 3,750 combat reserve. Seven aircraft factories should also be built, only two of which would be activated. And, he said, the United States had to defend the Western hemisphere "from the North Pole to the South Pole."[71]

Support of the White House was now the pacing factor in the rise of American air power, although other events also proved fortuitous. With the death of General Westover in a crash, Brig. Gen. H. H. Arnold became chief of the Air Corps in September 1938. He recruited officers to staff the air portions of the president's budget and message for Congress in January 1939 (Cols. Carl Spaatz and Joseph McNarney, Majs. Ira Eaker and Muir Fairchild, and Capts. George Kenney and Laurence Kuter). In November 1938, the army's new assistant chief of staff was Brig. Gen. George C. Marshall, who proved to be the point man in getting Arnold into the Combined Chiefs of Staff.[72] By late 1938, as Dr. Joseph Ames, chairman of the NACA, wrote Charles A. Lindbergh in France, a new atmosphere pervaded Washington. It was a state of "peacetime war." Lindbergh had urged NACA to aim for the development of a 500-mph airplane, and Ames, in reply, invited him to help NACA obtain additional laboratories for its aerodynamic and engine research and flight cleanup work on all new navy and army aircraft.

The take-off of American air power began on 12 January 1939, in President Roosevelt's State of the Union Message to Congress. Responding to Munich, he declared that "our existing forces are so utterly inadequate that they must be immediately strengthened," and he sought $300 million (less than he had said he wanted) for Army Air Corps aircraft procurement.[73] Within three months, Congress authorized upwards of a three-fold expansion of the Air Corps to 5,500 planes, 3,203 officers, and 45,000 enlisted men. This was a sharp contrast to the existing "utterly inadequate strength" of 1,700 tactical and training planes, 1,600 officers, and 18,000 enlisted men. As events would show, this was only the first expanded blueprint for the Army Air Corps. Planners had first to program for twenty tactical combat groups in the spring of 1939, and then reprogram for forty-one groups by May 1940, fifty-four groups by July 1940, and eighty-four groups by the fall of 1941. These goals could not be instantly attained, and many growing pains would be experienced.[74]

A major early problem soon stemmed from the purchase of first-line aircraft by

Britain and France at the cost of the buildup of the American forces. By the end of 1939, the two countries had ordered 2,500 aircraft of all kinds and by April 1940, 2,500 combat aircraft. Obsolete planes such as the P-36 and its water-cooled offspring, the P-40, had to be produced until the more advanced P-38s and P-39s could be built. B-25s and B-26s were ordered to replace B-18s and A-20s, and the P-47 design was pushed. For FY 1940, seventy B-17s were ordered as well as sixteen four-engine Consolidated B-24s on a second production line. Contract civilian flying schools expanded pilot training to 7,000 per year in 1940. Bases for air defense and training had to be built "boom-style" in many places, including Alaska, Puerto Rico, and Panama. In March 1939, the GHQAF became a responsibility of the chief of Air Corps, not the army general staff.[75]

In May 1939, Colonel Lindbergh returned from Europe to attend meetings of the NACA. He met with General Arnold at West Point, briefing him on European aviation, particularly the German Air Force. Lindbergh agreed to serve on the Kilner Board, with Colonels Spaatz and Earle Naiden, to determine the technical characteristics of all military aircraft. The Kilner Board also recommended that first-line aircraft sold to England and France carry with them a responsibility on the part of the purchasers to report on their combat effectiveness. With the help of Lindbergh, Arnold, and BuAer chief Adm. John Towers, the NACA put together its requirements for an additional laboratory.[76] In July 1939, production models of the B-17 arrived at GHQAF at Langley Field. Seven made a "goodwill flight" to Argentina, at an average speed of 260 mph.

On 1 September 1939, the day the German *Wehrmacht* lunged into isolated Poland and World War II began, George C. Marshall became acting chief of staff of the U.S. Army and Roosevelt declared the neutrality of the United States. U.S. Navy ships and PBY flying-boats began a Neutrality Patrol over the Carribbean and Atlantic sea approaches. Roosevelt also promptly dispatched an appeal to Germany, Italy, France, Britain, and Poland to refrain from "ruthless bombing from the air of civilians in unfortified centers of population." Hitler replied that FDR's request "corresponds completely with my own point of view."[77] American airmen noted that the German Air Force seized command of the air by destroying in one day the Polish Air Force on its airfields. The bombing of Warsaw to end the final resistance was the largest such bombardment of a city to date. The Polish campaign ended quickly and demonstrated a new word for the textbooks—*Blitzkrieg*. Screaming Stuka Ju-87 dive-bombers were shown in newsreels around the world and shocked Americans. The "phony war" began in Western Europe, but there was no air war yet.

American airmen also noted that Polish opposition to the Luftwaffe in the air had been virtually nil, although the Stuka, however stable a bombing platform, was indeed vulnerable without air superiority or greater defensive firepower. There was some soul-searching. Major Harold George, commanding the 94th Bombardment Squadron, advised the GHQAF commander that "today American bombardment groups could not truly defend themselves against American pursuit groups." Early outcomes of the continuing bomber versus fighter debate were an increase in the defensive armament of the war-improved B-17s and arguments for "pursuit escorts," even before the effectiveness of radar was fully appreciated.

General Arnold dispatched hand-picked observers to the "phony war." Lieutenant Colonel George Kenney reported from Paris that observation balloons were worthless and most reconnaissance planes were slow and vulnerable. Just before the German attack in the west, Gen. Delos Emmons and Colonel Carl Spaatz were sent to Europe. From May to September 1940, Spaatz observed the fall of Belgium, Holland, and France, and the Battle of Britain. Once the British Fighter Command's system of integrated radar-fighter sector control was appreciated, even General Arnold began to think that night operations might be essential for B-17 forces. Spaatz's diary and his reports consistently maintained that England, fighting alone, would survive. The Luftwaffe, he argued, would not win daylight air superiority over England or otherwise achieve a decision during the ill-coordinated bombing campaign against the city of London.[78] The German bombers could not defend themselves from the Hurricanes and the Spitfires, and Spaatz clearly agreed with Winston Churchill, who most persuasively stated (what has been quoted only partially ever since) that Bomber Command was a part of the Battle of Britain. On 20 August 1940, Churchill said, lest anyone continue to ignore it:

> Never in the field of human conflict was so much owed by so many to so few. All hearts go out to the fighter pilots, whose brilliant actions we see with our own eyes day after day, but we must never forget that all the time, night after night, month after month, our bomber squadrons travel far into Germany, find their targets in the darkness by the highest navigational skill, aim their attacks, often under the heaviest fire with serious loss, with deliberate, careful discrimination, and inflict shattering blows upon the whole of the technical and war-making structure of the Nazi power. On no part of the RAF does the weight of the war fall more heavily than on the daylight bombers who will play an invaluable part in the case of the invasion and whose unflinching zeal it has been necessary in the meanwhile on numerous occasions to restrain.[79]

Bomber Command's first raid on Berlin, a modest one, caused Hitler to alter the outcome of the entire Battle of Britain, the first turning point for the Allies in World War II.

While the German *Blitzkrieg* raced through Belgium and Holland and poured over France, President Roosevelt began his secret correspondence with the other former naval person, Winston Churchill, now just made prime minister. This tie led to the Atlantic Charter and coalition planning for the defeat of the Axis. On 16 May 1940, before Dunkirk and the Battle of Britain, Roosevelt called upon Congress for a further expansion of annual American aircraft production to 50,000 airplanes. He emphasized the importance of the army and navy air arms in hemispheric defense, but also planned to make available to the RAF new bombers from among the 50,000.[80] Months before Roosevelt's request, General Marshall had approved the "First Aviation Objective" of the Air Corps, a force of 12,835 planes by April 1942. But by July 1941, the army and the navy had authorization for 50,000 airplanes. Arnold and Marshall also agreed that the buildup of the army air forces did not mean their complete independence; they still depended on the army's supporting services. In the process, however, the Army Air Corps became the Army Air Forces (AAF).[81]

The tragedy of France in May 1940 masked other decisions by President Roosevelt which influenced the future history of air power, including his unpublicized decision to develop an atomic bomb before the Germans could. In June 1940 he created the National Defense Research Committee, later called the Office of Scientific Research and Development (SRD), in response to a proposal by Vannevar Bush of Massachusetts Institute of Teachnology, chairman of the NACA and the first director of the SRD. The new office established working committees, similar to NACA's technical committees, to develop high technology armaments such as the proximity fuse, computers, radars, and rockets, all without putting scientists into uniform. NACA retained responsibility for aerodynamic research, though most of its work in the war would focus on solving the problems of existing equipment.[82]

One of the remarkable documents in the history of American air power was the first product of the new War Plans Division of the Air Staff of the Army Air Forces. "AWPD-1" appeared just before Pearl Harbor in response to a White House request through the secretaries of war and of navy on 8 July 1941 for an estimate of the "overall production requirements required [sic] to defeat our potential enemies." Arnold placed Col. Harold George in charge of an Air War Plans Division, to which was assigned a bevy of non-Ph.D.s who were products of the Air Corps Tactical School.[83] The plan was drafted, approved by Marshall and the secretary of war, and submitted to the president on 1 September 1941 as the "Air Annex" to the estimate of overall production requirements.

AWPD-1 became the blue print for the procurement and deployment of the rapidly expanding Army Air Forces, particularly for the European theater. In concept, it was a "synthesis" of the doctrine of strategic air power as it had evolved at the Air Corps Tactical School and as it related to a global war outlined in the joint Rainbow 5 Plan modified by the Anglo-American Combined Chiefs of Staff. AWPD-1 proposed "Possible Lines of Action" which called for the defeat of the Luftwaffe and the support of an invasion of Nazi-held Europe by targeting bombing on electric power, oil, and transportation systems, in that order of priority. These priorities were remarkable forecasts, confirmed by events and the postwar testimony of the members of the U.S. Strategic Bombing Survey, Albert Speer, and scholars.[84] The only postwar complaint would come from the airmen authors who yet believe that AWPD-1 could have been carried out much sooner at less cost in blood and energy. But that document's projection that strategic air power could not be built up in England to conduct decisive operations until mid-1944 proved right on the mark.

The major contribution of the AAF, according to AWPD-1, was for simultaneous war against Germany and Japan by strategic bombing. The AAF would need 239 combat groups and 108 support squadrons, 63,467 planes of all types, and 2,164,916 men. By April 1944, if the concerted effort was instituted immediately, the air offensive against Germany would reach effective strength. This forecast by anonymous staff officers ranks among the most valid in modern military history, since the Army

Air Forces eventually had a peak strength of 2,400,000 men, 243 combat groups, and nearly 80,000 aircraft.[85]

In the fall of 1941, the navy air arm, like the army air forces, was not yet prepared or deployed for a global war in Europe or Asia. The navy could muster eight aircraft carriers, seven large and one small, five patrol wings and marine aircraft wings, 5,900 pilots, and 21,678 enlisted men. The naval battles to come in the Pacific would be fought most often in the air where surface fleets never saw one another.[86] For openers, Japanese naval air forces struck Pearl Harbor on 7 December 1941.

Notes

1. The author has previously discussed the historical evolution of air power in "The Impact of Air Power," *Air University Quarterly Review* 2 (Winter 1948); "The Meaning of National Air Power," *National Air Review* (NAA) (January 1950), 13; "Some Fallacies Concerning Air Power," *Annals of the American Academy of Political and Social Science* 299 (May 1955); "The Evolution of Air Power," *The Impact of Air Power* (Princeton, N.J.: Van Nostrand, 1959), 5–18; "Technical Change and Western Military Thought, 1914–1918," *Military Affairs* 24 (Spring 1960); "The Contemporary Spectrum of War," in M. Kranzberg and C. Pursell, eds., *Technology in Western Civilization*, 2 vols. (New York: Oxford University Press, 1967), 2:576–90; and, "Introduction," *Two Hundred Years of Flight in America* (San Diego, Calif.: Univelt, 1977).

2. Most useful accounts of early U.S. military aviation are I. B. Holley, *Ideas and Weapons* (New Haven, Conn.: Yale University Press, 1953), and Alfred Goldberg, ed., *A History of the United States Air Force, 1907–1957* (Princeton, N.J.: Van Nostrand, 1957), 2–11.

3. Charles H. Gibbs-Smith, *Aviation: An Historical Survey* (London: H.M.S.O., 1970), 129–34.

4. Benjamin D. Foulois and C. V. Glines, *From the Wright Brothers to the Astronauts: The Memoirs of Benjamin D. Foulois* (New York: McGraw-Hill, 1968), 43; John F. Shiner, "The Army Air Arm in Transition: Gen. Benjamin D. Foulois and the Air Corps, 1931–1935" (Ph.D. diss., Ohio State University, 1975), 7.

5. B. D. Foulois, interview with author, Andrews AFB, Md., 23 Aug. 1960.

6. Best concise history of lighter-than-air flight in the United States is Richard K. Smith, "The Airship in America," *Two Hundred Years of Flight in America*.

7. Francis T. Miller, *The World in the Air*, 2 vols. (New York: Putnam's, 1930), 2:155.

8. Peter Supf, *Das Buch der deutschen Fluggeschichte* (Berlin: Verlagsanstalt Hermann Klemm, 1935).

9. H. G. Wells, *The War in the Air* (New York: Dover, 1950), 69; Edward Mead Earle, "H. G. Wells," *Nationalism and Internationalism* (New York: Columbia University Press, 1950), 116–18.

10. Walter T. Bonney, *The Heritage of Kitty Hawk* (New York: Norton, 1962), 141–42.

11. Quoted in E. L. Jones, "History of the U.S. Air Arm," a looseleaf chronology prepared over many years, microfilm in National Air and Space Museum Library, M-186, 294; cf. H. H. Arnold, *Global Mission* (New York: Harper, 1949), 31–47.

12. Thomas D. Milling, interview with Thomas Greer, January 1952, *Development of Air Doctrine in the Army Air Arm, 1917–1941* (USAF Historical Study No. 89, Air University, 1955), 3.

13. Cf. J. F. C. Fuller, *A Military History of the Western World*, 3 vols. (New York: Funk & Wagnalls, 1953), 3:215f.; Theodore Ropp, *War in the Modern World* (Durham, N.C.: Duke University Press, 1959), 222–24, 248, 317.

14. Holley, *Ideas and Weapons*, 33–34.

15. Cf. Stanton A. Coblentz, *From Arrow to Atom Bomb: The Psychological History of War* (New York: Beechhurst, 1953), 389–92.

16. On early navy aviation, see A. D. Turnbull and C. L. Lord, *History of United States Naval Aviation* (New Haven, Conn.: Yale University Press, 1949); A. D. Van Wyen and L. M. Pearson, *U.S. Naval Aviation, 1910–1960* (Washington: NAVWEPS 00-80P-1, 1960); William Armstrong, "Aircraft Go to Sea: A Brief History of Aviation in the U.S. Navy," *Aerospace Historian* 25 (June 1978); T. Roscoe, *On the Seas and in the Skies: A History of the U.S. Navy's Air Power* (New York: Hawthorn, 1970).

17. Robert Sherrod, *History of Marine Corps Aviation in World War II* (Washington, D.C.: Combat Forces Press, 1952), 1–18; E. C. Johnson, *Marine Corps Aviation: The Early Years, 1912–1940* (Washington, D.C.: History and Museums Division, Headquarters, U.S. Marine Corps, 1977).

18. Jerome C. Hunsaker, "Forty Years of Aeronautical Research," *Smithsonian Report for 1955* (Washington, D.C.: Government Printing Office, 1956), 241–71; Daniel J. Kevles, *The Physicists* (New York: Knopf, 1978), 104–5; George Gray, *Frontiers of Flight* (New York: Knopf, 1948), 9–15; Arthur Levine, "U.S. Aeronautical Research Policy, 1917–1958" (Ph.D. diss., Columbia University, 1963).

19. Statement dated 16 April 1915, in E. L. Jones, "Chronology on American Flight" (no pagination), microfilm in NASM Library (M-104), NACA secretariat shared offices with the Signal Corps Aviation Section its first year.

20. Cf. Holley, *Ideas and Weapons*, 106–112; Goldberg, 11.

21. Kevles, 112f, 133–34; Michael Keller, "History of the NACA Langley Laboratory, 1917–1947" (Ph.D. diss., University of Arizona, 1969).

22. Alfred F. Hurley, *Billy Mitchell: Crusader for Air Power* (Bloomington: Indiana University Press, 1975), 20.

23. W. F. Craven and J. L. Cate, eds., *The Army Air Forces in World War II*, 7 vols. (Chicago: University of Chicago Press, 1948), 1:5.

24. Raymond H. Fredette, *The Sky on Fire: The First Battle of Britain* (New York: Holt, Rinehart & Winston, 1966), 256; E. M. Emme, "German Air Power, 1919–1939" (Ph.D. diss., University of Iowa, 1949), 2.

25. Fredette, 57.

26. Ibid., 196–99.

27. The best dissection of the U.S. aircraft program remains Holley, *Ideas and Weapons*, 118–46.

28. Hurley, 22–27; Craven and Cate, 1:12–13.

29. Andrew Boyle, *Trenchard* (New York: Norton, 1962), 168, 180–81, 185–88.

30. Ibid., 298–99.

31. Craven and Cate, 1:10–16; Goldberg, 23–27.

32. The L-12 on 14 May and L-14 on 14 June 1917. Douglas H. Robinson, *Giants in the Sky: A History of the Rigid Airship* (Seattle: University of Washington Press, 1973), 132–33; Turnbull and Lord, 139.

33. Van Wyen and Pearson, 29; Armstrong, 83; Robert L. Perry, "Trends in Military Aeronautics, 1908–1976," *Two Hundred Years of Flight in America*.

34. Ropp, 250–92.

35. Gerald Wheeler, *A Sailor's Life: Admiral William Veazie Pratt* (Washington, D.C.: Naval History Division, Department of the Navy, 1974), 354–56; Arnold, 121.

36. Craven and Cate, 1:17–24; Goldberg, 29–30.

37. Hurley, 64–66.

38. Hurley, 40–42, 56–109: Goldberg, 30–31.

39. William Mitchell, "Airplanes in National Defense," *Annals of the American Academy of Political and Social Science* 131 (May 1927).

40. James L. Cate, "Development of U.S. Air Doctrine, 1917–1941," *Air University Quarterly Review* 1 (Winter 1947); Emme, *Impact of Air Power*, 186–91; Robert F. Futrell, *Ideas,*

Concepts, Doctrine: A History of Basic Thinking in the United States Air Force, 1907–1964 (Maxwell AFB, Ala.: Aerospace Studies Institute, Air University, 1971), 17–30.

41. J. L. B. Atkinson, "Italian Influence on the Origins of the American Concept of Strategic Bombardment," *Air Power Historian* 4 (July 1957); Colonel Edgar S. Gorrell (USAF), "An American Proposal for Strategic Bombing of Germany [report of 28 November 1917]," *Air Power Historian* 5 (April 1958).

42. Arnold, 157–58.

43. Clark G. Reynolds, *The Fast Carriers* (New York: McGraw-Hill, 1968), 14–15.

44. Van Wyen and Pearson, 31.

45. Elting E. Morison, *Admiral Sims and the Modern American Navy* (Boston: Houghton Mifflin, 1942), 506, as quoted in Reynolds, 14; Wheeler, *Sailor's Life,* 192–93.

46. U.S. Navy, Administrative Histories, World War II, 48, "Bureau of Aeronautics," 1:41.

47. Van Wyen and Pearson, 39–41.

48. Gerald Wheeler, "Mitchell, Moffett, and Air Power," *Air Power Historian* 8 (April 1961); idem, *Sailor's Life,* 199–201.

49. Richard K. Smith, *First Across* (Annapolis, Md.: Naval Institute Press, 1973).

50. Van Wyen and Pearson, 39–43; I. B. Holley, "An Enduring Challenge," Harmon Memorial Lectures in Military History No. 16 (USAF Academy, Colo.: U.S. Air Force Academy, 1974), 1–2.

51. Cf. Wheeler, *Sailor's Life,* 170–79.

52. R. K. Smith, "The Airship," *Two Hundred Years of Flight in America.*

53. U.S. Navy, Administrative Histories, 36, "Aviation in Fleet Exercises, 1911–1939," vols. 15 and 16; Reynolds, 4–6; cf. Samuel E. Morison, *The Two Ocean War* (Boston: Little, Brown, 1963), 13–16.

54. Wheeler, *Sailor's Life,* 356; Sherrod, 29–33; Johnson, 64–82.

55. Ira C. Eaker, "Major General James E. Fechet, 1927–1931," *Air Force Magazine* 61 (September 1978); Goldberg, 37–38.

56. Futrell, 35.

57. Futrell, 31–33; Emme, "Emergence of Nazi *Luftpolitik,* 1933–35," *Air Power Historian* 7 (April 1960).

58. *New York Times,* 11 Nov. 1932, 4.

59. Goldberg, 37–38; I. B. Holley, Jr., *Buying Aircraft: Materiel Procurement for the Army Air Forces* (Washington, D.C.: Office of the Chief of Military History, Department of the Army, 1964), 124–49.

60. Goldberg, 40–41; Arnold, 155.

61. Futrell, 36–37.

62. Ibid., 37.

63. Goldberg, 41–43.

64. Reynolds, 19–20; Craven and Cate, 1:101f.

65. Curtis E. LeMay, *Mission with LeMay* (Garden City, N.Y.: Doubleday, 1965), 183–92.

66. Goldberg, 42–43.

67. Charles A. Lindbergh, *Autobiography of Values* (New York: Harcourt, Brace, Jovanovich, 1978), 163–75; idem, *Wartime Journals of Charles A. Lindbergh* (New York: Harcourt, Brace, Jovanovich, 1970), 48–82; cf. Leonard Mosley, *Lindbergh: A Biography* (Garden City, N.Y.: Doubleday, 1976), 214–45; J. W. Wheeler Bennett, *Munich* (New York: Duell, Sloan & Pearce, 1948), 99–100, 159, 416.

68. Emme, "German Air Power," 357–409, and idem, *Impact,* 10–12, 58–68.

69. Henry H. Adams, *Harry Hopkins* (New York: Putnam, 1977), 140; Arnold, 177.

70. Robert E. Sherwood, *Roosevelt and Hopkins,* rev. ed. (New York: Harper, 1950), 99–100.

71. Craven and Cate, 1:105; Futrell, 50.

72. Laurence S. Kuter, "George C. Marshall: Architect of Airpower," *Air Force Magazine* 61 (August 1978).

73. Futrell, 48–49; Arnold, 168–71.
74. Craven and Cate, 1:105; Futrell, 50f.
75. Holley, *Buying Aircraft,* 169–171.
76. Lindbergh, *Wartime Journals,* 183–89, 194, 208, 214, 229, 233, 248; Arnold, 188–89, 91.
77. Text in Emme, *Impact,* 68.
78. Carl Spaatz, "Evolution of Air Power," *Military Affairs* 11 (Spring 1947); "Leaves from my Battle-of-Britain Diary," *Air Power Historian* 4 (April 1957); Futrell, 52–53.
79. Quoted in Emme, *Impact,* 78–79.
80. Holley, *Buying Aircraft,* 229–46.
81. Futrell, 55.
82. J. P. Baxter, *Scientists Against Time* (Boston: Little, Brown, 1947); Vannevar Bush, *Pieces of the Action* (New York: Morrow, 1970).
83. Original members were Lt. Cols. Orvil A. Anderson and Kenneth Walker, and Maj. Haywood S. Hansell, but George acquired detail of Cols. M. R. Schneider and A. W. Vanaman, and Majs. Samuel E. Anderson, Laurence S. Kuter, and Hoyt S. Vandenberg. Best coverage is in Futrell, 50–62.
84. U.S. Strategic Bombing Survey, *Overall Report,* 30 September 1945, submitted to President Harry S. Truman, 39–50; Haywood S. Hansell, *The Air Plan That Defeated Hitler* (Atlanta, Ga.: Haywood S. Hansell, 1972), Craven and Cate, 1:50–52, 146–50, 131–32; Emme, *Impact,* 188–91; Futrell, 59–62; Holley, *Buying Aircraft,* 237–38; D. MacIsaac, *Strategic Bombing in World War Two: The Story of the United States Strategic Bombing Survey* (New York: Garland, 1976); D. MacIsaac, ed., *The United States Strategic Bombing Survey,* 10 vols. (New York: Garland, 1976); Arnold Krammer, "Fueling the Third Reich," *Technology and Culture* 19 (July 1978); Albert Speer, *Inside the Third Reich* (New York: Macmillan, 1970); and idem, *Spandau: The Secret Diaries* (New York: Macmillan, 1977), 48–49.
85. Goldberg, 49.
86. Van Wyen and Pearson, 85.

The Rise and Fall of the Imperial Japanese Air Forces

Alvin D. Coox

Unique Considerations[1]

Twentieth-century Imperial Japan exhibited a number of characteristics which differentiated it from other Great Powers. Japan emerged from 250 years of self-imposed feudalistic isolation only in the 1850s, and then under considerable Western duress. The end of the Shogunate and the founding of a modern state occurred only with the ascension of the Emperor in 1868. No national military establishment appeared until 1873, when the first draft laws were introduced. From this date until the Japanese attacked Pearl Harbor, a mere sixty-eight years intervened.

In addition to the relative brevity of Imperial Japan's modern experience, mention must also be made of the country's geographic and intellectual isolation from Western currents, including scientific and technological know-how, and from the ongoing Industrial Revolution. Vis-à-vis the industrialized world of Europe and the United States, Meiji Japan began its existence as an underdeveloped, have-not member of what today would be called the Third World. One could not have predicted with certainty, a century ago, that Japan would escape the colonial fate of other large and ancient Asian countries such as India and China.

Prewar Allied Ignorance and Underestimation of Japanese Aviation[2]

Against this unique backdrop a formidable military establishment was created. Yet, so far as Japanese military and naval aviation was concerned, prewar Allied knowledge was dismal. Complacent, chauvinistic, and arrogant, Western intelligence services generally regarded Japanese airmen as inferior to their European or American counterparts.

Japanese pilots were thought to be myopic, night-blind, poor at dive bombing, and accident-prone. After all, was it not well known that Japanese babies were strapped papoose-style on their mother's back, and thus they suffered from twisted vision and a wobbly sense of balance? In addition, the Japanese diet of rice caused vitamin and protein deficiencies which in turn contributed to vision problems, particularly in thin air and in darkness. When Japanese pilots sank the British capital ships *Repulse* and *Prince of Wales* off Malaya at the end of 1941, certain incredulous Western observers insisted that it was German pilots who had flown this special mission.

Just as Western intelligence denigrated Japanese flying personnel, so were Japanese aircraft deemed to be fairly numerous but "mostly derivative and emphatically second rate." Japanese warplanes were shoddy, made of plywood and glued paper. An Australian combat pilot, one of the very few who survived the Malayan campaign, recalls that Allied intelligence "experts" constantly issued assurances that the best Japanese fighter aircraft, "were old fabric-covered biplanes which couldn't stand a chance against the [Brewster] Buffaloes."[3]

That this ignorance of Japanese aviation was not entirely accidental, we know from the Japanese sources. The Japanese displayed only obsolete or obsolescent weaponry and equipment to the public. They downplayed their specifications and capabilities, and they kept the better materiel in the homeland. As one source said, thus were they "free from prying eyes, and we led the world seriously to underestimate the combat strength of our naval aviation."[4] Imperial Japan, of course, was also a tightly controlled totalitarian state by the 1930s, and its laws on military secrecy were taken very seriously and enforced very vigorously.

Japanese Army Air Force: Early Developments

Japanese military aviation history began with the construction of primitive balloons in 1877–78 and the purchase of a French model. The first successful manned Japanese military balloon went aloft in 1901. During the Russo-Japanese War in 1904, a provisional Army balloon unit operated against Port Arthur.[5]

As early as 1909, the Army General Staff (AGS) was showing interest in three-dimensional warfare. Within a year, a joint Army-Navy-civil research committee sent a civilian engineer to Europe. During 1910, the Japanese Army tried to assemble domestic aircraft; neither of two models was successful. Thereupon, it was decided to dispatch two officers to Europe and America to investigate the possibilities of buying aircraft and to study flying techniques. By the end of 1910, Captain Hino had flown a German (Grade) plane over a training field in Tokyo for 1 minute 20 seconds, a distance of 1,000 meters at 20 meters altitude. Captain Tokugawa flew a French Farman for 3 minutes at 70 meters altitude for a distance of 3,000 meters. This was the beginning of the Japanese military air force.[6]

During World War I, the Japanese Army engaged in minor operations to clear the Germans from the Far East in 1914. All operational Japanese aircraft (three Farmans, one Nieuport, and one balloon, with a total of eight pilots and three observers) went to attack Tsingtao in North China. The first Imperial Japanese Army Air Force (IJAAF) air-to-air combat experience occurred when the Nieuport engaged a pesky German plane without result. The first Japanese bombing operations also took place at Tsingtao, against the city and against shipping (fifteen sorties, forty-four bombs or mountain artillery shells). In December 1915, an Army air battalion was formed at Tokorozawa near Tokyo, built around one aviation and one balloon company, under the transportation corps belonging to the Imperial Guards Division.[7]

After the Bolshevik Revolution, thirty-one Japanese Army planes took part in the Siberian Expedition of 1918–22. Nothing much was learned from the campaigns at Tsingtao or in Siberia, except that much better aircraft were needed. Because of their continuing interest in the Manchurian theater, however, the Japanese forces in Siberia did devote considerable attention to weather and terrain conditions on the Asian continent.[8]

Cut off from European and American aviation sources in World War I, the Japanese Army Air Force made little progress. In 1914, Japanese airplane production totaled six; in 1915, four. In 1918, donations to the Army by the president of a Japanese shipping line enabled that service to purchase twenty Sopwith bombers, six Spad scouts, and three Nieuport trainers. With the imminent arrival of numerous modern aircraft, the IJAAF set up the first flying school in March 1918 at Tokorozawa.[9]

As the war wound down in Europe, the Allies began to release surplus materiel for sale. In 1918, the French sold the Japanese thirty Salmson bombers and three balloons; in 1919, forty Nieuport trainers and one hundred Spad fighters. From Britain, the Japanese bought fifty Sopwith fighters. At this point the IJAAF ceased purchases.[10]

A French offer to send military aviation advisers to Japan, all expenses paid, was welcomed by the IJAAF (July 1918). Their specialties were in flying, gunnery, propulsion, photography, and communications. In January 1919, fifty-seven French advisers arrived under Colonel Faure.[11] Among the first consequences was the reorganization of the IJAAF flying schools, at the time located at Akeno and Shimoshizu. Several years later, during the progressive war ministership of Gen. Ugaki Kazushige, an independent IJAAF was established on 1 May 1925. The Air Battalion was upgraded to Air Regiment, and an Army Air Force Headquarters was created.[12]

IJAAF Combat Experience

The escalation of Japanese troubles with China led to the outbreak of the limited Manchurian Incident (1931) and the all-out China incident, an undeclared war (1937). The IJAAF saw its first real combat during the latter protracted conflict. Since the Army Air Force achieved early air superiority over the Chinese, the IJAAF essentially used the China theater as a training zone, much more suitable than the constricted Japanese homeland. The IJAAF stressed ground support, although bombers did strike Chinese interior cities such as Chungking.[13]

The most important IJAAF battle experience before the Pacific War took place over the frontier between western Manchukuo (Japanese occupied Manchuria) and the Soviet Russian satellite state, the Mongolian People's Republic (MPR), between May and September 1939—the little-known but fiercely fought Nomonhan or Khalkhin-Gol Incident. A greatly outnumbered Kwantung Army Air Force slaughtered the Soviet Far Eastern Air Force, largely novices at first, in swirling dogfights until later summer.[14] Then the tide turned, especially after the Soviet ground offensive of 20 August and the attrition of Japanese planes and crews by Marshal Georgii Zhukov's aces, newly arrived from Spain. Zhukov mentions reinforcement by twenty-one Heroes of the Soviet Union.[15]

Whereas few Soviet pilots are known to have made kill-claims of any size, more than fifty IJAAF pilots became aces against the Russians. Of them, fourteen Japanese pilots claimed twenty kills or more. Before his death in August, Second Lieutenant Shinohara shot down fifty-eight Soviet planes. The conflicting data are irreconcilable: skeptics scoff at Japanese claims to have shot down or destroyed 1,389 Soviet planes, using a total available inventory of 574, and losses of 192. For their part the Russians said they brought down 660 Japanese planes, lost 207, and committed only 450. The Japanese are convinced that as many as 3,000 Soviet planes saw action on the Nomonhan front at one time or another.[16]

To the lore of air warfare, *à la japonaise,* must be added the exploit of Second Lieutenant Kanbara (nine victories plus three probables). On 7 August, Kanbara shot down a Soviet fighter plane, landed beside it on the grassy steppes, leaped from the cockpit, drew his samurai sword, and chopped down the hapless Russian pilot on the spot.[17]

Rise of the Japanese Naval Air Force

Like the Army, the Japanese Navy began its aviation history by building unsuccessful observation balloons circa 1877. After the creation of the Naval Aeronautical Research Committee in 1912, six Imperial Japanese Navy (IJN) officers were dispatched to France and the United States. They were to recommend purchase of seaplanes and were to learn to fly and maintain them. Two of these IJN officers flew Farman and Curtiss seaplanes from the new naval air station at Oppama in November 1912. Domestic training of IJN pilots proceeded on a small scale.[18]

In 1913, the first Japanese seaplane tender, *Wakamiya Maru,* entered service. The tender saw combat against the Germans at Tsingtao in 1914, when four IJN seaplanes flew reconnaissance and bombing sorties. In 1916, the first IJN air corps was activated at Yokosuka; in March 1918, the second naval air corps was established at Sasebo. The first operational IJN plane designed domestically was turned out in 1917 by the Yokosuka Naval Arsenal.[19]

During the period of diminished activity immediately following World War I, the tender *Wakamiya Maru* was the scene of an Imperial Japanese Naval Air Force (IJNAF) success. In June 1920, an officer took off in a British plane from a deck specially installed on the seaplane tender. Meanwhile, the world's first true aircraft carrier, *Hosho,* had been laid down in December 1919 and would be completed three years later.[20]

The tempo of Japanese naval aviation activity accelerated with the arrival in Japan in 1921 of a team of ten British experts, including the head of Sopwith. To meet the requirements of carrier operations, a deck fighter along the lines of the Sopwith was developed in Japan. With the appearance of the aircraft carrier *Hosho* in 1922–23, it was time to try deck takeoffs and landings. A former British naval lieutenant was hired as the first test pilot. After his success, takeoff and landing adventures by IJN pilots soon followed.[21]

Like foreign counterparts, Japanese airmen after World War I sought to demonstrate the vulnerability of surface vessels to air power. The Japanese Navy decided to sacrifice tonnage already earmarked for the scrap heap by the terms of the naval limitation agreements reached at the Washington Conference (1921–22). A captured Tsarist Russian battleship was picked for the successful bombing experiment by Navy and Army planes on 9 July 1924, in Sagami Bay. Several years later, in April 1927, the Japanese Navy Air Force was split from the IJN battleships; the new chief reported directly to the navy minister.[22]

British influence on the IJNAF continued. Three British inspectors came out in 1929 to conduct training in aircraft inspection. In 1931, two more British officers came to Yokosuka to provide instruction in tactics and strategy, flying, gunnery, and ordnance. Few will remember that the famous Genda Minoru (captain, IJN; later general, Japanese Self Defense Force) received advanced training from these British air officers. Genda later found the British lectures on offensive tactics for carrier aviation highly useful in his own planning assignments against Pearl Harbor.[23]

After the signing of the London naval accord in 1930, the IJNAF secretly pursued four build-up plans. In 1932 the Naval Air Establishment was created. Early development was disappointing. When Adm. Yamamoto Isoroku became chief of engineering at IJNAF HQ in 1931, he already had discerned the weakness of Japanese naval fighters. The validity of his criticisms—"Fit the aircraft carrier to the fighter plane, not the plane to the carrier"—was demonstrated during the Shanghai Incident of 1932. One American-piloted Chinese Boeing demonstration biplane mauled formations of three and six IJNAF aircraft before being shot down. Although the IJNAF managed to achieve its first air-to-air kill in history during the six-to-one encounter, the Japanese Navy Air Force also lost its first pilot in combat. The poor technical performance of its fighters was particularly disturbing.[24]

Meanwhile, new Japanese aircraft carriers were steadily joining the fleet. *Hosho* was followed by *Akagi* (converted from a battlecruiser) in 1927 and the next year by *Kaga* (converted from a battleship). Both new carriers, of 29,600 tons, carried sixty planes. The small *Ryujo,* carrying forty-eight planes, was completed in 1933. Thus, with four carriers available in the early 1930s, the IJNAF devoted intensified attention to aircraft design. The Japanese aviation industry began to produce excellent carrier fighter, attack bomber, and flying boat designs in the middle of the decade. These modern planes would stand the Japanese Navy in very good stead when enormous operational offensive needs arose in the later 1930s and early 1940s.[25]

When full-scale fighting commenced in China in July–August 1937, carrier- and shore-based planes went into action immediately. At first the Chinese Air Force chewed up the IJNAF's weak biplane fighters. In August, *Kaga* lost eleven of twelve unescorted attack bombers in one raid. Subsequently, the untried Claude monoplanes arrived to sweep the skies clear over China by early December. In one action, the Claudes downed ten new Russian I-16s.[26] The IJNAF did not participate in the huge air battles at Nomonhan in 1939, but, to the Japanese Army's immense chagrin, the IJAAF was beginning to think of calling for naval air assistance toward the end of that costly warfare.[27]

The tactic of firing torpedoes from carrier planes was first attempted in 1930. Two years later a torpedo designed exclusively for aerial use was developed. In 1934, IJNAF planes began practicing the release of torpedoes in the period between night and dawn. Between 1932 and 1935, dive bombing came to be regarded as especially useful in surprise attacks on aircraft carriers. From about 1936, IJN thinking veered toward the decisive air battle prior to decisive combat between the main surface fleet.[28]

Between 1937 and 1939 the carriers *Soryu and Hiryu* were completed. Several months before the outbreak of the Pacific War *Shokaku* and *Zuikaku* were added. *Shoho* and *Zuiho* were being converted from high-speed oilers. Thus, drawing on the new carrier inventory, the IJN was able to form the 1st Air Flotilla in April 1941. This flotilla, including six carriers, would implement Yamamoto's grand design by striking Pearl Harbor. From the heavy carriers' decks would fly 360 torpedo and dive bombers, horizontal and high-level bombers, and fighters—about 25 percent of the frontline IJNAF strength in 1941.[29]

Levels of Experience and Training

Japanese Army and Navy airmen gained much combat experience from the fighting over China and Manchuria/Mongolia after 1937. Their fighter pilots were among the best, if not the best, in the world by 1941. Around 50 percent of IJAAF pilots had had combat experience against the Chinese and the Russians. Ten percent of land-based Navy pilots saw action in China. Eight IJN pilots were in the ten to fourteen kill category; four IJA pilots in the seven to ten bracket.[30]

As long as they could, the Japanese stressed high levels of training. By the time of the Pacific War, IJAAF flying schools were turning out pilots at the rate of 750 per year, while Navy training units were graduating 2,000 per year. Army training emphasized pilots; the IJNAF focused on training aircrewmen. Until the Pacific War, two out of three IJAAF flying cadets were enlisted men, one out of three were regular officer candidates. Navy trainees were 90 percent enlisted personnel.[31]

Japanese pilots received about 300 flying hours in training units before assignment to a tactical unit. (The comparable figure for primary, basic, and advanced training given to a U.S. Army Air Corps cadet was 200 hours.) In 1941 the Japanese Navy gave special tactical training to the carrier airmen preparing for torpedo attacks in the shallow waters of Pearl Harbor and to the airmen based in the Marshall Islands who were to hit Wake Island at long range over water.[32]

By 1941, 3,500 IJNAF pilots had been graduated from flying schools or training units. The figure for the Army was 2,500 pilots. About 600 of the best Navy pilots, averaging 800 flying hours, were attached to the carriers. Army and Navy pilots operating in Malaya and the Philippines averaged 500–600 hours; squadron and flight commanders had much more experience.[33]

Throughout the Pacific War, the Japanese remained convinced of the need for a large number of pilots even if quality had to be sacrificed. To the very end there were enough men to fly the planes produced. The last figure for 1945 was 18,000 pilots to operate 10,700 effective planes (4,800 Army, 5,900 Navy). But while the Japanese turned out 5,400 new pilots in 1943, the Americans were producing 82,714 that year.[34]

At the end of the war, the average flying time of regular IJNAF and IJAAF pilots entering combat had declined to around 100 hours. Few pilots were left with more than 600 flying hours, and they were usually assigned as instructors, staff officers, or escorts for suicide forces. Half of all remaining pilots had less than 100 hours; they were to be employed in the first suicide assaults against the expected Allied invasion. "The caliber of the pilots produced during the wartime years," said IJNAF ace Saburo Sakai, "was at best questionable."[35]

Japanese Aircraft Production

The Big Three aviation firms in Japan were Mitsubishi (1918), Nakajima (airframes, 1917; engines, 1924), and Kawasaki (1919). Smaller manufacturers included Aichi, Tachikawa, Kawanishi, Hitachi, and Nippon Hikoki (Japan Aircraft). The Japanese government encouraged, protected, controlled, guaranteed, and, to a certain extent, subsidized the domestic aviation industry, which went over to exclusive military-naval production about 1939. Cooperation between aircraft manufacturers was as nonexistent as that between the Army and the Navy. The armed forces developed their own separate aeronautical design and minor production facilities (four IJN air depots, one IJA air arsenal).[36]

Total annual military aircraft production rose from 445 in 1930, to 1,181 in 1936, and to 4,768 in 1940. In 1941, a total of 5,088 military planes were produced. (The comparable figure for Germany that year of 1941 was 11,706; for the United States, 19,433.) Between 1942 and 1945, Japan produced 58,822 planes which tended to increase in weight and to be improved in performance. (The comparable figure for Germany was 92,656; for the United States, 261,826.) The peak year for the Japanese was 1944 (28,180 planes). German and American production peaked that year, too, at 39,807 and 100,752, respectively.[37]

If one Japanese wartime plane must be singled out, it would have to be the IJN's beloved carrier-based, single seat Mitsubishi A6M Zero (Zeke). In all, the Japanese Navy took delivery of 10,938 Zeros. It was the feeling of IJN pilots that the Zero fighter was "about equal" to the Curtiss P-40 and Grumman F4F Wildcat but no match for the Vought F4U Corsair and Grumman F6F Hellcat, which the Japanese naval pilots "particularly disliked."[38] No original Japanese jet fighter design attained the production stage by war's end.[39]

When the Pacific War broke out, the Japanese inventory was 7,500 planes. Wartime production added 65,000, for a total of 72,500. At war's end the inventory was 18,500. Thus 54,000 planes must have been lost: combat losses, 20,000; training, 10,000; other noncombat losses, 20,000; ferrying, 4,000.[40]

In the main campaigns, Japanese plane losses to all causes totaled 40,000. The history of the air war in the Pacific can be traced through the following data:[41]

Campaign or Phase of War	Planes Lost
December 1941–April 1942	1,000
Dutch East Indies	1,200
Midway-Aleutians	300
China-Manchuria	2,000
Solomons-Bismarcks-New Guinea	10,000
Central Pacific	3,000
Southeast Asia	2,200
2nd Philippines campaign, 1944–45	9,000
Ryukyus	7,000
Defense of Japan	4,200

During the air war, numerous Japanese pilots became aces. The ace of aces was a Navy pilot, Warrant Officer Nishizawa, with a confirmed score of 87 before he was killed in action. If unofficial kills are included, he downed at least 102 enemy aircraft. The highest IJAAF score (apart from Shinohara's 58 at Nomonhan) was 51 achieved by M/Sgt Anabuki.[42]

Kamikaze: The Divine Wind Expedient

The Japanese *kamikaze* (divine wind) or *tokko* (special attack) operations of desperation deserve a separate paper. Not only did the *kamikaze* warriors (including one-man Baka guided missiles) attack ships, they also rammed B-29s and crash-landed on Allied airfields. For suicide missions, the Army decided that at least seventy flying hours were necessary for pilots. The Navy deemed thirty to fifty hours sufficient if training planes were used for the attacks. During winter 1944–45 and spring 1945, all regular training was halted in favor of suicide pilot preparation. Expendable, low-powered trainers proved maneuverable, cheap to build, and fairly easy to fly. Since the planes carried bomb loads of only 50–250 kilograms, however, they were often loaded with extra gasoline, and hand grenades were sometimes heaped around the pilot in the cockpit.[43]

Suffice it to say that the *kamikaze* effort against warships was massive and occasionally spectacular. In the second Philippines campaign, there were 650 suicide missions, with 26.8 percent achieving hits or damaging near-misses and 2.9 percent sinkings. This experimental offensive cost the Japanese only 14 percent of the 4,000 planes they lost in combat. During the Okinawa campaign, the *kamikaze* lost 63 percent of the 3,000 planes shot down, the IJNAF having flown 1,050 suicide sorties; the IJAAF, 850. Twenty-five Allied ships were sunk; Allied vessels were hit 182 times, suffering damaging near-misses 97 times. The damage rate was only 14.7 percent effective (1.3 percent sinkings).[44] The war ended before the *kamikaze* could unleash their ultimate death-defying operations.

Why the Japanese Lost the Air War

The reasons for the destruction of the IJAAF and the IJNAF in World War II cannot, of course, be separated from the larger geostrategic, economic, scientific, technological, demographic, and psychological reasons for the defeat of Imperial Japan. Nevertheless, so far as the Japanese air forces were concerned, a number of specific explanations can be adduced:

(1) Early successes lulled the Japanese into a false sense of security. When they "woke up" toward the end of 1943, it was too late for them to recover.

(2) The Japanese doctrinal approach to air power was narrow and uncoordinated. The IJAAF was chronically subordinated to ground forces in a tactical role. Neither the IJAAF nor the IJNAF (which had a slightly broader conception) could ever mount sustained and heavy strategic attacks at long range against economic targets or rear zones.[45]

(3) It was the belief of the IJNAF that its Army counterpart would cooperate only if operations were conducted over land. "The Army flyers," says Genda, "didn't like to fly over the ocean." The AGS "acted as though they didn't realize the importance of the control of the seas."[46]

(4) The Japanese did not exploit the advantages of interior lines of communication. They frittered away their best air units in piecemeal fashion around their far-flung defensive perimeter—the consequence of envisaging a relatively short and victorious war.[47]

(5) With the isolation of Japan from the continent and Southeast Asia, the importation of oil and other natural resources dwindled seriously. By the end of the war, training time for IJNAF pilots engaged in other than suicide operations had to be reduced to fifteen hours per pilot per month. Substitute aviation fuels were introduced or tested. Some bordered on desperation and included alcohol, pineroot oil, camphor oil, isopropy, and ether.[48]

(6) Not only did the IJAAF and IJNAF fail to cooperate effectively, but the Army and Navy competed viciously for allocations of Japan's limited supplies of raw materials and production facilities. This "civil war" far transcended the usual connotation of rivalry, jealousy, or competition. Realistically speaking, unification of the separate military and naval air forces was an impossibility.[49]

(7) The Japanese lavishly expended the veteran, highly trained pilots with whom they started the war. When they escalated the replacement training program, they underestimated the difficulties and emphasized quantity. The new pilots were a poor match for the improved Allied air forces and, indeed, for their own seniors. The most advanced Japanese wartime planes proved too "hot" for the inexperienced new men to handle. It was the weakness of orthodox air operations which gave birth to the wasteful tactic of the *kamikaze*. The Japanese did not rotate seasoned pilots from combat nor did they try to preserve them by developing air-sea rescue techniques.[50]

(8) The qualitative edge of Japanese aircraft was lost as the war went on. Japanese planes suffered chronically from weak landing gear and poor brakes. This was attended by a general deterioration in maintenance, repair, supply, engine and flight testing, workmanship, components, and dispersal. Here, too, Army-Navy coopera-

tion was almost nonexistent. Spare parts for their varied and complex aircraft types were in constant short supply.[51]

(9) Air-ground communications and the absence of good fighter-bombers hampered air support of ground operations. The Japanese never developed an air transport capability for troop carrying or supply drops. Indeed, the Japanese armed forces despised and neglected logistical considerations. Japanese combat officers called logistics "boring" and neglected it whenever they could.[52]

(10) Although they produced some excellent aircraft, the Japanese eventually could not match the Allies in developing new kinds of aircraft and engines in quality or quantity, especially after the Japanese aviation industry began to suffer from enemy air attacks as well as from the consequences of hasty wartime attempts at expansion and dispersal. Duplication of effort and useless secretiveness were rife. New Japanese airplane output suffered from material deficiencies, compounded by ineffective substitute components and inferior workmanship, insufficient testing (many trainers received no test flights), clumsy flying, and costly ferrying losses caused by navigational mistakes, mechanical failures, poor maintenance, and pilot error. The IJNAF found itself rejecting 30 to 50 percent of the planes produced late in the war; corrections might take a precious month. Cdr. Nomura Ryosuke admitted after the war that IJNAF pilots became "convinced in their own mind that they were flying greatly inferior aircraft," and they "had a horror of American fighters."[53]

(11) Japanese operations in defense of the homeland were essentially ineffective. Their early warning radar was poor in quality and inadequate in quantity. Their collection and analysis of warning data were inefficient; their control of interceptor units was haphazard; and their night-fighter techniques were primitive. As usual, the Army and the Navy failed to cooperate; many of their radar installations were operated side by side. When IJN picket boats broadcast reports of visual observation of approaching enemy planes, the IJA would have to monitor that particular channel to obtain intelligence for its own use. In addition, as a percentage of the total Japanese fighter force, no more than 26.5 percent (450 planes) were ever assigned to defend the homeland. The figure was only 16.8 percent as late as December 1944. At war's end, 535 fighters were assigned, but this amounted merely to 16.5 percent of the 3,250 planes available. Interceptor attacks per B-29 sortie amounted only to 0.02 in July 1945, 0.04 in August.[54]

Conclusion

It is difficult to believe that the combat life of the powerful Imperial Japanese air forces lasted little more than eight years, from 1937 to 1945. During that brief but momentous period, Japanese warplanes blackened the skies from China to Mongolia, from Hawaii to Alaska, from Malaya to Burma and India and Ceylon, from Australia to the Philippines. For the nation which bewailed its scanty natural resources, its limited land area, and its demographic inferiority, Imperial Japan certainly waged military operations on a grand and ferocious scale. Nor were the Japanese frugal with their manpower. According to incomplete data, in the worst months of the war IJNAF flight personnel were dying at the rate of 1,500 to 1,800 per month. The Japanese Navy alone had lost 17,360 airmen by May 1945.[55]

Those of us who can remember the old red-ball (*hi no maru*) insignia certainly did not react as benignly or as calmly as we do today when we see very similar markings on Japan Air Lines passenger planes or Japan Air Self Defense Force (JASDF) military aircraft. Our survey of Japanese military and naval aviation has thus come full circle.

Notes

1. For background, see Alvin D. Coox, "Chrysanthemum and Star: Army and Society in Modern Japan," in David MacIsaac, ed., *The Military and Society: Proceedings of the Fifth Military History Symposium, USAF Academy, 1972* (Office of Air Force History, Headquarters USAF and United States Air Force Academy, 1972).

2. Unless otherwise noted, this section is based on Otto D. Tolischus, *Through Japanese Eyes* (New York, 1945), 144; John Deane Potter, *Yamamoto: The Man Who Menaced America* (New York, 1967), 47–48; Walter D. Edmonds, *They Fought with What They Had: The Story of the Army Air Forces in the Southwest Pacific, 1941-1942* (Boston, 1951), 5–6; S. Woodburn Kirby, *The War against Japan* (London, 1957–59), 116–17, 147, 166–67, 240.

3. Martin Caidin, *The Ragged, Rugged Warriors* (New York, 1966), 268.

4. Masatake Okumiya and Jiro Horikoshi, with Martin Caidin. *Zero! The Story of Japan's Air War in the Pacific: 1941–45* (New York, 1957), 12.

5. Boeicho boeikenshusho senshi shitsu (Office of Military History, Institute for Defense Studies, Defense Agency), vol. 87 *Rikugun koku no kaihatsu seisan hokyu* (Army air ordnance: Development, production, supply) (Tokyo, 1975), 3–7 (hereafter cited as BBSS, 87).

6. Ibid., 7–14.

7. Shibuya Atsushi, *Hikoki 60-nen* (Sixty years of aviation) (Tokyo, 1972), 110–23; BBSS, 87:14–19.

8. BBSS, 87:23–26.

9. Vice-Adm. Kuwabara Torao, *Baigun koku kaisoroku* (Recollections of naval aviation) (Tokyo, 1964), 144; BBSS, 87:23.

10. BBSS, 87:20–25.

11. Ibid., 35–36.

12. Naito Ichiro, Izawa Yasuho, and Hata Ikuhiko, eds., *Koku joho, besatsu* (Air intelligence, special edition), *Nihon rikugun sentokitai* (Japanese army fighter units) (Tokyo, 1973), 256; BBSS, 87:37–42.

13. Naito Ichiro, *Nippon hikoki monogatari* (Stories of Japanese aviation), 3 vols. (Tokyo, 1972), 1:278–82; BBSS, 87:87–90, 160, 168, 172–220.

14. BBSS, 87:168–72.

15. Georgii K. Zhukov, *The Memoirs of Marshal Zhukov* (New York, 1971), 150.

16. Naito, 1:280.

17. Naito et al., 329.

18. Kuwabara, 23–56. For survey bearing on this entire section, to the London Naval Conference of 1930, see Boeicho boeikenshusho senshi shitsu, vol. 95 *Kaigun koku gaishi* (General history of naval aviation) (Tokyo, 1976), 1–29 (hereafter cited as BBSS, 95).

19. Kuwabara, 77–94.

20. Ibid., 160, 154.

21. Shibuya, 169–72; Kuwabara, 145, 184, 190–93.

22. Kuwabara, 203–10.

23. Genda Minoru, *Kaigun kokutai shimatsuki* (Record of the particulars of naval air units), 2 vols. (Tokyo, 1961), 1:60–68.
24. Section based on BBSS, 95:chap. 2; U.S. Strategic Bombing Survey (hereafter USSBS). (Pacific), Military Analysis Division. *Japanese Air Power* (Washington, D.C., 1946), 4.
25. S. L. Mayer, ed. *The Japanese War Machine* (Secaucus, N.J., 1976), 76; R. J. Francillon, *Japanese Aircraft of the Pacific War* (New York, 1970), 39.
26. Okumiya and Horikoshi, 13–15.
27. Capt. Ohmae Toshikazu, IJN, interview with the author.
28. Potter, 52; USSBS, *Japanese Air Power,* 4.
29. Mayer, 82–85; USSBS, *Japanese Air Power,* 6; USSBS, (Pacific) Naval Analysis Division. *Interrogations of Japanese Officials* (OPNAV-P-03-100), 2 vols. (Washington, D.C., 1946), 1:23 (Capt. Fuchida Mitsuo, IJN).
30. Christopher Shores, *Fighter Aces* (London, 1975), 120; USSBS, *Japanese Air Power,* 35.
31. USSBS, *Interrogations,* 2:533 (Cdr. Terai Yoshimori).
32. USSBS, *Japanese Air Power,* 6, 35, 38.
33. Ibid., 5.
34. Ibid., 35–39.
35. Alvin D. Coox, *Japan: The Final Agony* (New York, 1970), 95, 96; USSBS, *Japanese Air Power,* 40; Saburo Sakai, with Martin Caidin and Fred Saito. *Samurai!* (New York, 1958), 26–27.
36. Kuwabara, 168–74; BBSS, 87:31–58 (licensing arrangements), 58–86 (domestic production); USSBS, Aircraft Division. *The Japanese Aircraft Industry* (Washington, D.C., 1947), *passim.*
37. USSBS, *Japanese Air Power,* 28–29, 32; USSBS, *Japanese Aircraft Industry,* 1–5, 115–57.
38. Shibuya, 199–207; Yanagida Kunio, *Zeroshiki sentoki* (The type Zero fighter plane) (Tokyo, 1977), *passim;* Okumiya and Horikoshi, 2–3; USSBS, *Interrogations,* 2:532 (Cdr. Nomura Ryosuke); William Green, *War Planes of the Second World War,* vol. 3 *Fighters* (Garden City, N.Y., 1964), 47–50.
39. Francillon, 404–7.
40. USSBS, *Japanese Air Power,* 30, 33; USSBS, *Interrogations,* 2:374–75 (Comdr. J. Fukamizu).
41. USSBS, *Japanese Air Power,* 34, Exh. D.
42. Izawa, *passim.* There are numerous errors of transcription and identification in Shores, 124, and Mayer, 135.
43. USSBS, *Interrogations,* 1:60–64 (Capt. Inoguchi Rikihei, IJN); USSBS, *Japanese Air Power,* 36, 60, 72–73.
44. USSBS, *Japanese Air Power,* 23–24, 60–80; USSBS, *Interrogations,* 1:24 (Capt. Fuchida Mitsuo, IJN).
45. USSBS, *Japanese Air Power,* 1.
46. USSBS, *Interrogations,* 2:496–97 (Genda).
47. USSBS, *Japanese Air Power,* 3.
48. Coox, 94–95; USSBS, *Interrogations,* 2:533 (Cdr. Terai Yoshimori); USSBS, *Japanese Air Power,* 24–25.
49. USSBS, *Interrogations,* 2:329 (Adm. Yonai Mitsumasa), 2:496–97 (Genda). For historical background, see BBSS, 95:74–75, 80.
50. USSBS, *Japanese Air Power,* 2.
51. Coox, 55; USSBS, *Japanese Air Power,* 31, 36.
52. Colonel Imaoka Yutaka, IJA, interview with the author; USSBS, *Japanese Air Power,* 2; USSBS, *Interrogations,* 2:531–32 (Cdr. Nomura Ryosuke).
53. Coox, 95; USSBS, *Japanese Aircraft Industry,* App. 5; USSBS, *Interrogations,* 2:532 (Comdr. Nomura Ryosuke).
54. USSBS, *Japanese Air Power,* 2, 26–27. 45–59; USSBS, Military Analysis Division. *Air Campaigns of the Pacific War* (Washington, D.C., 1946), 52–53.
55. BBSS, 95:Chart 3.

Soviet Air Power in World War II

Kenneth R. Whiting

When the Germans attacked the USSR on 22 June 1941, the Soviets had more combat aircraft than did the Luftwaffe and its allies on the Soviet border. Within a few days, however, the Soviets had lost much of their air force, and the Luftwaffe roamed the skies over Russia with almost complete impunity. Four years later, the pitiful remnants of the once-mighty German Air Force were unable to put up even a token opposition against the thousands of Soviet planes swarming over Berlin. Why were the Soviet airmen sitting ducks in June 1941, and what happened to reverse the situation by 1944–45? These are the questions bound to arise in the mind of anyone examining Soviet air power in World War II.

Before describing the Russo-German air war, however, it would seem appropriate to assess what the Soviet air forces (VVS)[1] brought into the war in the way of equipment, combat experience, and doctrine. In other words, we need to understand how the Soviet VVS had evolved slowly from a mixed bag of foreign aircraft inherited from the Tsarist regime into an organization quantitatively and qualitatively capable of first holding and then defeating the Luftwaffe.

Early Military Aviation Developments

Until Josef Stalin's Five-Year Plans for the industrialization of the Soviet Union at a forced tempo began to produce results in the early 1930s, the Red Army was largely a mass of infantry militia. Aviation, armor, and technical personnel made up only 10 percent of the armed forces. The expansion of the Soviet aircraft industry called for in the First Five-Year Plan began from an extremely small base. In 1929, the air force of the Red Army consisted of "a thousand combat aircraft of old construction," and there was little hope of any improvement until Soviet industry could provide the basis for a modern aircraft industry.[2] During the rehabilitation of the national economy in the New Economic Program (NEP) period (1921–28), the aircraft industry could progress only at a snail's pace because of the lack of a machine-tool industry, the shortage of metals of any quality, and the scarcity of technically trained manpower. During that period, the USSR was importing most of its aircraft and just about all of its engines. The 1929–32 period, however, witnessed a real expansion of the aircraft industry; old plants were expanded and modernized, and new ones were built. According to an official Soviet source, between 1928 and 1932 the labor force in the aviation industry increased by 750 percent and the number of engineers and technicians by 1,000 percent.[3]

Just how many aircraft plants there were in 1928 and how many were built in 1932 was the subject of confused guessing by Western observers. The estimates of the number of aircraft plants in 1932 vary from six airframe and twelve engine plants to a total of over forty (with 150,000 personnel). In the Second Five-Year Plan (1933–37), the output of aircraft quadrupled, going from 860 in 1930 to 3,578 in 1937.[4] By 1938, there were probably around seventy plants (twenty-eight airframe, fourteen engine, and thirty-two for other components).[5] Exact figures are difficult to find for, as the authors of *The Soviet Aircraft Industry* point out, the Soviet habit of referring to aircraft plants by numbers, or by the honorary designation *imeni* (in the name of), makes for a good deal of confusion. For example, the Fili plant built by Junkers, to produce all-metal planes during the early part of the Red Army–Reichswehr honeymoon was redesignated Plant No. 22 *imeni* Gorbunov when the Russians took it over.[6] Although the specifics are lacking, the evidence as a whole indicates a rapid expansion of the Soviet aircraft industry during the first two five-year plans.

During the 1930s, Soviet aircraft designers were under intense pressure to overcome the country's dependence on foreign aircraft. In 1929, N. N. Polikarpov got the jump on the next decade with his R-5 reconnaissance plane. By 1934, the R-5 was equipped with a new 680-hp engine and had a top speed of 135 mph. Produced in a number of versions, the R-5 is best known as a reconnaissance aircraft, but also served as a two-seat fighter (DI-2) and as a dive-bomber. It was produced in large numbers and was used in the Russo-German War up to 1944.[7] A. N. Tupolev produced a new bomber in late 1930, the TB-3 (*tyazhelyy bombardirovshchik,* or heavy bomber), a four-engine monoplane with a top speed of 110 mph. Over 800 were produced and were used as night bombers early in the struggle against the Germans.[8]

In 1933, Polikarpov came to the fore as the preeminent Soviet designer of fighter aircraft when he produced two outstanding fighters in the same year. The I-15, a single-seat sesquiplane with a gull upper wing, was powered by a 700-hp M-35 engine (Wright Cyclone built under license). Entering service in late 1934, the I-15 had a top speed of 230 mph and a ceiling of 32,000 ft. The I-16, developed in the same year, was a single-seat, single-engine, monoplane powered by a 480-hp M-22 engine (a Bristol Jupiter under license) which gave it a top speed of 220 mph.[9] In 1934 it got a new engine, a 775-hp M-25B (Wright Cyclone 9) which increased its speed to 280 mph.[10] The best of the Soviet fighters in the late 1930s, the I-16 remained in production until 1941 and was used during the early part of the Russo-German War.

In 1934, Tupolev turned out the SB-2 (*skorostnoy bombardirovshchik,* or fast bomber), a monoplane powered by two 860-hp M-100 engines (Hispano-Suiza under license) which gave it a top speed of 230 mph. Carrying four machine guns and a bomb load of 2,000 lbs., the SB-2 went into series production in late 1935, and over 6,000 were produced.[11] Vladimir M. Petlyakov's team, under Tupolev's management, designed the TB-7 bomber in 1936. First flown in December of that year, the TB-7 did not go into serial production until 1939 and into service until 1941. Production of the Pe-8, as it was called by then, was phased out in 1944. It was a four-engine monoplane-bomber powered, in its earliest production version, by four Mikulin-designed

AM-35A (1,350 hp each) engines. It had a top speed of 250 mph, a range of 3,000 miles, increased in 1941 to 4,800 miles when it got M-30B engines, and it carried a bomb load of 4,400 lbs.[12]

By the late 1930s, the VVS had a sizeable inventory of combat aircraft, about 2,500. The best machines were the I-15 and I-16 fighters, the SB-2 and TB-3 bombers, and the R-5 reconnaissance plane. This was the stable of planes available for the subsequent Soviet adventures in Spain, China, and Mongolia in the late 1930s.

One of the problems involved in the creation of a bigger and more potent VVS in the 1930s was that of training the pilots and the technicians needed to keep the aircraft operational. Although a Soviet calculation in 1933 that it would take more than a hundred trained men to keep one airplane serviced and operational was certainly an exaggeration, the demand for trained manpower was bound to increase exponentially as aircraft poured off the assembly lines and were sent into operational units. Several developments, however, helped to solve the manpower problem. For one thing, the ongoing mechanization of agriculture resulting from collectivization was developing a reserve of potential mechanics among the peasants tinkering with and operating tractors and trucks. In addition, the Society for the Promotion of Defense, Aviation and Chemical Warfare (called *Osoaviakhim* in its Russian acronym), a voluntary organization dedicated to training young people in those skills needed by the armed forces, taught tens of thousands how to operate, maintain, and repair engines, radios, and motor vehicles. *Osoaviakhim* also saw to it that the young enthusiasts learned how to shoot straight and even had its own aircraft in which future pilots could learn the rudiments of flying. After studying theory for a year in the evenings and on weekends, the aspirant began a fifty-hour flight training program in a U-2, UT-1, UT-2, or I-5.[13] "By the end of 1940 the clubs had almost achieved their target of 100,000 trained pilots."[14] Thus, it was not too surprising that the Soviets were able to replace the pilots lost in the early part of the Russo-German War. But the help of *Osoaviakhim* notwithstanding, the VVS had to become one great technical training institution with academics and flying schools mushrooming up all over the country.

Combat Experience in the 1930s

While Stalin was pushing industrialization and building up a formidable military force in the late 1920s and early 1930s, he was trying to follow a relatively nonbelligerent foreign policy. The disaster that befell the Soviet intervention in Chinese affairs between 1924 and 1927, the absence of any chance of revolution in Western Europe in the early 1930s, plus the concentration upon the "building of socialism in one country," contributed to a semi-isolationist period in Soviet foreign policy. There was, however, some apprehension about Japanese objectives in the Far East, apprehensions that led to the formation of the semi-autonomous Special Far Eastern Army, with accompanying air forces, in August 1929.

The only Soviet military action in the Far East in this period was a conflict with the Chinese over the Soviet rights to co-management of the Chinese Eastern Railway, which crossed Manchuria. Chang Hsüeh-liang, warlord of Manchuria, had begun a campaign to drive the Russians out of the Chinese Eastern Railway administration. Marshal V. K. Blyukher, the commander of the new Special Far Eastern Army, a force of about 100,000 men supported by tanks and aircraft, hit the Chinese along the Sungari River and also encircled a large force near Manchouli. It was all over in six weeks. The Soviets used thirty-two aircraft in the first blooding of the VVS.[15]

By 1934, it was obvious that Stalin was going to have to shelve his "semi-isolationist" policy and seek foreign allies, since, in addition to the expansionism of the Japanese in the East, the Germans under Adolph Hitler were becoming a threat in the West. Thus, the Soviets faced potential enemies in both East and West, the perennial nightmare of Russia's policy-makers, Tsarist or Communist. In September 1934, the Soviet Union joined the League of Nations, and the Seventh (and last) Comintern Congress in 1935 adopted a "united front" policy which called on all Communist parties to cooperate with any anti-fascist party, whatever its leanings otherwise.

Hardly had Stalin moved toward collective security when his partners, England and France, began to be pushed around by Hitler and Benito Mussolini. Then the Spanish Civil War put him in a dilemma: he had either to let down his leftist allies in the popular fronts or support the Loyalist government against Gen. Francisco Franco. He solved the dilemma by intervening in a very cautious manner through the device of the International Brigades. Beginning in October 1936, the Soviets provided aircraft, tanks, and artillery, as well as the skilled personnel needed to operate the weapon systems and to act as instructors.

Just how many men Stalin sent to Spain is unknown. According to an official Soviet account, he sent "557 Soviet volunteers including 23 military advisors, 49 instructors, 29 artillery experts, including anti-aircraft specialists, 141 aircraft pilots, 107 tank drivers, and 29 sailors; communications specialists, engineers, and doctors totalled 106, and there were 73 interpreters and other specialists."[16] Other sources, equally unreliable, give much higher figures.

In aircraft, the Soviets provided about 1,500 machines, although in any one month not more than a third of that number was operational. Of the thousand or so fighter aircraft, around 500 to 600 were I-15s or I-15Bs; the rest were I-16s.[17] There were over 200 SB-2 bombers; the remainder were R-5 reconnaissance planes. Soviet aircraft made up over 90 percent of the Republican air force by early 1937, and the Republicans had air superiority until late that year. At that point, the Nazis equipped the Condor Legion in Spain with Me-109 fighters and Ju-87 dive-bombers, which were superior to the Soviet I-15s and I-16s. The obvious inability of the Soviet fighters to cope successfully with the Germans led Stalin to begin phasing out the Soviet air force in Spain in mid-1938, so that by the end of the year all Soviet aircraft were out of the country.

Although Soviet fliers gained valuable combat experience in Spain, the concepts derived were mostly negative. For example, the VVS came to the conclusion that strategic bombing was an ineffective use of pilots and machines, a conclusion the Germans also drew from their Spanish experience. In retrospect, considering the modesty of the bombing effort in both cases plus the rather primitive equipment involved in that effort, it is not surprising that neither the Luftwaffe nor the VVS was impressed with the results obtained in the Spanish adventure. The Soviet pilots were also made painfully aware of the inferiority of their machines in combat with the German Me-109s. All in all, the Soviet involvement in the Spanish Civil War, especially in the air war, was far from successful.

The Soviets, while trying to consolidate a "united front" against Germany and Italy in Europe and to aid the Spanish Loyalists in their own peculiar way, did not neglect the threat posed to their Far East region by Japanese expansion into Manchuria, Northern China, and Inner Mongolia. Every effort was put forth to make the Special Far Eastern Army as self-sufficient as possible so that Soviet forces would be able to face Germany and Japan simultaneously if worst came to worst. The Japanese estimated that Blyukher's army east of Lake Baykal quadrupled its strength between 1931 and 1936 and in the latter year consisted of nearly 20 rifle and four cavalry divisions plus 1,200 aircraft and an equal number of tanks. When the Japanese began their all-out attempt to conquer China in July 1937, the Soviets added still more men and equipment to their Far Eastern forces; and the number of aircraft rose to nearly 2,000 by 1938.[18]

Stalin believed that by helping the Chinese, he could keep the Japanese so busy in China that they would not be tempted to make any incursions into Soviet territory. The Soviets not only delivered aircraft to the beleaguered Chinese, but also set up and maintained depots and assembly plants, trained Chinese pilots, and sent "volunteer" Russian pilots. For example, on 29 April 1939, in the air battle over Wuhan, over half the sixty-five Soviet-built fighters were flown by Russians. The Japanese ambassador in Moscow, Shigemitsu Mamorn, protested the Soviet involvement, stating that some 500 Soviet aircraft and 200 pilots had entered China (up to May 1938).

The Soviet aircraft used in China were I-15 and I-16 fighters and SB-2 and TB-3 bombers. The I-15s and I-16s, which had shown up badly against the Me-109s in Spain, did much better against the less effective Japanese fighters of that period, and the Soviet pilots in China were able to evaluate the ability of the Japanese pilots, to study their air tactics, and to observe Japanese equipment. China was the ideal area for testing Soviet aircraft and trying out air tactics under actual combat conditions. It was in China that the Soviets realized that the 7.62mm machine gun was a very inadequate weapon for making bomber kills; as a result, they began to install the 12.7mm gun, the equivalent of the American 50- caliber. Japanese control of the air, easily acquired following the near annihilation of the Chinese Air Force (CAF) in 1937, was increasingly difficult to maintain in 1938 when Soviet aircraft and pilots entered the fray.

Although Soviet assistance to China was somewhat erratic because of trouble with the Japanese at Lake Khasan in 1938 and Khalkhin-Gol in 1939 as well as the war with Finland in the winter of 1939–40, the Russians were the main external contributors to China's defense between 1937 and late 1940. In September 1939, the American ambassador in Moscow claimed that the Soviets had sent at least 1,000 aircraft and 2,000 pilots to China.[19]

While the Soviets were engaging the Japanese indirectly in China in the late 1930s, they found themselves in direct confrontation on two occasions: at Changkufeng, or Lake Khasan, in 1938, and at Khalkhin-Gol in 1939. On both occasions the Japanese seemed to be feeling out Soviet resolve and capability, and in both confrontations the Soviets demonstrated an ability to defend their borders.

The Lake Khasan engagement, to use the Soviet terminology, began in early July 1938 when the Japanese protested the Soviet fortification of Changfukeng Hill, a point between Lake Khasan and the Tumen River on the disputed Korean-USSR border.[20] This frontier skirmish, which began on 29 July, rapidly escalated into a "limited war," as both sides put in more forces until the Soviets had twenty-seven infantry battalions plus several artillery and tank regiments. At that point, the Japanese thought the engagement was getting out of hand, and hostilities ceased on 11 August with the Soviets in control of the disputed territory.

The VVS component of the 1st Independent Red Banner Far Eastern Army which confronted the Japanese at Lake Khasan was commanded by P. Rychagov, who was executed in 1941 as one of the scapegoats for the massive Nazi destruction of Soviet aircraft. His airmen, facing light Japanese opposition, were able to penetrate the enemy positions in depth and demonstrated a considerable capability. But Soviet aviators found that aviation was not very effective against an enemy well entrenched with his artillery well dug in.[21]

The next clash with the Japanese was, according to Soviet terminology, the Khalkhin-Gol Incident, which took place on the border of Manchukuo and Mongolia between the little Khalkhin-Gol River and the village of Nomonhan.[22] It began on 11 May 1939 and lasted until 16 September; again both sides kept increasing their commitments. During June the main activity was in the air, some attacks involving over a hundred planes. In July, Georgi K. Zhukov took command of Soviet forces and launched a decisive offensive in late August. He insisted on very close air-ground cooperation and had his pilots study the terrain jointly with their infantry and tank colleagues. His successful employment of some 500 Soviet aircraft went a long way toward insuring victory, especially by inhibiting enemy reinforcement of the battlefield.

The Red Air Force on the Eve of the Great Patriotic War

As a result of their experiences in Spain, in China, and against the Japanese, the Soviets acquired a great deal of expertise. Some of the Air Force leaders had become quite knowledgeable about strategy, tactics, and the like. But then Stalin unleashed the purge. Marshal M. N. Tukhachevsky, along with six other top commanders, was shot on 12 June 1937 to begin the senseless purge that wiped out about four-fifths of the top commanders of the Red Army. No military force could stand a blood-letting of that magnitude without suffering pernicious anemia in its command system.[23]

Soviet aviation was especially hard hit by the Purge. Ya. I. Alksnis, who had succeeded P. I. Baranov as commander of the VVS in 1931, was arrested in 1937 and probably died in 1940. His deputy, V. V. Khripin, also disappeared in 1937. Alksnis was succeeded by a nonentity named A. D. Loktionov, who in turn gave way in September 1939 to Ya. I. Smushkevich. The latter was a veteran of the Soviet air activities in Spain and had made quite a reputation as an air commander in the Far East. He was destined to be shot as a scapegoat during the debacle of June 1941. About 75 percent of the senior officers in the VVS were eliminated by the end of 1939. The purge also extended to the aircraft industry, to the research organizations, and to some of the design bureaus. Even Tupolev was under arrest for a short period. To an undetermined extent, therefore, the wisdom that had been accumulated in the various campaigns between 1936 and 1939 was thrown away. It would seem fair to say that the poor showing of the VVS in the Winter War with Finland and in the early phase of the Great Patriotic War could be at least partially attributed to Stalin's blood lust in the late 1930s.[24]

While Zhukov was thumping the Japanese in Mongolia, Stalin surprised the world with the Soviet-German Non-Aggression Pact on 23 August 1939, an act tantamount to Soviet acquiescence to the German invasion of Poland. With 1,600 aircraft at its disposal in Poland, the Luftwaffe went after the enemy's airfields at the outset and destroyed the Polish Air Force within a week.[25] In short, the air war in Poland was a romp for the Luftwaffe and a dress rehearsal for the attack on Russia twenty-two months later.

For the Soviet Union, any euphoria engendered by the victory at Khalkhin-Gol or the easy entry into Poland was chilled in the Winter War with Finland. When the Finns refused to relinquish territory, Foreign Minister V. M. Molotov stated that it would be up to the military to clarify the situation; and on 30 November the Red Army attacked. The Finnish campaign was not the VVS's finest hour. Since the Finnish Air Force had only about 145 obsolete planes, the Soviet VVS had a 15-to-1 advantage. Nevertheless, coordination with the ground forces was extremely poor, bombing accuracy was mediocre, and even the fighters in air-to-air combat were unimpressive. One set of figures gives the Soviet losses as 684 aircraft compared to Finnish losses of 62.[26]

The miserable performance of the Red Army and its VVS in the Winter War, Zhukov's candid appraisal of Soviet shortcomings at Khalkhin-Gol, and close observation of German efficiency in Poland alerted Stalin to the need to reorganize the Red Army and to equip it with better weapons.[27] K. E. Voroshilov was replaced as commissar of defense by Marshal Semyen K. Timoshenko, and Georgi Zhukov became chief of the General Staff. The new bosses immediately began reequipping the VVS with new types of planes better able to stand up to the Luftwaffe. But, as Zhukov says in his memoirs, when the Germans attacked, the VVS was in the midst of its reorganization, its pilots were not yet fully trained in the new aircraft in the inventory, and only 15 percent of them were trained for night flying.[28]

In January 1940, A. I. Shakurin replaced M. M. Kaganovich as head of the

Aviation Industry Commissariat and went to work with a will. According to Shakurin, his job was to accelerate the output of better aircraft at literally breakneck speed, instructions he got from Stalin himself. His job was helped by the completion of the new TSAGI (research facility), replete with laboratories and wind tunnels, an expansion and modernization that had been under way since 1935.[29]

While the VVS was demonstrating its strengths and weaknesses in action on the Khalkhin-Gol, over the plains of Poland, and in the cold and fog of Finland, Soviet aircraft designers proceeded to come up with new types of aircraft ranging from heavy bombers to dive-bombers to fighters.

Vladimir M. Petlyakov's Pe-2 dive-bomber went into series production in 1940, and over the next five years 11,426 were turned out. It carried five machine guns and a 3,300-lb bomb load and had a top speed of 335 mph at 16,000 ft.[30] Sergei V. Il'hushin's DB-3F (*dal'nyy bombardirovshchik,* or long-range bomber; the F stood for *forsirovannie,* or supercharged), later redesignated Il-4, was in service by 1940. It had a top speed of 265 mph at 20,000 ft and a range of 2,000 miles. The Il-4 became the backbone of Soviet long-range aviation during the war; 5,256 were built between 1940 and 1944.[31]

In 1940, three new fighters went into series production: the MiG-3, the Yak-1, and the LaGG-3. And in 1941 the famous Il-2 *Shturmovik* began to come into the service. These four aircraft were to be produced in large numbers during World War II. The MiG-3, some 3,322 of which were built in the 1940–41 period, was a product of the team of Artem Mikoyan and Mikhail Gurevich. This new fighter had a top speed of slightly over 400 mph at 22,000 ft. A match for the German Me-109 above 16,000 ft, it was at a disadvantage below 13,000 ft.[32] Aleksandr S. Yakovlev's Yak-1 was influenced by the British Spitfire and the German Me-109, both of which he had seen in visits to England and Germany. A low-wing monoplane, it had a top speed of 400 mph at 20,000 ft. The Soviets produced 8,721 of them during World War II.[33] The LaGG-3, produced by the team of Lavochkin, Gorbunov, and Budkov, went into series production in early 1941. A low-wing monoplane with the same short fuselage that characterized most of the fighters of that period, the plane had a top speed of 335 mph; total production was 6,529.[34]

Il'yushin's famous Il-2 *Shturmovik* dive-bomber completed its state acceptance trials in March 1941 and went into series production immediately. A few were being sent to VVS units by July. Its armament consisted of two 23mm cannons, two 12.7mm machine guns, and eight rockets or 13,000 lbs of bombs. It was so heavily armored that it was called a "flying tank." The Il'yushin *Shturmovik* became one of the most celebrated planes of World War II, a tank-destroyer par excellence. The Soviets produced 36,163.[35]

According to Shakurin, in the second half of 1940 production ended on all the old fighters. Since by that time there were twenty-eight aircraft, fourteen engine, and thirty-two aircraft component factories in operation, Shakurin had every right to anticipate a VVS adequately equipped with modern machines within the next few years.[36] Unfortunately for the VVS, the overwhelming bulk of the aircraft it received up to mid-1941 were obsolete since the newer types did not begin to flow into combat units until early 1941, just before the Nazis destroyed most of them on the ground.[37]

The German Onslaught, 1941–42

As Soviet aircraft designers were coming up with better products in 1940–41, Hitler was even then readying his forces for war with Russia. Convinced that a *blitzkrieg* against Russia was feasible, on 18 December 1940, he issued Directive No. 21, or "Case Barbarossa," which described in general terms the strategy for the attack on the Soviet Union set for mid-May 1941. The Soviet VVS got short shrift in "Barbarossa:"

> The enemy will be energetically pursued and a line will be reached from which the Russian Air Force can no longer attack German territory. The final objective of the operation is to erect a barrier against Asiatic Russia on the general line Volga-Archangel. The last surviving industrial area of Russia in the Urals can then, if necessary, be eliminated by the Air Force. The effective operation of the Russian Air Force is to be prevented from the beginning of the attack by powerful blows.[38]

Because of German involvement in the Yugoslavian-Greek campaign, "Barbarossa" had to be delayed until 22 June. When the Luftwaffe did strike, however, the "powerful blows" Hitler had called for were powerful indeed. Caught parked on their airfields, the Soviet planes were sitting ducks for the German fliers, and as many as 2,000 Soviet aircraft may have been destroyed in the first forty-eight hours of the war—even the Soviet admitted to 1,500 lost in the first twenty-four hours.[39]

The Red Army and its VVS were caught flat-footed. Rather than use territory acquired in Poland, Romania, and elsewhere between 1939 and 1941 as a buffer through which the Germans would have to penetrate, Stalin had moved his forces to forward areas in this new territory. The VVS had moved a number of airfields close to the new frontier created in September 1939, and most of the other airfields in the western military districts were being reconstructed by the NKVD in a leisurely fashion. The Soviet fighters, concentrated on a few fields, presented an ideal target for the Luftwaffe's surprise attack. Furthermore, the Aircraft Observation, Warning, and Signal Service was poorly organized and provided little, if any, early warning.[40]

The German pilots, having the advantages of complete surprise plus favorable weather, were able to fly continuous high-altitude and low-level attacks on Soviet airfields. Luftwaffe bombers flew up to six missions a day, while dive-bombers and fighters flew up to eight. Since the Soviet aircraft were lined up on the airfields in rows with no protection, the German pilots had perfect targets to aim at. The few Soviet planes that did manage to get into the air were immediately shot down.

The destruction was almost unbelievable. According to a German account, the first wave of 637 bombers and 231 fighters was directed at Soviet airfields,[41] and even the official Soviet account of the war has 1,000 German bombers attacking sixty-six Soviet airfields with a loss of 1,200 aircraft, 800 of that number on the ground.[42] The German High Soviet Command reported the destruction of just over 4,000 Soviet aircraft by 29 June, that is, a week after the start of the offensive.[43] The Soviet and German figures for kills and losses are unreliable at best, and the discrepancies sometimes border upon the ludicrous, but even the Soviets admit the almost unbelievable havoc wrought by the Luftwaffe in the opening days of the German offensive.

Poorly organized anti-aircraft defense, inferior planes, inexperienced pilots, and utter confusion in the upper echelons of command combined to make Soviet efforts to counter the Luftwaffe onslaught futile. The Soviet I-15Bs and I-16s were not in the same league with the German Messerschmitt 109s. About all the Soviet fighters and dive-bombers could do in the summer of 1941 was to try to give some assistance to the Soviet ground forces. Furthermore, the Soviet DBA, or Long-Range Bombardment Aviation, equipped with Il-4s and obsolete TB-3s, was unable to hamper the German offensive by striking deep behind the lines. Because of the dreadful situation on the ground, Long-Range Aviation was used primarily for close-support operations, which was hardly the most efficient use of the DBA.[44]

According to Field Marshal Kesselring, commander of the Luftwaffe's 2nd Air Fleet, German pilots achieved "air superiority" two days after the opening of hostilities. (In his book, he also describes the massacre of Soviet medium bombers as they arrived over German targets at regular intervals and were shot down with ridiculous ease by the German fighters.[45])

Without air support, either tactical or strategic, the Red Army was at the mercy of the Luftwaffe, and the German Panzer Groups could operate deep behind the Russian fronts with little hindrance from the VVS, while calling upon their own air units when they got into a tight spot. The German *blitzkrieg* proceeded as Hitler expected. In the first three months of the war, Field Marshal Ritter von Leeb's Army Group North pushed through the Baltic states and began the siege of Leningrad; Fedor von Bock's Army Group Center trapped three Soviet armies and four mechanized corps for a total of 287,000 prisoners; and Gerd von Rundstedt's Army Group South achieved one of the greatest "round-ups" of the war when it captured in the Ukraine some 665,000 prisoners, 3,178 guns, and 884 armored vehicles.[46]

The Soviets had almost as many aircraft as the enemy's 1,150 planes for the Ukraine, but 75 percent of the Soviet aircraft were obsolete. The VVS, according to the Russian version, flew over 26,000 sorties during the August–September fighting in the Kiev and Black Sea area.[47] Plocher, however, points out that the Luftwaffe had air superiority during the whole of the Ukrainian campaign and was able to prevent "any serious Soviet air interference with German ground forces"; the Luftwaffe carried out "virtually undisturbed attacks against Russian troops and materiel in the pocket."[48] The magnitude of the German victory would seem either to support Plocher's version or demonstrate the ineffectiveness of 26,000 Soviet sorties.

By the end of September, Hitler became enthusiastic about a renewal of the drive on Moscow by Army Group Center. The campaign began well with a great double encirclement of the Soviet forces in the Vyazma-Bryansk pocket.[49] While the infantry mopped up the Russians trapped in the pocket, the panzers pushed ahead. Then came the rains, and the German advance was stopped dead in its tracks—not so much by the Russians as by mud, as roads became bottomless bogs that could not be negotiated by wheeled vehicles or even tanks. There was nothing the invaders could do but wait until cold froze the ground.

The figures given in the official Soviet account of the air war in Russian for the first four months—that is, up to the October pause in the German drive on Moscow—are 250,000 sorties, mostly against German tank and motorized troops, and the destruction of 3,500 enemy aircraft.[50] The Soviet estimate of German losses is undoubtedly on the high side; but even using the Soviet figures, the Luftwaffe comes out very well in comparison with the VVS losses in the summer and fall of 1941.

The Luftwaffe, however, was down to 2,000 aircraft by early November, since some units had been withdrawn for rest and repair after four months of intensive effort and other units had been transferred to the Mediterranean and West European fronts to cope with the growing U.S.-British threat.[51] Furthermore, as the Luftwaffe's strength on the Russian front began to thin out in late 1941, the Soviets were getting new and better planes. The battle of Moscow in November–December was to demonstrate that the days of overwhelming Luftwaffe air superiority were numbered.

Stalin, stunned at first by Hitler's surprise attack, recovered quickly and consolidated control of the war in his own hands. He had already, in May 1941, made himself chairman of the Council of People's Commissars, or SOVNARKOM (*Soviet Narodnykh Komissarov*), thus combining control of both party and government. A week after the Germans struck, he created, with himself at the head, the State Defense Committee, or GKO (*Gosudarstvennyy Komitet Oborony*), which had absolute control of the government and the armed forces. GKO administered military matters through *Stavka* of the Supreme High Command (*Verkhovnogo Glavnokomandovaniya*), or *Stavka VGK*. *Stavka,* in turn, worked through either the General Staff or through its own *Stavka* representatives at the various fronts.[52] Early in July 1941, *Stavka VGK* coordinated the air forces of three fronts (Northwestern, Western, and Southwestern), including units of the 7th PVO (national air defense) interceptor force and some DBA units. In August, a number of VVS units were coordinated with units from the Reserve and from DBA on the combined Bryansk and Central fronts. To carry out these coordinated operations, *Stavka VGK* dispatched its own representative to the headquarters of the combined front to be responsible for air actions. The *Stavka* representative for aviation used the staff and communications of the front commander to control air operations and reported to both the front commander and to *Stavka* on the results. The system worked so well that the functions of the aviation representative of *Stavka VGK* were gradually increased.[53]

Having been immobilized by the *rasputitsa,* the "season of bad roads," from mid-October to mid-November, the Germans finally got moving again when the weather cleared and frost hardened the ground enough for aircraft, tanks, and wheeled vehicles to operate.[54] But the Germans were in for an unpleasant surprise as they neared Moscow. During the late summer and early fall, Stalin had pulled in toward Moscow well-trained troops and aircraft from the Far East and Trans-Baykal commands as well as forces from Outer Mongolia and Central Asia. These reinforcements for the defense of Moscow included over a thousand planes.[55]

The cold that ended the *rasputitsa* became much more intense in late November and early December, and freezing weather reduced the Luftwaffe to a semimobile force of frozen planes. The Soviet aviators, on the other hand, knew how to care for their aircraft in the cold and fared much better. According to the official account, they flew 51,300 sorties during the two-month battle for Moscow, 86 percent of them in close support missions, while "the enemy lost about 1,400 planes in the Moscow sector."[56] Leaving aside the validity of the statistics, the authors do go on to point out, and correctly, that the increased activity of the VVS was the result of the availability of good airfields with good technical services in the Moscow area plus the fact that cover was provided by the Moscow *PVO Strany* interceptors. Furthermore, for the first time, Frontal, Long-Range Bomber, and PVO fighter aviation were unified under the single control of the commander in chief of the VVS, thus facilitating an economy of effort and a higher degree of flexibility.[57]

On 30 November, Zhukov and the General Staff got Stalin's approval for a counterattack which involved all three Soviet fronts defending Moscow. The counterattack got under way on 5 December, and over the next three weeks the Soviet offensive rolled the Germans back from the capital.

Soviet air was a vital element in the counterattack, as Frontal Aviation, the Moscow PVO, the *Stavka* Reserve, and Long-Range Aviation supported ground operations. (Incidentally, "Long-Range Aviation" would seem to be a misnomer for an outfit that, to quote Zhukov, "bombed and strafed his [German] infantry marching formations, tank and truck columns.")[58] The Soviet counterattack in December 1941 was the first time the Luftwaffe had been on the defensive since September 1939, and the VVS had even gained air superiority in some localities by December. Moreover, new aircraft were beginning to arrive on the fronts in respectable quantities from the new eastern factories by the end of 1941.

When it had become obvious to the GKO shortly after the German onslaught began that the enemy would in all probability overrun much of the heavily industrialized region of Russia, the Soviets decided to transfer as many plants as possible from the Ukraine and Russia east of the Volga River. Many of them went to Central Asia, the Urals, and Siberia—all regions out of bomber range for the short-legged German Luftwaffe. It was during this period that the absence of a German long-range bombing capability was so crucial. According to the official Soviet account, they had moved 1,360 large plants and ten million people by the end of December, a total of 1.5 million tons of freight and humans.[59]

The evacuation of most of the aircraft industry to the east caused a severe drop in output in the second half of 1941 and the first three months of 1942 (1,039 aircraft in January, 915 in February, and 1,647 in March). After that, however, the production rate accelerated swiftly: over 25,000 for 1942, 35,000 for 1943, 40,000 in 1944, and 20,900 for the first half of 1945. Counting the 15,735 produced in 1941, the total Soviet output during the Great Patriotic War was about 137,000 planes of all types. Half of the total were single-engine Soviet fighters, and about 40,000, or nearly one-third of the total, were Il-2 *Shturmoviks*.[60]

Combat operations in 1941 had revealed serious shortcomings in the Soviet Air Force organizational structure, the main weakness being an inability to concentrate air in massive operations. Air power was being used in a piecemeal fashion, partly because of the way in which it was parcelled out to front commanders, to army commanders, and to *Stavka*. In April 1942, General A. A. Novikov replaced Zhigarev as commander in chief and began to restructure the VVS by creating "air armies." The 1st Air Army, formed on 5 May 1942, was made up of two fighter and two composite divisions, a U-2 night bomber regiment, a reconnaissance squadron, and a liaison squadron. Eventually there were seventeen air armies, formed from frontal and army aviation, and varying in size according to the importance of the theater and the availability of aircraft. In the 1942–43 period they averaged 900 to 1,000 aircraft, in the 1944–45 period around 1,500; for certain operations, some air armies had 2,500 to 3,000 aircraft.[61] Furthermore, the old composite divisions (combining fighter, attack, and sometimes tactical bomber aircraft) gave way to divisions made up of a single type of plane. The creation of the air armies was a giant step forward in mobility, concentration of forces, and central control of the Soviet air forces. In addition, General Novikov was devoted to the task of building up the *Stavka* Reserve, a force that could be shunted from one front or theater to another with some speed.[62]

Stalin, greedy for victories to offset the ignominious routs of 1941 and buoyed up by the successful defense of Moscow, pushed his generals into a series of ill-advised offensives in early 1942. As a result, the Soviets suffered several setbacks in the Crimea and at Kharkov, forcing them once again back onto the defensive.

Hitler, in a state of euphoria as a result of the Soviet fiasco at Kharkov, came up with an overly ambitious schedule for 1942. The main emphasis was on drives through the Don Bend and along the Volga with a simultaneous drive through the Kuban to the oilfields at Grozny and Baku. By 23 August, the German tanks reached the Volga just north of Stalingrad. The Luftwaffe then proceeded to reduce the city to rubble, and from mid-September to mid-November the men of the Red Army and the Wehrmacht fought tooth and toe-nail in the wreckage of the city strung out along the right bank of the Volga for thirty miles. The main brunt of the cellar-to-cellar fighting fell on Marshal V. I. Chuikov's 62nd Army. The Germans referred to this war in the rubble as the *Rattenkrieg* (war of the rats).[63]

VVS's main tasks in the defense of Stalingrad were close air support, reconnaissance, and very short-range bombing. As the authors of the Soviet official history put it: "Ground-attack-planes and fighters operating with infantry and artillery attacked the enemy right on the front line, and aircraft of the front and long-range bombers struck against reserves, artillery and troops located 2 to 5 kilometers from the front line."[64] The commander in chief of the VVS, General Novikov, remained at Stalingrad to see to it that his troops did their job right, as did the ADD commander, General A. Ye. Golovanov. Novikov, as the *Stavka VGK* representative to coordinate air at Stalingrad, was involved in the planning of the counteroffensive being prepared by Zhukov while the battle for the city raged. The importance of the VVS's role can be seen in the following account of how the counteroffensive was planned. When Novikov informed Zhukov that his aviation was not yet ready, the latter informed *Stavka*. On 12 November, Zhukov received a reply informing him that it would be better to postpone operations until air support was ready. As *Stavka* put it: "The experience of the war shows that operations against the Germans can be successful only if carried out with superiority in the air."[65]

The Soviet Offensives, Stalingrad to Berlin

The Stalingrad counteroffensive which began on 19 November 1942, marked the end of the first period of the Great Patriotic War.[66] By then, the VVS had a superiority in numbers and, on occasion, even superiority in the air. At the end of the first period two new fighters came into the inventory, the La-5 and the Yak-9. The La-5 was an adaptation of the LaGG-3 and went into series production in July 1942, thus making it available for action at Stalingrad by September. The 287th Fighter Division was equipped with La-5s, which, according to the Soviets, were faster in level flight than the German fighters.[67] The Yak-9, a modified Yak-7, entered combat over Stalingrad. It had a top speed of about 360 mph and was armed with a 37mm cannon and two 12.7mm guns.[68] Furthermore, the strength of the fighter regiments was increased from twenty-two to thirty-two aircraft. Experience during 1941 and most of 1942 had proved the desirability of making the basic battle unit the *zveno*, or flight of four aircraft, subdivided into two pairs (*para*). A relative abundance of new aircraft, better organization (especially the creation of the air armies and bolstering of the *Stavka* reserves), and sharper tactics—partly derived through those of the opponent—all meant a large step forward in VVS's drive for air superiority.

The counteroffensive launched on 19 November worked like a charm. By 23 November, Gen. Friedrich von Paulus's 6th Army, some 250,000 men, had been encircled in the "cauldron," or as the German has it, the *Kessel*, an area about the size of Connecticut.

Once the trap was closed, it would have seemed logical for von Paulus to fight his way out while his troops still had some vigor. The *Fuehrer* had lost his grasp of reality, however, and began to clutch at straws. He accepted Herman Goering's promise that the Luftwaffe could supply the 6th Army and decided to keep the 6th Army rolled up in a "hedgehog" before Stalingrad, to await the 1943 offensive that could rescue them.

Colonel-General von Richthofen, commander in chief of the Luftwaffe 4th Air Fleet, although proclaiming the plan "stark staring madness," proceeded to put it into effect.[69] He had around 320 Ju-52 and Ju-86 transports at Tazinskaya, or "Tazi," and about 190 He-111 bombers at Morosovskaya, or "Moro." The Ju-52, the lumbering "good old auntie," had long been the transport workhorse of the Luftwaffe. It was a three-engine monoplane with a cruising speed of 150 mph and a range of 250 to

800 miles, depending on the load-fuel ratio. The Ju-86 carried an even smaller load. Since it was 140 miles from "Tazi" to the Pitomik airfield in the *Kessel,* neither transport could trade off much fuel for freight. The He-111 was a twin-engine bomber which cruised at 225 mph and could haul two tons of freight 760 miles.

Right from the start, the resupply operation was the victim not only of the shortage of adequate transport but of the weather as well, and the planes had to stand down for days on end. Although Goering had promised to deliver 600 tons a day, the high-point of the airlift came when 700 tons were delivered between 19 and 21 December—that is, 700 tons for all three days together! Then the Russians took both "Tazi" and "Moro," and the German transports had to travel 200 miles between their new bases and Pitomik. During the whole operation the VVS made life miserable for the lumbering transports, forcing them to fly in formations of forty or fifty with fighter escort, which made loading and unloading on the tiny Pitomik field a real problem.[70] The VVS even sent *Shturmovik* formations against the German airfields to destroy transports on the ground . One such raid, on 9 January 1943, hit the Sal'sk airfield and destroyed seventy-two aircraft.[71]

The Red Army overran Pitomik on 16 January, and the auxiliary airfield at Gumrak was seized on 21 January. The remnants of von Paulus's 6th Army were taken prisoner by the end of January. Between 24 November 1942, and 31 January 1943, in the space of a little over two months, the airlift had cost the Luftwaffe 266 Ju-52s, 165 He-111s, 42 Ju-86s, 9 Fw-200s, 7 He-177s, and 1 Ju-290—a total of 490 planes, which includes only transport losses.[72] Even worse, the image of the Luftwaffe as an irresistible force was shattered irreparably.

The Soviet claims are much higher. They have the Luftwaffe in the defense of Stalingrad up to 23 November 1942 losing 2,100 planes and between 19 November 1942 and February 1943, losing 3,000 more. Between 22 June 1941 and 30 June 1942, the German losses in aircraft came to 14,700, if one believes the Russians.[73] Needless to say, German figures are quite different, some 2,951 planes lost and 1,997 damaged between 22 June 1941 and 8 April 1942.[74]

In spite of the enormous disparities in claims, there can be little doubt that by February 1943, the VVS was the mightier of the two air forces. The number of air armies had been increased; *Stavka* had ten air corps in its reserve; and the air effort was now synchronized—General Novikov, head of VVS, as a representative of *Stavka* coordinated the activities of Frontal Aviation, ADD, and the fighter element of PVO. During 1941–42 the Soviet aircraft industry delivered 33,857 planes to the VVS, while the German aircraft industry, including plants in the satellite nations, came up with only 20,857.[75]

Stalingrad was not only a definite watershed in the relationship between the Luftwaffe and the VVS, but the turning point in the Great Patriotic War as a whole. The German military machine in the east was on the defensive after the catastrophe on the Don and Volga. The German counterattacks later in the war were feeble things compared to the *blitzkrieg* encirclements of 1941 and 1942. Nevertheless, the Germans had by no means reached the end of their rope. Soviet offensives after Stalingrad outran their logistical support and resulted in Red Army units becoming dispersed as well as exhausted. Provided with this opportunity, the German commander, Field Marshal Erich von Manstein—for once given a relatively free hand by Hitler-launched counteroffensives that stunned the Russians and regained Kharkov. Von Manstein's counterstroke, however, was the last demonstration of the German free-wheeling use of armor and air power in deep penetration and envelopment, for the battle of Kursk in the summer of 1943 would destroy the Wehrmacht's initiative and most of its aircraft and tanks.

The Kursk salient, a protrusion of the Soviet front north of Kharkov and south of Orel, was a tempting target for a German offensive in the summer of 1943. The Russians, aware through their intelligence of the German plans, filled the bulge with guns and tanks, and *Stavka* sent Zhukov and Marshall A. M. Vasilevsky to coordinate the defense of the area. The 2nd, 5th, and 16th air armies, plus two PVO fighter divisions (about 3,000 aircraft), were assigned to the Voronezh and Central fronts in the Kursk area. In addition those fronts could call upon the Aircraft of the four adjacent fronts and upon *Stavka* reserves. Two-thirds of the Luftwaffe's aircraft on the Russian front were allocated to the Kursk offensive, which the Germans called *Operation Zitadelle,* some 2,000 planes in all (1,200 bombers, 600 fighters, 100 dive-bombers, 600 fighters, 100 dive-bombers, and 150 reconnaissance machines.[76]

As both sides built up their forces through May and June, the main activity was in the air. In early May, Soviet aircraft attacked German airfields in an effort to destroy Luftwaffe planes on the ground, a strategy so well taught them by the Germans in 1941. While they were assaulting German airfields, German bombers were running almost nightly missions against the Soviet military-industrial plants at Gorki, Saratov, and Yaroslavl. This "modest campaign was to remain the only German attempt at attacking Soviet industry.[77]

Kursk is best known as the greatest tank battle of World War II, but it was also an air battle of no small proportions. The two sides together fielded some 5,000 aircraft and at one stage in the battle, the German offensive in the Belgorod area against the Voronezh front, over 2,000 aircraft were operating in an area of 12 by 37 miles, and air battles often involved 100 to 150 planes.[78] Soviet numerical superiority prevailed. As one German writer puts it:

> The German efforts to regain air superiority during the summer 1943 offensive had no continued or full successes. After the last German attacks in the Kursk salient had failed in the autumn of 1943, the Russians definitely ruled the air.[79]

The difficulty the historian faces in trying to deal with aircraft losses on either side can be illustrated by several examples of so-called "official" figures on the battle of Kursk. According to Novikov and Kozhevnikov, the Soviet airmen made 118,000 sorties and destroyed in the air and on the ground some 3,700 German aircraft.[80] But General Plocher, citing "official" figures, has the Luftwaffe 1st Air Division alone at the battle of the Orel bulge flying "37,421 missions, achieving 1,733 aerial victories, of which 1,671 were accomplished by fighters alone, with the loss of only sixty-four German aircraft."[81]

The Russian counteroffensive drove the Germans back over the Dnieper River, and in the late fall of 1943 von Manstein had a 450-mile front to protect, with some of his infantry and panzer divisions down to regimental strength. To make things worse, Hitler was back to his old policy of "yield not an inch." With the VVS in control of the air and with the German ground forces frozen into an immobile defensive stance, the chief German asset, flexibility in strategy and tactics, was gone.

During the last half of 1943, the German Army in the east was deteriorating as rapidly as the Soviet forces were building; they now out-manned and out-gunned the Germans by large ratios, had a superiority in tanks, and had gained control of the air.[82] For example, in the three months immediately after the defeat at Kursk, von Manstein's forces received only 33,000 men to replace the 133,000 casualties in the Ukraine.[83] An overwhelming superiority enabled the Red Army to push forward on all fronts, from the Baltic down to the Balkans.

With regard to the air war, the VVS was not only getting more planes, but also better ones. The Yak-9, which made its first appearance over Stalingrad in the winter of 1942–43, was being used in 1944 not only as an interceptor, but also as a ground attack plane and a fighter-bomber. In mid-1943, Yakovlev increased its fuel capacity, giving the Yak-9D (*dal'niy,* or long range) a range of 870 miles. Its range was extended even further in 1944 as the Yak-9DD (*dal'niy deystviya,* or long-range operations) could fly from the Ukraine to Italy, a distance of 1,120 miles. This plane, with a top speed of 380 mph, was used as an escort for the American B-24 and B-17 bombers in their shuttle-bombing runs.[84] The Petlyakov Pe-2 also underwent improvements throughout the war. When the new German Me-109G appeared on the Russian front in early 1943, the Pe-2 was souped up with an M-105PF engine which could develop over 1,200 hp.[85]

The Yak-3 (replacing the Yak-1 on the production lines in the summer of 1943) poured into the VVS inventory in 1944. A 400-mph fighter, it was a match for the Me-109G and the Focke-Wulf Fw-190. The Lavochkin La-7, which went into series production in the summer of 1944, had a top speed of 420 mph and was especially designed to cope with the Fw-190.[86]

Even the German advantage of skilled and combat-hardened pilots had been dissipated by 1944. The murderous losses suffered by the Luftwaffe necessitated the use of newly fledged fliers. The VVS, however, was fairly wallowing in trained pilots by 1944. An even more important factor helped the Soviets gain control of the air in 1944, and that was the diversion of the best German interceptors to Western Europe to try to cope with the growing intensity of the Anglo-American air raids on the Reich and the invasions of Fortress Europe. Obviously, the Luftwaffe's resources were stretched too thin to be effective on any of the many fronts that had developed by 1944. The German bombers had to confine their activities to night operations since they had practically no fighter cover for daytime activities. Attempts to regain the initiative on the Eastern front, either on the ground or in the air, were bound to fail.

By early 1945 the Russians were poised to administer the *coup de grace* to their

Nazi foes. On the Soviet-German front they had 11 air armies with a total of over 15,000 aircraft against the Luftwaffe's 1,875 planes.[87] The VVS's overwhelming edge over the Luftwaffe was dramatically illustrated when Rudenko's 16th Air Army grew to over 2,500 aircraft in January 1945, giving him more than a 20-to-1 superiority over his opponent, while Krasovsky's 2nd Air Army grew to 2,588 aircraft.[88] In January 1945, the Red Army smashed into Poland and began its march on Berlin at the rate of twelve to fourteen miles a day. In the attack on Berlin in April 1945, the VVS was able to concentrate 7,500 of its 15,540 aircraft against the pitiful remnants of the once-proud Luftwaffe. The Soviet claim of 1,132 German planes shot down in the battle for Berlin may be dubious, but there can be no doubt about who controlled the air over that city.[89]

War against Japan, 1945

Once Germany had surrendered, the Soviets were free to enter the conflict against Japan. Until the Yalta Conference in February 1945, Stalin wanted no part of a two-front war, since the Russo-Japanese Neutrality Pact of 13 April 1941 allowed him to concentrate his forces in the west and draw down on forces in the east. With Germany on the ropes, however, Stalin at Yalta agreed "that in two or three months after Germany has surrendered and the war in Europe has terminated the Soviet Union shall enter the war against Japan."[90]

The build-up of Soviet forces in the Far East began soon after the Yalta meeting. According to Japanese intelligence, by June a daily average of ten troop trains and five munitions trains arrived in the Far East. The Japanese estimated that between April and the end of July, the Soviets increased their strength in the Far East from 850,000 to 1,600,000 troops, 1,300 to 4,500 tanks, and 3,500 to 6,500 aircraft.[91] General John Deane, military attaché to Moscow, gives slightly different figures: 1,500,000 men, 3,000 tanks, and 5,000 aircraft;[92] while the Soviet figures for their forces in that area on 9 August 1945 were 1,577,725 troops, 3,704 tanks, and 5,368 aircraft, of which 4,807 were combat planes.[93] These forces faced a total Japanese opposition in Manchuria, Inner Mongolia, Korea, and the Kurile Islands of about 1,000,000 men, 1,215 tanks, 1,800 aircraft, and 6,700 guns and mortars.[94] The Japanese forces and their Mongolian and Manchukuoan allies were the residue left behind when the Japanese high command pulled out the best cadre to send to other fronts.

The Soviet offensive, commanded by Marshall Vasilevsky, began on 9 August and called for all three fronts to push into Manchuria, with the main thrust plunging through the Greater Khingan Mountains toward Changchun and Mukden. Soviet tank forces penetrated some 250 miles into Manchuria by 15 August, their greatest problem not Japanese resistance but fuel for their machines. By 19 August, the Japanese Kwantung Army had arranged surrender terms with Vasilevsky.

Air operations played a minor role in the August campaign in the Far East. The VVS flew only 14,030 combat sorties and 7,427 noncombat missions, partly because the weather was so awful between 11 and 20 August. About a fourth of the sorties were reconnaissance, but the most important contribution of the air force to the campaign was the hauling of supplies and men. The transports carried 2,777 tons of POL, 16,497 men, and 2,000 tons of munitions and other materiel.[95] The Japanese planes opposing the VVS were obsolete, the best having been siphoned off to oppose the American drive across the Pacific. The Japanese fighters, Type 97 and Type 1 (Nakajima fighters "Nate" and "Oscar") were 60 to 100 mph slower than the Soviet La-9s and La-7s, while the Mitsubishi bombers were 100 mph slower than the Pe-2s and Tu-2s.[96]

Despite the fact that the Red Army was attacking a badly demoralized Kwantung Army, the speed of the armored and motorized forces, the closely synchronized air support, and the business-like way in which the whole operation was carried out, all testify to lessons well-learned on the German front during four years of hard campaigning. The comparison between the smoothly running military machine that plunged into Manchuria, Northern China, and Korea on 9 August 1945, and the bewildered Red Army that had faced the Germans on 22 June 1941, was a vivid demonstration of how well Soviet commanders had been trained in the murderously effective school of combat in four years.

An Analysis of Soviet Air Power in the Great Patriotic War

If the statistics of the air war in Russia are debatable, evaluations of how well or how poorly the VVS and the Luftwaffe fought the air war are even more at variance. But for all that, the writer dealing with the Russo-German air war must try to show that all the trees he has described do total a forest.

Although the VVS took a murderous licking in the summer and fall of 1941, probably losing around 10,000 planes, a high percentage of these were destroyed on the ground and thus did not entail the loss of pilots and navigators. This factor was to loom large in favor of the Soviets when aircraft did become available in respectable numbers in 1942 since it was easier to replace a plane than a trained pilot. By the spring of 1942, the Soviet aviation industry was rolling out enough aircraft to put the VVS back in business. In addition, by November 1942 the Allies had delivered 3,000 planes to the Russians.[97]

During the Great Patriotic War, the Soviet aircraft industry turned out 125,000 planes, while the Germans produced only 100,000 between 1941 and mid-1945. The Soviets, moreover, had only one front to supply, while the Germans were using large numbers of their aircraft in the Mediterranean theater and in defending the Reich against British and American bombers. By 1943 the Luftwaffe was reducing the number of aircraft in Russia to supply the needs of the Mediterranean and home fronts. This left the Eastern front with a relative scarcity of planes, many of which were obsolete.

Some German historians of the air war in Russia regard the decision not to build, or at least to give a low priority to the building of, a four-engine bomber as a fatal mistake. As early as the Battle of Britain in the summer and fall of 1940, the lack of a long-range bomber was one of the deciding factors in the outcome. If German aircraft had been able to range far and wide over all of the United Kingdom and also out to sea along the supply routes, the Royal Air Force would have had to disperse its interceptors and radar so widely as to be almost ineffective. The limited range of the German aircraft restricted their attacks to a definite area, one that could be adequately covered by British radar and interceptors.

The situation in Russia in 1941 and 1942 is grist to the mill of these *ex post facto* students of strategic air warfare. In 1941 the target, in their opinion, should have been the railroads crammed with eastbound trains loaded with dismantled aircraft plants and skilled workers. But Hitler's Barbarossa directive forbade the diversion of aircraft for the destruction of Soviet industry until the battle was won on the ground with close air support. In 1942, the ideal target was the Soviet aircraft industry, newly established in its eastern locale, but within range of the Luftwaffe planes since the Wehrmacht was still pushing forward. The best way to shut off the flow of aircraft to the VVS would have been to hit the source of supply, the aircraft plants.

It was not until June 1943, however, that Luftwaffe bombers began any strategic bombing of Soviet industrial targets. Between 4 and 21 June the "strategic" bombing force of the Luftwaffe, Air Corps IV, flew 993 sorties and released 1,538 tons of bombs on a tank factory in Gorki, a synthetic rubber plant in Yaroslavl, and an oil refinery in Saratov.[98] Although the Germans came up with some grandiose plans in 1944 and 1945 for hitting Soviet industry, especially the power plants in the Volga-Moscow region, the June 1943 raids constituted the only serious German "onslaught" against the Soviet defense industry.

The main reason for the poor bombing performance of the Luftwaffe in Russia was the lack of a decent strategic bomber. The bombers used in Russia, the He-111 and the Ju-88, had a combat radius of around 600 miles with a 1-ton bomb load. Both were too slow for other than night operations even in the Eastern theater.[99] Furthermore, German bomber strength in Russia never exceeded 600 planes, and by early 1943 many of those had been expended on close support of the ground forces, as well as being used as a transport force in the airlifts into the Demyansk "pocket" and the Stalingrad "Kessel." The great hope of German strategic bombing enthusiasts was the He-177 four-engine bomber. But it never lived up to its advance billing, and the dozen or so used in the Stalingrad airlift marked its only appearance in the Russo-German War.[100]

The Soviet ADD, although entitled "long-range aviation," did very little strategic bombing. Of the nearly four million sorties flown by all components of Soviet aviation, less than 7 percent could be termed "strategic bombing," even after stretching that term outrageously.[101] Because of the nature of the war in Russia—enormous forces engaged in ground operations—close support of those operations and very short-range bombing was the order of the day. "Strategic bombing" usually referred to attacks a few miles beyond the Forward Edge of the Battle Area (FEBA).

Instead of the long-range bomber, the dive-bomber was the air weapon par excellence on the Eastern front. In the German case, as early as 1936 the emphasis was on the Ju-87 *Stuka,* and all German bombers were to be designed with a dive-bombing capability (even the 30-ton He-177), a requirement that precluded any effective strategic-bomber design. As long as the Luftwaffe was carrying out *blitzkrieg* operations in restricted areas such as Poland and the Low Countries against feeble opposition, the slow and lumbering *Stukas* were effective, especially against armored forces, communications, and even streams of refugees. But when air superiority went to the enemy, as in Russia after the middle of 1943, the Ju-87 became a sitting duck for the faster Soviet aircraft, particularly when the *Stuka* was coming out of a dive.

The Soviets were as enthusiastic about the dive-bomber as were the Germans. The Il-2 *Shturmovik* was a better assault plane than the *Stuka.* As Yakovlev put it, the fact that the best Soviet planes were designed in the late 1930s and early 1940s rather than in the early 1930s, as were the German planes, meant that they had more potential for improvement since the state of the art was developing rapidly. The addition of a rear gunner in the *Shturmovik* took care of its main weakness-attack from the rear when in a dive or coming out of one. The Il-2 was probably the best assault plane in World War II.

The large role played by the *Stuka* and the *Shturmovik* is proof positive of the main emphasis in the Russo-German air war—close support of the ground forces. As one German historian put it: "strategic air warfare played no role in Germany's campaign against Soviet Russia."[102] But he might well have added that the Soviet campaign was just as weak on the strategic air side as was the German. According to Soviet statistics, 47 percent of the sorties flown by VVS and ADD were for close air support. There was no urgent need, however, for the Russians to go in for strategic air warfare. The partisans were busy interdicting a good deal of the German rail and truck traffic in Russia, and the American and British bombers were working over the Reich home front itself by 1943. This allowed the Russians to concentrate their air on supporting the ground forces.

Another Luftwaffe deficiency helped the Russians enormously, namely, the shortage of transport planes. The main German transport, from the involvement in the Spanish Civil War in 1936 until the end of World War II, was the Ju-52, a three-engine relatively light and definitely slow monoplane. Even that was not produced in sufficient numbers to serve all the Luftwaffe air fleets. As it was also the main trainer in German flying schools, the output of pilots was constantly hampered when the air fleets requisitioned both Ju-52s and instructors to fly them during the frequent emergencies.[103] The fact that the Ju-52 was the transport workhorse for such a long period would seem to demonstrate an obtuseness on the part of the Luftwaffe high command about the value of air transport, especially in an area as vast as Russia. The disastrous attempt to airlift supplies into the Stalingrad *Kessel* seems to have had little influence on the thinking of the Luftwaffe high command during the last three years of the war.

Without going all the way with the favorite gambit of Germans writing about the air war in Russia—namely piling most of the blame of Hitler from 1942 on—there is some fire amidst all that smoke. Hitler was a ground force oriented leader and, with some exceptions, left aviation to the commander in chief of the Luftwaffe, Reichsmarschall Herman Goering. Goering, in turn, because of his "supinity" and "frivolous insouciance," left most of the direction to successive incumbents of the office of Chief of the General Staff of the Luftwaffe, especially Generaloberst Hans Jeschonnek, who held the job between February 1939 and 19 August 1943.[104] Jeschonnek was incapable of questioning an order or an opinion expressed by Hitler, however potentially disastrous it might be. Moreover, as Goering's stock with the Fuehrer declined, the more the Reichsmarschall acquiesced in the latter's decisions and even tried to placate him by promising more than he could deliver, the Stalingrad airlift being a case in point.

Stalin, however, was an aviation buff, taking an intense interest in design and production even before the war. His interest extended to the VVS's command structure and the procurement of its machines, and one of his outstanding designers, Yakovlev, gives him high marks as knowledgeable in things aeronautical. Like his top commanders, Stalin grew as a military figure during the war; and, although prone to botch things up in 1941 and early 1942, he eventually assembled a capable staff in the *Stavka,* whose advice he listened to before making decisions. Despite Khrushchev's claim that Stalin plotted strategic operations on a schoolboy's globe, most of the testimony of those close to him on the *Stavka* portray him as keenly interested in, and knowledgeable about, the military situation at the front. It is hard to visualize Stalin as relying on his "intuition" or consulting an astrologer.

Soviet aviation in the Great Patriotic War was sustained by an expanding industrial infrastructure and transportation system able to operate without German interference.[105] It had the advantage of fighting on only one front, while the Luftwaffe was forced to siphon off its best aircraft to put a defensive roof over the Reich from 1943 forward and to furnish air support for a doomed effort in the Mediterranean area. Due to good prewar planning and the efforts of the civilian-party organization, the *Osoaviakhim,* the VVS never suffered from a shortage of pilots as did the Luftwaffe in the last two years of the war. Over and above all else, however, it was the productive capacity of the Soviet aviation industry that enabled the VVS to gain air superiority in the second half of the war—it simply swamped the Luftwaffe under a flood of first-rate aircraft.

Notes

1. In 1941, the Soviet air forces (*Voenno-vozdushnye sily,* or VVS), came in five varieties: Long-range aviation (*dal'nebombardirovochnaya aviatsiya,* or DBA); Frontal aviation (*VVS fronta*); Army aviation (*VVS armii*); Corps aviation (*korpusnye aviaeskadril'i*); Reserve aviation (*aviatsionnie armii reserva*). DBA was controlled by the High Command Frontal aviation by the *front* commander, VVS of the Army units were attached to each army, while both Corps and Reserve aviation were controlled directly by the High Command and could be shifted about as needed. Most writers refer to the lot as the VVS. In addition, Soviet air power included the interceptor component of PVO *Strany* (the national air defense force), Naval aviation, and the

Civil Air Fleet.

2. *Istoriya velikoy otechestvennoy voyna Sovetskogo Soyuza 1941–1945* (The history of the Great Fatherland War of the Soviet Union, 1941–1945), 6 vols. (Moscow: Veonnoe Izdatel'stvo Ministerstva Oborony Soyuza SSR, 1960–65), 1:90 (hereafter cited as *IVOV*). This is the nearest thing to a definitive work that the Soviets have produced on World War II.

3. *Aviatstroitel'* (June 1933), 1–2, as cited in *The Soviet Aircraft Industry* (Chapel Hill, N.C.: Institute for Research in Social Science, 1955), 6.

4. *IVOV*, 1:65.

5. *Fortune*, August 1937, 70–77.

6. *Soviet Aircraft Industry*, 13–14.

7. *Aviatsiya i Kosmonavtika* (October 1973), 23.

8. Ibid. (November 1973), 22, and ibid. (December 1973), 24.

9. Soviet designations for engines are as confused as those for aircraft. The "M" stands for the Russian "motor," the same as in English; "AM" for Aleksandr Mikulin, "Sh" or "ASH" for Arkadiy Shvetsov, and "VK" for Vladimir Klimov, In the early period only the "M" was used and it applied to foreign as well as indigenously produced engines.

10. *Aviatsiya i Kosmonavtika* (January 1974), 23.

11. H. Nowara and G. Duval, *Russian Civil and Military Aircraft, 1884–1969* (London: Fountain, 1971), 107–10; *Aviatsiya i Kosmonavtika* (February 1974), 22–23.

12. *Aviatsiya i Kosmonavtika* (May 1974), 22–23.

13. The U-2 (*uchebnyy*, or trainer) was a 1928 Polikarpov product and over 40.000 were produced. The UT-1 and UT-2 (*uchno-trenirovochnyy*, or basic trainer) were designed by Yakovlev in 1938, and the I-5 (*istebitel'*, or fighter) another Polikarpov product, a stubby little monoplane of 1930 vintage.

14. W. Generalleutnant Schwabedissen, *The Russian Air Force in the Eyes of German Commanders* (Air University: USAF Historical Division, 1960), 26.

15. A good brief account in John Erickson, *The Soviet High Command* (New York: St. Martin's, 1962), 241–44.

16. *IVOV*, 1:113.

17. The I-16 had many nicknames applied to it during the Spanish Civil War. It was called *Rata* (Rat) by the Franco forces, *Mosca* (Fly) by the Loyalists, while the Soviet fliers referred to it as *Ishak* (Donkey). With its short, barrel-like configuration it was an easy plane to identify and everyone in Spain got to know it.

18. *Japanese Special Studies on Manchuria* (Washington, D.C.: Office of the Chief of Military History, Department of the Army, 1956), 13:53.

19. An excellent account of the Soviet involvement in the Sino-Japanese War is Gordon Pickler, "United States and the Chinese Nationalist Air Force, 1931–49" (Ph.D. diss., Florida State University).

20. Soviet historians refer to the engagement as the Battle of Lake Khasan. A detailed account is given in *IVOV*, 1:231–37. The Japanese call it the Chengkufeng Incident.

21. *Japanese Studies on Manchuria*, vol. 11, pt. 3, book A. "Small Wars and Border Problems: The Changkufeng Incident," 120–21. The Japanese claim that the Russians employed 27 infantry battalions, 100 pieces of artillery, 20 tanks, and a sizeable number of aircraft. Ibid., 115–17.

22. The Japanese refer to the conflict as the Nomonhan Incident, while the Soviets call it the Khalkhin-Gol Incident. The best account representing the Japanese point of view is in the two-volume work: *Japanese Studies in Manchuria*, vol. 11, pt. 3, books A and B, "Small Wars and Border Problems: The Nomonhan Incident." This work also includes an English translation of a Soviet account: S. N. Shishkin, *Khalkhin-Gol* (Moscow: Military Publishing House, 1954). The most recent Soviet effort is in *IVOV*, 1:236–45. Also see Erickson, *Soviet High Command*, 517–23, 532–37.

23. Erickson, *Soviet High Command*, 505-6.
24. Ibid., 500-501.
25. Asher Lee, *The German Air Force* (New York: Harpers, 1946), 45-48; *IVOV*, 1:201-3.
26. Eloise Engle and Lauri Paananen, *The Winter War* (New York: Scribners, 1973), 62.
27. Georgii Zhukov, *The Memoirs of Marshal Zhukov* (New York: Delacorte, 1971), 203.
28. Ibid.
29. A. L. Shakurin, "Aviatsionnaya Promyshiennost' Nakanune Velikoy Otechestvennoy Voyny" (The aviation industry on the eve of the Great Fatherland War), *Voprosy istorii* (February 1974), 81-99.
30. Jean Alexander, *Russian Aircraft Since 1940* (London: Putnam, 1975), 295-304; *Avialsiya i' Kosmonavtika* (October 1974), 27.
31. Alexander, 86-93.
32. Ibid., 193-95; the figures giving the total production of Soviet aircraft during the Russo-German War found in Alexander coincide perfectly with those given by Aleksandr S. Yakovlev in a table in his *Fifty Years of Soviet Aircraft Construction* (translated for NASA by the Israel Program for Scientific Translations) (Washington, D.C., n.d.), 55.
33. Alexander, 421-24; *Aviatsiya i Kosmonavtika* (November 1974), 24-25.
34. Yakovlev, *Fifty Years*, 55.
35. Ibid., 45-46.
36. Shakurin, 83.
37. Zhukov, 201, says the Red Army received 17,745 combat planes, including 3,719 of the latest types, between January 1939 and 22 June 1941, a little over 7,000 aircraft a year.
38. Text of "Case Barbarossa" in H. R. Trevor-Roper, ed., *Blitzkrieg to Defeat: Hitler's War Directives, 1934-1945* (New York: Holt, Rinehart & Winston, 1964), 49-52.
39. Erickson, *Soviet High Command*, 593; in *IVOV*, 2:16, it is stated that in the first day the Luftwaffe attacked 66 airdromes along the frontier on which were parked the newest types of fighters and some 1,500 were destroyed either on the ground or in air combat.
40. Aleksandr Yakovlev, *The Aim of a Lifetime* (translation of *Tsel' Zhizni* by Vladimir Vezey) (Moscow: Progress, 1972), 133-34. Yakovlev, as a leading aircraft designer and as deputy commissar of the aircraft industry, was in a good position to observe the VVS during the Great Patriotic War.
41. Generalleutnant Hermann Plocher, *The German Air Force versus Russia, 1941* (Air University: USAF Historical Division, 1965), 41.
42. *The Soviet Air Force in World War II*, ed. Ray Wagner and trans. Leland Fetzer (New York: Doubleday, 1973), 35.
43. Plocher, 41.
44. Up to March 1942, the Soviet long-range bombing force was called *Dal'nyaya bombardirovochnaya aviatsiya* (Long-range bombardment aviation), or DBA; up to December 1944 it was named *Aviatsiya dal'nego delvstviya* (Aviation of long-range operations), or ADD; from then until 1946 it was the 18th Air Army, and from 1946 on it has been designated simply *Dal'nyy aviatsiya* (Long-range aviation), or DA. The "long-range" part of the designation, whether for DBA, ADD, or the 18th Air Army, was until 1945 a relative term since the inventory included more medium- and short-range bombers than long-range ones. See A. Tsykin. "Taktika Dal'ney Bombardirovochnoy Aviatsii v LetneOsenney Kampanii (1941 goda)" (Tactics of long-range bombardment aviation in the summer-fall campaign [1941]), *Voenno-Istoricheskiy Zhurnal* (December 1971), 65.
45. Plocher, 42.
46. Paul Carell, *Hitler Moves East, 1941-1943*, translated by Ewald Osers (Boston: Little, Brown, 1964), 127: in *IVOV*, 2:111, the Soviet authors claim that they had only 677,085 troops on the Southwestern Front, after the battle some 150,541 remained, thus making the German figure absurd.

47. *Soviet Air Force in World War II,* 60–63.
48. Plocher, 127.
49. The German figures are probably exaggerated; the various Soviet sources differ considerably on the extent of the losses in the Vyazma-Bryansk debacle, but all give much lower figures than do the Germans. See Alexander Werth, *Russia At War, 1941–1945* (New York: Dutton, 1964), 230–31, for Russian estimates. John Erickson, *The Road to Stalingrad* (New York: Harper & Row, 1975), 1:219, points out that since the Russians could only muster 90,000 men in the Mozhaisk defense sector, the main defense line after the debacle, the Soviet losses must have been desperate as they had 800,000 men when the battle began.
50. *Soviet Air Force in World War II,* 67.
51. Asher Lee, *The German Air Force,* p. 117.
52. G. Mikhaylovskiy and I. Vyrodov, "Vysshie organy rukhovodstva voynoy" (Higher organs of wartime command), *Voenno-istoricheskiy zhurnal* (April 1978); see chart of command structure for World War II on p. 25.
53. M. Kozhevnikov, "Koordinatsiya deystviy VVS predstavitelyami Stavki VGK po aviatsii" (Coordination of VVS operations by the representatives of Stavka VGK for aviation), *Voenno-istoricheskiy zhurnal* (February 1974), 32.
54. Albert Seaton, *The Battle for Moscow, 1941–1942* (New York: Stein & Day, 1971), chap. 4.
55. J. T. Greenwood, "The Great Patriotic War, 1941–1945," in R. Higham and J. Kipp, eds., *Soviet Aviation and Air Power: A Historical View* (Boulder, Colo.: Westview, 1977), 21. Erich von Manstein, *Lost Victories* (Chicago: Regenery, 1958), 632, puts the aircraft reinforcements from the East at 1,500.
56. *Soviet Air Force in World War II,* 79.
57. Ibid.
58. Zhukov, 337–38.
59. *IVOV,* 6:45–46.
60. Alexander, 4, 7.
61. See M. Kozhevnikov, "Rozhdenenie vozdushnykh armiy" (Birth of the air armies), *Voennoistoricheskiy zhurnal* (September 1972) for details. A translation of this article by J. Waddell can be found in *Aerospace Historian,* June 1975.
62. Greenwood, 89.
63. The best account of the fierce struggle within the city is in Marshal V. I. Chuikov, *Nachalo puti* (Moscow, 1959), English translation by Harold Silver, *The Battle for Stalingrad* (New York: Holt, Rinehart & Winston), 1964. Marshal A. I. Yeremenko, Chuikov's superior, as commander of the Stalingrad Front, describes the battle as seen from headquarters in his book, *Stalingrad* (Moscow, 1961).
64. *Soviet Air Force in World War II,* 103–4.
65. A. Novikov and M. Kozhevnikov, "Bor'ba za strategicheskoe gospodstvo v vozdukhe" (The struggle for strategic command of the air), *Voennoistoricheskiy zhurnal* (March 1972), 27.
66. Soviet historians are addicted to "periodization," and World War II is no exception. They divide it into the "imperialist" period from 1 September 1939 to 22 June 1941, and then divide the Great Patriotic War into three periods: 22 June 1941 to 18 November 1942; 19 November 1942 to 31 December 1943; and from 1 January 1944 to the end of the war.
67. Novikov and Kozhevnikov, 26; Alexander, 168–70.
68. Alexander, 426–29.
69. Cajus Bekker, *The Luftwaffe War Diaries,* trans. F. Ziegler (London: Macdonald, 1964), 278.
70. Ibid., 283–85.
71. Novikov and Kozhevnikov, 28.
72. Bekker, 294.

73. *Soviet Air Force in World War II,* 110.
74. Bekker, 377, app. 14.
75. *Soviet Air Force in World War II,* 114–17.
76. Ibid., 164–65; Novikov and Kozhevnikov, 29; Generalleutnant Hermann Plocher, *The German Air Force Versus Russia, 1943,* ed. Harry R. Fletcher (Air University: USAF Historical Division, 1967), 75–83. Plocher gives 1,830 operational aircraft as the total employment for *Zitadelle.*
77. Oleg Hoeffding, *German Air Attacks against Industry and Railroads in Russia, 1941–1945* (Santa Monica, Calif.: Rand, RM-6206-PR, 1974), v.
78. *Soviet Air Force in World War II,* 174.
79. Schwabedissen, 168.
80. *Soviet Air Force in World War II,* 29.
81. Plocher, *German Air Force versus Russia, 1943,* 105.
82. According to the Soviets, *IVOV,* 5:467, the Red Army only increased 11 percent in manpower during 1943, but increased 80 percent in guns, 33 percent in tanks, and 100 percent in aircraft.
83. Alan Clark, *Barbarossa: The Russo-German Conflict 1941–45* (New York: Morrow, 1965), 370.
84. Alexander, 426–29.
85. Ibid., 299–300.
86. Ibid., 430–33, 172–73.
87. From January 1944 to January 1945, Russia's inventory of aircraft went from 8,800 to over 15,000. Greenwood, 118–19.
88. Ibid., 119.
89. *Soviet Air Force in World War II,* 361.
90. Text of the "Agreement Concerning the Entry of the Soviet Union into the War Against Japan, signed at Yalta February 11, 1945" in Max Beloff, *Soviet Policy in the Far East, 1944–1951* (London: Oxford University Press, 1953), 25.
91. *Japanese Special Studies on Manchuria,* vol. 13, "Study of Strategical and Tactical Peculiarities of Far Eastern Russia and the Soviet Far Eastern Forces," 111–12.
92. John R. Deane, *The Strange Alliance* (New York: Viking, 1947), 248.
93. Raymond Garthoff, "Soviet Intervention in Manchuria, 1945–1946," *Orbis* 10 (Summer 1966), 527.
94. Ibid.; *IVOV,* 5:548.
95. Garthoff, 531.
96. *Soviet Air Force in World War II,* 368.
97. Robert A. Kilmarx, *A History of Soviet Air Power* (New York: Praeger, 1962), 184. Soviet historians tend to downgrade U.S. Lend-Lease in general and aircraft in particular. During the Great Patriotic War, the United States delivered 14,018 aircraft to the USSR. Robert H. Jones, *The Roads to Russia: United States Lend-Lease to the Soviet Union* (Norman: University of Oklahoma Press, 1969), app. A, tab. 11.
98. Hoeffding, *German Air Attacks,* 25–28.
99. Ibid., 16.
100. Ibid., 17, 18–21.
101. Oleg Hoeffding, *Soviet Interdiction Operations, 1941–1945* (Santa Monica, Calif.: RAND, R-556-PR, 1970), 5; Greenwood, 130–31.
102. Richard Suchenwirth, *Historical Turning Points in the German Air Force War Effort* (Air University: USAF Historical Division, 1959), 77.
103. Ibid., 20–27; Suchenwirth, 3–4, says the two old fighter pilots, Goering and Jeschonnek, had a distaste for transport pilots.
104. Hoeffding, *German Air Attacks,* 8.
105. Schabedissen, 389.

Higher Command and Leadership in the Luftwaffe, 1935-1945

Horst Boog

It is not unusual for those who have lost a war to be more critical of themselves than those who came out of it as victors. In this context, some historians concentrate on the well-known attributes of the Luftwaffe, among which must be included its able application of such general operational principles as that of the interior line, of the concentration of forces at decisive places, and of surprise and successful cooperation with the land forces. Others recount the many acts of bravery and the generally high morale and fighting spirit of the Luftwaffe. But, in many ways, the important focus should be on a number of special facets of Luftwaffe command and leadership patterns or mind-sets which, arguably, proved to be decisive in the loss of the war in general and the air war in particular. This does not mean that without this faulty command thinking World War II would have been won by Adolf Hitler. But had it not existed, it would have been more difficult to defeat him and the Luftwaffe. It is important to first describe these patterns of command thinking, then analyze the consequences which they produced, and, finally, consider the origins of this method of thinking. These traits can be discerned most clearly in the attitudes and pattern of thought of Luftwaffe General Staff officers who were educated and trained at the *Luftkriegsakademie* (Air War Academy). They can also be found in Luftwaffe field manuals.

Five Special Traits of Luftwaffe Command Thinking

Concentration on Purely Military Matters

One of the most characteristic traits of Luftwaffe command thinking was its concentration on purely military matters. The humanistic model of the highly and universally educated individual able to reach decisions independently, as well as the principle of the universal assignability of the general staff officer continued to exist only in theory. About 70 to 80 percent of the available time in the curricula of the Air War Academy was reserved for military subjects. For the most part, technical subjects such as armament, economics, and industrial operations, factory organization, and mechanics, among others, had already been deleted in peacetime. They were not resurrected during the war. Also eliminated were subjects of a general nature, such as foreign aviation developments, foreign languages, and sciences. Military history was taught only to illustrate operational and tactical problems. It did not examine the interdependence among politics, economics, and warfare at the level of grand strategy—an interdependence which would have been obvious to officers studying the American Civil War. The *Wehrmachtakademie* (Armed Forces Academy), which was supposed to train a few select general staff officers in grand strategy, was closed completely three years after its establishment. Interestingly enough, the comprehensive subject of "air warfare" did not exist at the Air War Academy. Although in war such comprehensive knowledge was needed to determine the timely change from air attack to air defense, or vice versa, and to arrive at a realistic relationship between aims and means, air attack and air defense were taught separately.

During the war, this truncated course of study suffered further reduction. Instead of general or universal training and education, the Air War Academy concentrated on the elements of routine staff work, especially the method of issuance of orders. The original aim, to train future chiefs of general staffs, was expressly renounced in the last years of the war. This overall reduction of topics reached its climax in 1943 when the Luftwaffe leadership decided that a thorough introduction into the problems of higher command and higher operational thinking was no longer possible.[1] The understanding of the outside world with its various problems and of broad strategic issues became increasingly difficult for the Luftwaffe General Staff officer. Consequently, he had unclear conceptions about overall conditions overseas and about the potential of foreign war industries, and did not have the background to deal with questions exceeding the operational scope of the European theaters of war. For example, when Hitler asked his immediate entourage about the location of Pearl Harbor after it had been attacked by the Japanese, none of the officers, including the Luftwaffe representatives, knew exactly where it was situated.

Suffering under the stress of a continuous load of staff work, general staff officers who later rose to important command and staff positions did not develop a broad, strategic view of the war situation. Field Marshal Wolfram von Richthofen was a good example of this limited view. Von Richthofen had received the typical military technical and academic training and was a master in the field of close air support of the Army. Yet his personal diary contains hardly any indication that he attempted to understand the war situation as a whole. The dangers arising from this narrow-mindedness were recognized toward the latter part of the war, and the courses at the Air War Academy were extended to broaden the outlook of the general staff officer candidates. These endeavors came too late to have any effect.

There were shortcomings not only in the field of education, knowledge, and capabilities, but also with respect to the level of general experience necessary to support the principle of universal assignability. A shortage of time and personnel as well as the growing demand for hard-to-obtain specialized knowledge and capabilities blocked transfers between different occupational careers and prevented officers from becoming familiar with the other service branches and the problems of other theaters of the war. When the Luftwaffe curtailed the routine rotation between staff and troop assignments, it led first to an estrangement between general staff officers and troop officers and finally to an open critique of Hermann Goering and his general staff by highly decorated fighter commanders. Troop assignments of general staff officers proved to be too short, and transfers from the A2, or from signal communications, to another activity were well nigh impossible. Transfers from the A3 (operations) to the A4 (materiel) branch were greatly disliked for many reasons.

Specialization was the natural consequence of these problems and, in view of the pressure of time, certainly the most effective way of getting results from general staff officers quickly. This reduction of the scope of experience is illustrated in an order whereby the general staff candidates, after having passed the Air War Academy, were to be sent back to their original units for a probationary period. In practice, factors such as the urgent needs of these units sometimes determined these assignments. Provided only limited opportunity for reassignment, staff officers remained largely unaware that conditions and Luftwaffe missions in the various theaters of the war were different. When the Luftwaffe eliminated the requirement that a portion of the general staff's membership be combat pilots, it further reduced that staff's familiarization with the diverse dimensions of aerial warfare. This caused Field Marshal Erhard Milch to complain that the Luftwaffe High Command was not able to think in appropriate dimensions. The more the ideal of universal assignability became a fiction, the more the general staff officer candidate of the Luftwaffe became a specialist in a very limited field determined largely by his branch of service. As a result, the comprehensive view which was in such high demand became progressively harder to attain.

This development met the particular requirements of the individual service branches. Grand strategy being the exclusive domain of Hitler, they did not need the strategist, with whom they could do little; they needed the manager possessing special knowledge, even though he was no longer exchangeable. Under these circumstances, the training of general staff officers in the understanding of the overall interdependence among the economy, armament, enemy situation, technology, grand strategy, and warfare had to suffer.

Finally, another type of specialization occurred increasingly during the war years due to the criteria used for selection and promotion of officers. Hitler and Chief of Luftwaffe General Staff Hans Jeschonnek demanded young, higher commanders[2] for whom the general staff officer's career was to be nothing more than a stepping stone to advancement. In their view, a Luftwaffe General Staff officer should not so much distinguish himself by his education and knowledge, but rather prove his qualities by showing courage, bravery, and resolution. These attributes represented preindustrial values and were influenced by the Social Darwinism of the National-Socialist ideology. The growing importance of physical and psychological values corresponded to the general endeavor of the German armed forces, toward the end of the war, to mobilize the last mental and ideological energies in compensation for the lack of material and personnel strength. It is astonishing that the general staff officer training at the Air War Academy could be kept free from ideological indoctrination almost until the end of the war. Even after the attempt on Hitler's life on 20 July 1944, the commander of the Academy refused to introduce what he called mass-psychological indoctrination into the curriculum for the training of general staff officers.[3]

Overemphasis on Tactics and Operations

The Luftwaffe leadership's narrow view was encouraged by an education system that overemphasized tactics and operations at the expense of other fields such as logistics, intelligence, technology and signal communications, training, and transportation. This second trait was sometimes called *Ia-Denken,* or A3-thinking, since the German *Ia*—or operations officer position—corresponds roughly to the American A3 position. It was manifested in the distribution of available instructional time to the different subjects during the general staff training courses as well as in the assignments of general staff candidates during the probationary year. The four basic tactical subjects of air attack, air defense, land and sea warfare were allocated 38 to 50 percent of the instructional time. Together with military history—which was primarily the history of tactics—and war games, these subjects received 44 to 66 percent of available instruction hours. Only 12 to 21 percent of the hours were allocated to support functions, ranging from intelligence, quartermaster and signal communication services, to navigational, photo, and mapping services. So, at the most, the support services were given merely two-fifths of the time of the basic tactical subjects. Intelligence, in fact, disappeared completely from the curriculum of the Air War Academy during the war.

The preference given to the tactical and operational side becomes even more obvious on examining the assignments of the general staff candidates during their probationary year after leaving the Academy. Although from 1935 onward, the first chief of the Luftwaffe General Staff, Walter Wever, and his successors had repeatedly pointed out the necessity for an adequate knowledge of logistics by general staff officers and had warned against an underestimation of this field, assignments of general staff candidates to operations positions dominated until the end of the war. This was contrary to the practice in the Army. Usually more than one-half of the successful candidates at the Air War Academy, and above all the most qualified of them, were sent to assignments as operations officers. Assignments to the intelligence service were rare during the war and ranked far behind even those of the quartermaster service. General staff candidates strove for an operations career as the most distinguished of all the general staff careers because it could lead to the position of chief of general staff.

To a certain degree, the higher value given to the operations positions was justified. This was the place where all the results of the other command activities were transformed into command decisions. As the former chief of the German Army General Staff Franz Halder put it, this was "the brain that maintained the connections within the command sphere and secured the presence of adequate forces in the right place."[4] The greater importance attributed to A3 work, therefore, cannot be wholly condemned, but the excessive emphasis on it to the point that other command activities were neglected can be criticized.

There was still another reason for the preferential treatment of the A3 service. The operations branches in troop staff organizations of the Luftwaffe contained more positions than the other branches for general staff officers. In fact, the operations branches had even more positions than those to be found in comparable Army staffs. The share of general staff officer positions in the operations sections ranged from 50 percent in air fleet staffs to 100 percent in air division staffs as compared with only 36 percent in army group staffs and 50 percent in army division staffs. While the large operations role in Luftwaffe troop staffs was justified by the greater diversification of tasks to be solved in this part of an air force staff, the question must be asked whether the other fields of command activity did not also deserve a higher share of general staff officer positions based on the various tasks they had to accomplish.

Lower Priority of Technology Compared to Tactics

A third Luftwaffe characteristic was that of according technology a much lower priority than tactics. While Wever repeatedly underlined the equality of rank between tactics and technology, one of his successors, Hans Jeschonnek, in 1939 rejected the opinion of his engineers that the technical superiority of an air force would be decisive. Since all European nations found themselves on one and the same level technologically, he argued, it is hardly possible to reach technical superiority for any lengthy period of time. It would be better, therefore, to stress the development of tactics so as to give the Luftwaffe a unique advantage. Yet later, slight technical advantages decided the outcome of the war in the air. Technology and tactics should have been developed concurrently.

The first step toward the devaluation of technology in the Luftwaffe General Staff was the elimination of specialized training courses for future technical general staff officers in the spring of 1938. One of the reasons for this was the assumption that technology in the Luftwaffe could be mastered by normal versatile "tactical" general staff and troop officers who would have the assistance of Luftwaffe engineers. This assumption did not prove to be correct. A second step in this direction was the gradual reduction of technical subjects in the curriculum of the Air War Academy, until, during the war, they were no longer taught. This development took place despite the fact that the importance of technology continued to be stressed in Luftwaffe manuals and directives.

Technology was never in high favor with most of the officers. The situation was symbolized at the top by Goering, who bragged about his technical ignorance. In this respect, he had something to brag about. In the Luftwaffe General Staff, there was no engineer or technically and scientifically trained officer in a responsible position. As in the German Navy's earlier experience, technology and technicians were quite often treated with disdain. However, it must be asked whether the original intention of the Luftwaffe to create officers of both high tactical and technical competence was sound, or, whether, from the beginning, such an objective was fallacious because of the impossibility of any individual mastering both areas.

Overemphasis on the Offensive

A fourth trait which narrowed command thinking arose from employment doctrine. Offensive assumptions shaped the Luftwaffe doctrine of air war until nearly the end of the conflict. General Wever considered the bomber to be the decisive weapon in air warfare,[5] an idea which remained in the basic Luftwaffe manual on the conduct of air war, No. 16, until 1945. Of course, conditioning this idea was a conviction nursed by the doctrine of land warfare and by Germany's unfavorable geostrategic situation in the middle of the European continent.

Luftwaffe officials believed that the protection of the country against air attack could only be safeguarded by the possession of a sufficiently large buffer zone. By itself, this idea of a buffer zone did not involve aggression and should not be confused with Hitler's policy of aggression, even though both had the same effect in the end. Nevertheless, this concept implicitly required the conquest of sufficient territory once a war had broken out. It was advocated by Hitler, who was influenced by the geopolitical ideas of Professor Karl Haushofer as well as by responsible Luftwaffe commanders and general staff officers. The latter were not air-minded enough yet to imagine that the homeland could also be protected from hostile attack by building up a strong fighter defense force. As in other countries, a conviction that there was no effective means of defense against air raids also nourished the emphasis on offense. For example, at the Air War Academy, 16 to 21 percent of available instruction time was devoted to the subject of air attack while air defense was accorded only half that amount, a ratio which went unchanged until the last months of the war.

Furthermore, this offensive emphasis was clear in the selection of candidates for the general staff officer training courses. Until the end of the war representatives of the attack weapons (bombers) dominated the selection list. In fact, officers from bomber, dive bomber, and attack units constituted between 40 and 70 percent of all candidates from the flying service arms, a percentage that was 100 percent above their proportional share. This overrepresentation was the logical result of the emphasis on the subject of air attack in the curriculum. In contrast, fighter pilots, whose function was largely defensive, were underrepresented. Although they composed on the average about 40 percent of all Luftwaffe air crews, they generally received 17 percent of the staff officer training school assignments allocated to aircrew members. On many occasions their percentage was closer to zero. Although this detrimental situation was recognized late in the war, it could not be changed because of the heavy losses of the fighter arm. This unequal representation at the Air War Academy was a conscious policy of the general staff and not a matter of pure chance.

Narrow View of the Luftwaffe Mission

A final trait of Luftwaffe command thinking was the narrow view of its "mission." Although the idea of the necessity of strategic air warfare always existed in Luftwaffe doctrine—at least implicitly—the concept of a cooperative air force prevailed. This meant that offensive thought was not interpreted according to the theories of Giulio Douhet but was oriented toward land warfare. The experiences of World War I were a primary cause for this emphasis: the successes of German flying units were predominantly in ground support operations, whereas attempts to carry on strategic air warfare with dirigibles and giant bombers had proved futile. Since there was neither an independent air force nor an air arm in Germany after 1919, the concept of an air force as an auxiliary weapon of the army persisted. Further, in the small army left to Germany after the Treaty of Versailles, officers with flight experience filled only the lower ranks, largely because of their youth, while army officers filled the more responsible and influential higher staff officer and general officer positions. Moreover, most of the higher ranking officers with air experience in the later Luftwaffe had been commanders of fighter squadrons or reconnaissance units during World War I; none of them had experience in commanding larger air forces or in conducting strategic air warfare.

The knowledge that Germany's material resources were rather limited and that far-reaching bomber attacks would bring results only after a long and indefinite time (as proven in the Spanish and Sino-Japanese Wars of the 1930s) strengthened the preference for the so-called "operative" and cooperative air war. This approach promised to bring about faster successes in conjunction with armored thrusts and other operations of the Army. Considerations of economy and inadequate aiming devices for horizontal bombing furthered the development of the dive bomber. The increase in that airplane's weight caused by its extra equipment and fittings shortened its range and encouraged preparation for aerial warfare over medium distances in support of army operations.

Most interestingly and significantly, the concept of "strategic air warfare" always existed latently as an idea, but it did not exist in the official terminology of Luftwaffe doctrine. While a number of German journalists wrote as if the Luftwaffe were a Douhetean instrument, its leaders instead concentrated on the ground cooperation mission. Although they considered strategic bombing a legitimate air force task, they did nothing until 1943 to develop the concept because of bomber force attrition in land operations, especially in Russia. Moreover, on account of Hitler's inability to establish his continental empire, the time for a real strategic bombing war had not yet, and never was to come. During the 1930s, Luftwaffe planners emphasized interdiction, or indirect cooperation with ground forces. Because they initially thought close air support would be very difficult, they did not begin to refine the tactics for this mission until shortly before the war. Only after having learned from Allied strategic air operations that it was better to destroy enemy tanks and planes at the places where they were produced, rather than at the front, did the Luftwaffe undertake a belated and unsuccessful strategic air campaign against Britain and industrial centers and electric power plants in the Soviet Union in 1943–44.

Consequences of These Traits

Organization of the German Air Force High Command

The restriction of Luftwaffe command thinking to purely military matters and the preponderance of the operational sphere of command over the supporting sectors directly affected the organization of the German Air Force High Command. None of the other branches of the German armed forces changed its high command organization as frequently as the Luftwaffe. The reasons for this were manifold. At first, there was the lack of operational experience with large forces within this young branch, which had come into existence only in 1935. Another factor was the difficulty of combining in the best possible way tactics with technology, which was far more important in the Air Force than in the Army.

Other very important reasons for the frequent organizational changes include the personal, political, and functional rivalries at the top. Goering, the domineering and selfish, vain and indolent commander in chief of the Air Force, followed the example of Hitler in his use of the principle of *divide et impera* in order to secure his position. He remained at odds with Milch, his very capable and ambitious but "civilian" undersecretary of state for aviation (*Staatssekretar der Luftfahrt*). Furthermore, Milch struggled continuously with the Air Force General Staff, which declined to recognize this "political" superior. Milch's conflict with Ernst Udet, director-general of air armament (*Generalluftzeugmeister*), was part of the larger struggle between the former and Goering because Goering played Udet off against Milch. Hitler also played his part in the game, helping to prevent the enforcement of necessary organizational and operational measures and a clear separation of the military from the ministerial functions. In this way he was able to keep Goering, his old companion and designated successor, in his office as Reich minister of aviation and commander in chief of the Air Force in spite of the fact that his incompetence as a military leader soon became obvious. Hitler tolerated the on-going personal feuds in accordance with the *divide et impera* principle as a constituent element of his regime.

Above all, however, the top organization of the Luftwaffe had been streamlined to conform to the immediate requirements of a short war. Later, changes were required because this streamlined organization proved to be insufficient for the growing demands of the subsequent long war. The fact that the German Air Force had used up seven chiefs of its general staff in ten years demonstrates these conditions in the German Air Force High Command. So, too, do the suicides of Jeschonnek, the chief of the general staff, and Udet, the director-general of air armament.

The organization of the Luftwaffe High Command developed in four phases in peace and war. In the first, all branches of the Air Ministry were subordinated to Secretary of State Milch who correlated and controlled all the command functions necessary for the establishment of the Luftwaffe. This central control was necessary during the initial phase of the establishment of the Luftwaffe and was indeed very effective because Milch was an able and strong personality. Although he was not very easy to get along with, he took care of practically everything of importance in the Luftwaffe; and, unlike most Luftwaffe officers, he possessed a wide knowledge of economic, technological, and industrial matters. His abilities and the high esteem Hitler had for him aroused Goering's envy. Milch's less than satisfactory military knowledge, experience, and leadership (from the officers' point of view), and at times his rather high-handed manner of dealing with the Air Force General Staff, intensified Goering's opposition to him.

As a result, in a second phase, lasting from the summer of 1937 to early 1939, Milch temporarily lost his overwhelming influence within the Air Force as well as his position as deputy to Goering. At the same time, the Air Force General Staff became more influential and narrower in its outlook. Thus, purely military concerns began to prevail in the top organization even though the establishment of the Luftwaffe had not yet been completed. This was reflected in the reduction of the general staff to something akin to Goering's personal military operations staff. While the reputation of the general staff within the Luftwaffe command hierarchy was increased by its direct subordination under the commander in chief, the staff abandoned to the unreliable Goering its claim to comprehensive command responsibility. Since he lacked the steadfastness and determination of Milch, Goering commanded only nominally, and the centralized control of the Luftwaffe became increasingly weak.

In order to arrive at a more effective short war operational command structure, the general staff confined itself voluntarily to those command functions that were absolutely indispensable from a military point of view. The narrowness of the organization's perspective reached its zenith at the beginning of World War II, when Jeschonnek, the chief of the Luftwaffe General Staff, jettisoned as "ballast" and unnecessary for the immediate purposes of air operations the training, signal communications, and medical inspectorates as well as the civilian air defense staff. The other general staff inspectorates had previously been abandoned to the director of training (*Chef des Ausbildunswesens*), who was responsible to Undersecretary of State Milch. Jeschonnek also separated his headquarters from the office of the Luftwaffe Quartermaster General, a man whom he did not like. In addition to other problems, this meant a disruption between the Luftwaffe operations and logistics sections. The chief of the general staff now took over the office of the chief of the Luftwaffe Operations Staff and concentrated wholly on the tactical and operational side of the air war. For the anticipated short war, and in the brief campaigns of the first year of the war, this organization proved quite effective. For the long war that soon took shape, however, it proved a failure.

Yet no major changes took place in the senior organization of the Air Force during the third phase, which lasted from 1939 to 1943–44. Nor was there any change in the position of the general staff as Goering's personal staff of operational assistants. On the other hand, by 1939 Milch had regained the function of deputy to Goering and had become inspector-general of the Air Force. By the end of 1941 he had also become director-general of air armament, thereby further consolidating his position. Only in operational matters did the chief of the general staff report directly to the commander in chief of the air force. In all other respects, he first had to inform the undersecretary of state before seeing the commander in chief. These circumstances again stirred up the animosity between Milch and the chief of the general staff. In addition, Hitler's fundamental order of 11 January 1940, which forbade anybody from receiving more information than he needed for the execution of his orders, severed the vital collaboration between the chief of the general staff and the undersecretary of state for aviation/director-general of air armament. The order affected other sectors of Luftwaffe command even more significantly.

Meanwhile, the general staff became increasingly aware that a long war called for a command organization that suited the various demands of such a war, including economic, technological, and industrial requirements. Accordingly, the scope of responsibility of the chief of the general staff widened to become the germ of a comprehensive command organization of the Luftwaffe. Yet this organization took a long time to bear fruit because of Goering's ineffectiveness and lack of responsibility, and because of the division of the top organization between the two rivals, the chief of the general staff and the undersecretary of state.

Under Günther Korten, the chief of the Air Force General Staff, and Karl Koller, his chief of operations, the senior organization of the Luftwaffe finally adapted to the requirements of a long war of attrition in the fourth phase during 1944 and early 1945. Continual reorganizations characterized this period. Not before the last days of the war was an optimum scheme found, and then mostly on paper only.

On 5 February 1944, the chief of the general staff, who, unlike the undersecretary of state, had been signing "by order of" the commander in chief until then and had not been "acting for" him, received the authority of a deputy of the commander in chief for all military and operational matters. He was thus placed on an equal footing with Goering's other two deputies: Bruno Loerzer, his friend and newly appointed personnel chief or, as he was soon to be designated, chief of personnel and national socialist conduct of the Air Force *(Chef der Personellen Rüstung und Nationalsozialistischen Führung der Luftwaffe),* and Milch, the undersecretary of state for aviation and inspector-general of the air force. The latter's official duties had not yet been confined to ministerial matters alone and still comprised aviation training and the entire field of technology and air armament. The old comprehensive designation of the Luftwaffe High Command—"The Reich minister of aviation and commander-in-chief of the Luftwaffe"—had derived from the ideological National Socialist "leader principle" *(Führerprinzip)* and former Prussian administrative practice. This designation was now subdivided into "High Command of the Air Force" *(Oberkommando der Luftwaffe),* comprising the chief of the general staff and the chief of personnel, and "The Reich minister of aviation," under which title the secretary of state had to sign. Goering still opposed the concentration of all command functions of military relevance in the hands of either a "chief of air warfare" or the chief of the Luftwaffe General Staff. Hitler had recommended such an organizational reshuffling, but did not press Goering on the issue. As a result, the latter continued the practice of having the inspectors of the various arms report directly to him. These individuals, who played an important part in the development of tactics and aviation technology, became Goering's messengers and representatives with the troops. In effect, the chief of the general staff had no real authority over them. Goering, like the party itself during the last years of the war, also claimed the field of moral leadership and personnel management for himself as a high-ranking member of the National Socialist party. He withheld these responsibilities from the chief of the general staff by organizing them into a separate realm for Loerzer, his friend and deputy.

A clear separation of the military from the ministerial functions occurred only when Milch was deprived of the offices of undersecretary of state and director-general for air armament in June–July 1944, when his function as inspector-general of the Air Force was reduced to insignificance, and after Loerzer's position had been abolished on 8 December 1944. Only then was it possible to concentrate all of the military functions in the hands of the chief of the general staff. The chief of staff soon took over supervision of the development and procurement of aircraft, weapons, and materials, while Albert Speer's ministry supervised production.

In March 1945, the chief of the general staff finally became the principal deputy to Goering and won comprehensive authority over the entire Luftwaffe, including the right to issue directives to Goering's other two deputies, the commander of the replacement air force (*Befehlshaber der Ersatz-Luftwaffe*), and the new chief of aviation. The commander of the replacement air force and the chief of the general staff constituted the "High Command of the Air Force" and were also responsible for those aspects of training and personnel replacement not handled by the Air Force Personnel Office under the commander in chief himself. The new chief of aviation (*Chef der Luftfahrt*) was responsible for the ministerial part of Goering's duties. Now the chief of the general staff under Goering was the responsible officer for the direction of the entire air war and of the reduced air armament sector.

This seemingly optimum solution was achieved, however, within an air force whose striking power and reputation had already been broken. In addition, the new senior organization never had a chance to function properly because Goering's continued penchant for creating ad hoc offices directly under him produced unending confusion. Eroded by the influence of these ad hoc offices and beset with battlefield disasters, the new organizational scheme was largely stillborn. By this time, Luftwaffe High Command initiative had degenerated into mere reactions to enemy initiatives. The replacement of functional considerations by the "leader principle" and the overwhelming force of events played their parts in paralyzing the structure of the Luftwaffe High Command. Proper operational and organizational measures had been taken too late, after having originally been directed toward the "false" (short) war.

Effect of Overemphasis on Operational Thinking

Apart from this, note again how the upper level of the Luftwaffe reflected the organization traits of its command practice and operational thinking described above. This was especially apparent in the so-called "command orientation" or "A3 thinking," that is, the overemphasis on operations and tactics at the expense of logistics, technology, training, and other infrastructural and supporting spheres of command. Training, torn apart and removed from general staff responsibilities until near the end of the war, received little high-level attention. Furthermore, the air transport and the quartermaster services had been degraded organizationally, and technology was not connected with the operative command from 1937 to 1944.

While the Luftwaffe took a wider view of military operations than the other armed services, the fundamental trait of concentration on its own military business was evident in the lack of integration of the Luftwaffe into the organization of the

Supreme Command of the Armed Forces. Indeed, the Luftwaffe was much less integrated into the Supreme Command than the Army and Navy. This deficiency resulted, of course, from Goering's opposition to an effective supreme command of the armed forces by someone other than himself. Goering accepted directives only from Hitler and refused to subordinate the Luftwaffe to the chief of staff of the Supreme Command of the Armed Forces or to the chief of the armed forces operations staff. There was no liaison officer of the Supreme Command of the Armed Forces with the Air Force High Command. The Luftwaffe representatives at the headquarters of the Supreme Command of the Armed Forces fell considerably behind those of the Army and Navy in terms of rank and number because Goering was not interested in strengthening the authority of this institution by dispatching generals to it or in having his direct contact to Hitler disturbed by high-ranking air officers there. He would do nothing that might reduce the Luftwaffe's and his own independence in his capacity of commander in chief Luftwaffe. Furthermore, in the command structure of the Air Ministry, there were no top-level advisory and coordinating councils or agencies to tie together air operations with the other armed services, pertinent ministries, and the scientific and industrial establishment. Such high-level planning, advisory, and controlling bodies in England and the United States as the Air Staff, the Defense Committee, the Ministerial Committee on Military Coordination, and the Joint Chiefs of Staff did not have places in the German Air Force command organization. Instead, the primary function of the Luftwaffe command structure was to execute orders.

The preponderance of operational interests also diminished the chances of military success through their negative effects on the quartermaster service, on air transport, and on training, technology, and intelligence. Because it is impossible to describe these negative effects in detail here, several examples must suffice.[6]

Since specialization in the quartermaster business, for instance, could harm an officer's career, that service was not popular, and the best officers were not assigned to it. In 1942, the last chief of the Luftwaffe Operations Staff felt degraded when, as operations officer of the Fourth Air Fleet in Russia, he was appointed quartermaster of that fleet, a position which ranked higher than that of the operations officer. According to Field Marshal Albert Kesselring, the quartermaster service also had a low reputation in the general staff. The Luftwaffe commander with Heinz Guderian's tank army confided in his diary in 1941 how much he hated all the rear services and how foreign they were to him. Not surprisingly, awards to quartermaster personnel were much less numerous than to the fighting troops. Also, personnel replacements for the supply organization had lowest priority.[7] General Henry H. Arnold, reflecting on his adversary's shortcomings, concluded that the Luftwaffe made a grave mistake by never providing for sufficient replacement of aircraft and crews.[8] The chapter on supply and replacement in Fundamental Field Manual No. 16 on the Conduct of Air War was never written, although this field manual went through several editions.

The mentality described here had a profound impact on the air war, as well as on the war in general, which was conducted in accordance with operational and strategic aims and not on the basis of logistical considerations. Good examples of this were: the failure to occupy Malta, an omission which greatly disturbed the Axis supply convoys to Africa; the Luftwaffe's promise to supply the Sixth Army in Stalingrad by air, although past experiences had already proven the impossibility of an undertaking of such dimensions; and the way in which Erwin Rommel stormed forward in Africa without sufficient numbers of tanks and troops to occupy the British stronghold in the Nile Delta. Hitler fought and lost World War II with an inadequate understanding of logistical considerations.

The treatment of air transportation is another significant example of the neglect of logistics. No mention is made of air transport as a means of supply in the Handbook for the Luftwaffe General Staff Service of 1939. When air transportation was needed, the necessary aircraft and crews were formed ad hoc from the advanced flight training schools using Ju-52 planes. If air transportation had ranked sufficiently high organizationally, the promise of air supply for Stalingrad would not have been given so readily to Hitler by Goering and his chief of the general staff. Neither man understood the subject very well. An air transport command with a competent staff and sufficient authority appeared only after Stalingrad and Tunis, where the Luftwaffe had lost most of its transport planes. What is said here about air transport is also true of the signal communications service. Although the signal troops made up 20 percent of the Luftwaffe strength, their chief was only a three-star general, while the rest of the Luftwaffe included ten four-star and five five-star generals.

As Field Marshal Kesselring and the quartermaster general of the Luftwaffe confirmed after the war, training was the stepchild of the service which adhered to the principle that the surprise strike at the beginning of an operation had to be conducted with full strength to include the maximum number of troops drawn from schools and reserves.[9] As already mentioned, training was taken from the general staff and given to the undersecretary of state in 1939. The chief of the general staff was more interested in maintaining large numbers in operational front-line units and less concerned with securing a sufficiently broad base of thoroughly trained crews. The training establishment was already too small at the beginning of the war. The last chief of the Luftwaffe General Staff wrote after the war: "The number of flying units was increased at the price of a low training level and a lack of reserves."[10] General Jeschonnek once said, "First we have to beat Russia, then we can continue training."[11] The advanced training schools were frequently deprived of their flight instructors, who, along with their planes, were assigned to air transport duties. The training time was steadily shortened to the extent that a German pilot at the end of the war received less than one-third the flight training time of an American pilot. Since the training schools had too few modern combat aircraft, young pilots had very little time to become accustomed to them in their operational units. When fuel had to be saved, that saving began in the training sector. When aluminum was in short supply, the production of training aircraft was curtailed. In 1944, the recently reestablished training branch of the general staff stated that the quantity of trainees had had a higher priority than the quality of their training.[12] As a result, more than 50 percent of flight accidents in 1944 were due to inadequate training.[13] Since noncombat aircraft losses very often were higher than those caused by enemy action, this figure takes on added significance. Actually there was a *circulus vitiosus*: the low-quality training caused higher losses which increased the shortage of aircraft and diminished the allocation of combat aircraft to the schools.

Technological Blunders

The rather low esteem among the military for technology led to the appointment of incompetent people to important positions. The best known case is that of the amiable and valiant Ernst Udet, a Bohemian, an artist and a clown in the air, but not the capable manager needed in the position of director-general of air armament.[14] He rose to that post because Goering wanted to please Hitler by appointing this well-known man who was so beloved by the people. Udet appointed as his chief engineer a young man who had no experience in the mass production of aircraft and who eventually was fired. Later in the war, an unqualified officer headed the technical sector, and his deputy freely confessed that he did not understand anything about technology. Goering usually appointed highly decorated young officers as his technical consultants because he felt that bravery in combat counted more than technical knowledge and experience. He preferred officers rather than expert engineers in positions of technical importance, military persons who would not accept the word "impossible." They decided the technical questions.[15] The word of combat-experienced officers counted for more than that of the engineers, and this resulted in constant alterations of aircraft types and frequent delays in production in the decisive first years of the war.

The ignorance of responsible Luftwaffe leaders about the problems associated with aircraft development can be explained. In the simplest terms, they became convinced that technology could be directed in accordance with the military principle of order and obedience. This fact went hand-in-hand with the pressure exerted by Hitler, the circumstances of the war, and their sometimes blind dedication to their superiors. Goering was always greatly astonished and furious when he could not quickly get the technical advances he wanted.[16] Von Richthofen reacted in like manner during his tenure as chief of the development branch in the technical office of the Luftwaffe.[17] Series production of an aircraft began before the completion of its testing phase, requiring alterations at the front and preventing many planes from becoming combat ready. The best-known cases of this wasteful policy were those of the He-177 bomber and the Me-210 destroyer projects. The He-162 jet fighter is another example. That fighter was a brilliant engineering achievement; but because it was in mass production only three months after its conception, it could not possibly stand the test of combat.

Since tactics and technology were organizationally separate, the general staff paid very little attention to problems involving both areas and sent hardly any technical requirements to the technical office. Goering saw to it that his general staff concerned itself with operational matters while the technical office devoted its energies exclusively to matters in its own sphere. The lack of tactical requirements for the technologists was also due in part to Hitler's reluctance to inform the general staffs about his plans early enough, so that the latter simply were not always in a position to issue such requirements. Hitler's policy shows that he too believed that mere orders were enough to direct industry and to shift it quickly to new programs.

Although increasing amounts of money were spent on research, the percentage expended in relation to the sums put into aircraft production steadily decreased.[18] Goering and Milch did not take much notice of this; their interest was in increasing production. They did not understand much about research and therefore could not secure the proper direction and coordination of this vital activity. The Department of Aviation Research was steadily downgraded within the Air Ministry until it was dissolved in 1942, to be replaced by an extremely inefficient Aviation Research Council.[19]

Some of the biggest blunders in the technical field were Goering's and Hitler's directives of 1940 and 1941 which canceled all development projects that did not promise to yield results within one year. This and other technical difficulties delayed the development of the first jet plane, the He-178, and of the first operational jet fighter, the Me-262, and the bulk of Luftwaffe combat aircraft at the end of the war were outdated. Although aircraft factories clandestinely carried on developmental work, their technical personnel were mostly shifted to the production side. This halt to development stemmed from ignorance of the importance of technological continuity, the need to produce as many proven weapons as possible, and Milch's hesitation to embark on entirely new projects. In this way, the Luftwaffe lost its initial qualitative advantage over enemy air forces—its only real advantage.

In this connection, a word must be said about the preference line officers enjoyed over engineers in the Luftwaffe, who had the status of civilians although wearing uniforms. As in the Navy, the technologist or engineer was not considered to be on an equal footing with the line officer. Because the conception of the military value of the scientist and engineer was inadequate, most were drafted into the Army as ordinary infantrymen. This mistake went uncorrected until far too late. Many Luftwaffe engineers developed a mixture of inferiority and superiority feelings toward the line officers because the latter generally and socially counted more but at the same time were ignorant about technology. In contrast to the officers, the engineers had only limited opportunities for advancement despite their wide responsibilities. For example, the 18 highest-ranking engineers held ranks equivalent to a one-star general in 1945 when the Luftwaffe had 176 one-star, 101 two-star, 57 three-star, 7 four-star, 4 five-star generals, and one Reich marshal. Many young engineers left their corps to join the line officer corps at reduced rank in order to improve their careers.

Low Priority Accorded Intelligence

Some remarks must be made on the relationship between operations and intelligence.[20] The latter never enjoyed the same reputation in the Luftwaffe as the former, although it can be said that intelligence work was rated higher before than during the war. The easy initial successes in the various blitz campaigns fostered the neglect of intelligence work, as did the disappointments later in the war, when intelligence forecasts proved to be false and spies were discovered in the Intelligence Branch of the Luftwaffe Operations Staff. The best officers of the general staff were not assigned to intelligence work or to the related attaché posts. A number of other developments illustrate the disregard of the importance of intelligence. For example, radio intelligence remained a secret realm of the chief of signal communications; technical intelligence fell under the control of the director-general of air armament; and both had no organizational connection with the central Intelligence Branch of the Luftwaffe General Staff. Some seven offices in the Luftwaffe collected and/or evaluated intelligence. A comprehensive field manual for intelligence work did not even exist.

The low priority attached to intelligence resulted in fundamental blunders in assessing the intentions and capabilities of Germany's three main opponents—England, the United States, and Russia. The potentials of all three were substantially underestimated before and during the decisive initial years of World War II. Hitler arrived at false decisions concerning Great Britain, Russia, and the United States due in part to the false assessments of his Luftwaffe's intelligence service. The predominantly military training of the intelligence officers usually led to intelligence assessments which proved correct in the narrow military sense—that is, in regard to the location of units, the types of weapons, strength of troops, and so on—but which were mostly wrong concerning the economic, political, and moral war potential of the opponents. Intelligence officers had not received the necessary broad and pertinent education. Moreover, they tried to conduct intelligence work by themselves as a purely military matter without coordinating their efforts with economists, technicians, and scientists. The results of Luftwaffe intelligence corresponded with the low priority assigned to this field of activity.

Dominance of Offense over Defense

The belief that effective air defense was impossible shaped the doctrine of the offensive as it evolved in the 1920s and 1930s. Most of the world's air forces shared the conviction that, since the bombers would always get through, the best defense would be offensive operations against the centers of the enemy's war potential. This emphasis on the offensive, which had nothing to do with aggression as far as the military were concerned, appears in all pertinent manuals on air war. The decisive role of an independent fighter force as an effective means of defense and of gaining command of the air was not yet in the minds of Luftwaffe strategists. Thinking about military aviation conformed to the principles of land warfare. The unfavorable geostrategic situation of Germany encouraged military thinkers to view the conquest of sufficient buffer zones as the best defense against air attacks. The two-dimensional thinking of land warfare prevailed over the three-dimensional thinking best suited to an air force. The easy victories in the early years of the war confirmed the belief in the superiority of the air offensive over the defensive and were a major reason for the delay in creating an effective, centralized air defense system. It was not until 1943–44 that the various local defense systems were unified into one organization covering all of German air space. This centralized air defense was but a belated reaction to the Allied strategic air offensive.

Nevertheless, the idea of the greater value of the offensive prevailed in the heads of many a Luftwaffe leader including Koller, the chief of the operations staff. The land war–minded Hitler, who at that time exerted an overwhelming influence on the Luftwaffe, in contrast to his impact in the early stages of the war, was a staunch disciple of offensive air operations. He and Goering perverted the doctrine of the air offensive by repeatedly employing the fighter force in support of the land fronts (e.g., against the Allied invasion troops, the last time on 1 January 1945 in what was called operation *Bodenplatte*), although the precondition for attack, the control of the air over one's own territory, no longer existed. This false employment of fighters weakened the air defense of Germany, which, in the opinion of Adolf Galland, the General of Fighters, had to be strengthened to allow undisturbed industrial production. This offensive mentality produced fruitless bomber attacks on England when the air defense of the home country was at stake and prevented a timely shift in priority from bomber to fighter production. Likewise, Hitler's land war–mindedness prevented a shift in priority from the production of more and more ineffective antiaircraft guns and ammunition to fighters. For example, the aluminum for the fuses of the heavy anti-aircraft artillery (AAA) shells produced during the war would have sufficed for the production of about 40,000 additional fighters.[21] This does not mean that no AAA guns should have been constructed. They were, of course, necessary for local defense and anti-tank warfare. Yet, there were not enough of them to put up a curtain of fire, whereas perhaps 10,000 more air defense fighters would have greatly enhanced the deterrent capability of home defense. Although to a certain degree, land and sea operations required the production of offensive aircraft, these operations withdrew resources from air defense.

The Role of the Luftwaffe: Strategic Bombing or Tactical Support?

The above observations lead to the question of whether the Luftwaffe was meant to be primarily an independent strategic air force or a force cooperating with the Army and Navy. Although it is generally known that it finally turned out to be a cooperation force, opinions differ as to whether this was the Luftwaffe's primary purpose as originally conceived.

Communist historians tended to regard the Luftwaffe, as well as "capitalist" air forces, primarily as a strategic and terror instrument. Their intent was to demonstrate how "humane" the Soviet Air Force practice of cooperation with the Army was in contrast to the "barbarian" method of the "imperialist" air forces. Those historians tried to turn the deficiencies and ineffectiveness of the Soviet strategic bomber force into a virtue by saying that it was dedicated to the more "humane" mission of army cooperation.[22] This was, of course, just one facet of the ideological struggle against the "class enemy," also known as the policy of "peaceful coexistence."

According to Fundamental Field Manual No. 16, the Luftwaffe had three tasks:

first, to annihilate the enemy air force by attacks on its ground organization and industrial base rather than by fighting it in the air; second, to support the operations of the Army and Navy; and, third, to bomb the centers of war potential in the rear of enemy territory. Obviously, the Luftwaffe doctrine of employment had a strategic and a cooperative component. Although this enumeration did not imply a priority for the supporting role of the Luftwaffe over its strategic role, this was in fact the case since the strategic role was to be resorted to only when there was a standstill in the land war. In corroboration of this, it is worthwhile to remember what has been said previously about the concepts of "operative" and "strategic" air warfare. Although "operative air war" was a very unclear concept, it shows the tight linkage of air war thinking with land operations. The Luftwaffe order of battle was not structured in accordance with independent offensive or defensive functions, but was designed to cooperate with the Army in the ratio of one air fleet to each army group. The suffering caused by the Allied strategic air campaign finally made the Luftwaffe comprehend strategic air war and compelled certain of its operational thinkers belatedly to demand—or to deplore the nonexistence of—a strategic air force.

In contrast to this de facto, predominantly cooperative employment of the Luftwaffe, which corresponded with its doctrine, one could argue that as early as 1933 Hitler conceived the Air Force as a *Risiko-Luftwaffe* (risk air force) whose function, from the beginning, was to deter potential attackers by the menace of an indiscriminate strategic air offensive against their industries and population, thus protecting the growth of German war potential. It is quite true that in subsequent years the Luftwaffe did have the strategic task of helping Hitler extort political advantages from other countries. The plans to expand the Luftwaffe strength five-fold in 1938–39 and to give it a considerable strategic component is taken as proof of its strategic role in Hitler's attempt to dominate the world. Finally, the Battle of Britain is considered proof of the intrinsic purpose of the Luftwaffe. None of these arguments, however, proves that the Luftwaffe was actually designed as a strategic weapon. The politicians (including Goering) assumed the risk and blackmailing roles so as to use the Luftwaffe for bluffing; Milch, in 1936, thought of it more as a cooperative weapon.

In reality, responsible Luftwaffe commanders considered their service, even in 1939, unfit for a strategic air war overseas and did not intend to build up an air force composed of a large number of strategic bombers. In their view, the proper doctrine of employment, which was greatly conditioned by Germany's geostrategic position, called for an independent air force capable of assisting the advancing ground forces, which thus would remove the imminent air threat. The first chief of the Luftwaffe General Staff, Wever, ordered priority given to the fast medium bomber in May 1936, before his fatal accident.[23] The development of the heavy bomber had a very low priority although it was never canceled. The decision of Goering in April 1937 to scrap the existing prototypes of strategic bombers was just the belated execution of Wever's earlier decision and was in keeping with the general opinion within the Luftwaffe that the fast medium bomber was needed more urgently than the big bomber for its potential tasks.[24] Of course, Wever had been aware of the possibilities of the long-range strategic bomber. Under the influence of the successful Allied strategic air campaign against Germany, Luftwaffe officers working for the U.S. Historical Division glorified him after the war as the "father of strategic thought in the Luftwaffe." But Wever was also a realist who knew what the Luftwaffe needed first in the near future. So, although Luftwaffe air doctrine was double tracked, comprising a strategic and cooperative component, the accent lay on the latter component both in theory and, even more so, in practice. The impact of the experience of the Spanish Civil War, combined with the doctrine of combined operations, convinced the Luftwaffe to develop a close air support corps and not a strategic bomber command. The latter was vainly envisaged only late in the war.

In summing up, one arrives at the conclusion that it was the politicians, exploiting foreign propaganda, as in the case of the bombing of Guernica in 1937, who imposed a strategic disguise upon the Luftwaffe in order to deter potential enemies. On the other hand, responsible Luftwaffe commanders shaped their weapon with a view to its prevailing cooperative tasks, strategic employment being considered, but thought unnecessary in a future war against adjacent opponents whose territories were to be conquered and not destroyed. According to doctrine, the Luftwaffe was to fight the enemy's armed forces rather than its civilian population. It was not until the brilliant victories in Poland, Norway, and France that Luftwaffe leaders, in the wake of the enthusiasm and euphoria arising from these victories against weaker opponents, forgot the negative results of previous war games and came to believe in their ability to fulfill the strategic task of bringing England to her knees independently. They bluffed themselves by their own propaganda. Of course, this explains, at least in part, their high hopes. The geostrategic situation had meanwhile changed much to Germany's advantage, and no one as yet had experienced the difficulties associated with strategic bombing operations. Had the Luftwaffe been designed as a strategic force, and had the situation of an immovable frontline with England not come so unexpectedly early, that service certainly would have developed and employed the four-engine bomber. But all this is hypothetical and not historical. It suffices to repeat that the experiences of the Spanish Civil War indicated to the Luftwaffe leadership the importance of ground cooperation operations and that Army leaders in the late 1930s continued to pressure the Luftwaffe to maintain a strong ground support force.

German emphasis on the dive bomber also encouraged the Air Force to become a tactical rather than strategic force. The Luftwaffe developed the dive bomber because it lacked a suitable bomb-sight for horizontal attacks and because it needed an effective but cheap bombing aircraft. The dive bomber seemed to answer these needs, which led to the decision to make all bombardment aircraft dive bombers. As previously noted, the move to dive bombers increased aircraft weight and reduced range, thereby encouraging the Air Force to view its bombers principally as a ground cooperation force.

After the disappointments of the Battle of Britain, the Luftwaffe was employed mainly in the ground support role envisaged by its doctrine. The wide use of its medium bombers in close air support (especially in Russia) undermined the force. Only too late did the Luftwaffe High Command realize that the bomber force could achieve better results if used against strategic targets. But it was Hitler, the Supreme Commander of the Armed Forces as well as the Commander in Chief of the Army, who now prevented this insight from being put into practice by ordering increasing but very costly close air support. So, although the Luftwaffe doctrine for the conduct of air war provided for both a strategic and a cooperative function, the accent in doctrine and practice was on cooperation. It was this mode of employment from which the Luftwaffe could never free itself and which became its Verdun.

Origins of the Luftwaffe's Mode of Thinking

Having delineated the fundamental traits of Luftwaffe command thinking and some of their consequences in practice, a word should be said in explanation of the origins of this way of thinking.

Most of the fatal consequences for the Luftwaffe can be understood in terms of its policy of extensive rather than intensive armament. Hitler pressed the armed services to reach his rearmament goals within the shortest possible time. It was the number of soldiers and aircraft in the combat units that counted, and not the thorough and time-consuming construction of a durable infrastructure or the formation of reserves. Preparations were designed for a short war against not more than one weak opponent at a time. This was the kind of war Hitler expected. The Luftwaffe's attitude toward the support and infrastructural spheres of command as well as toward technical research, training, and reserves was entirely consistent with Hitler's views on military preparation. The fiasco came when the war turned into a world war of attrition which Hitler had not expected. He had hoped to achieve his goals in Europe before the big powers in the East and West had time to rearm.

Traditional German Military Attitudes

Hitler's short war mentality was only partly to blame for the demise of Germany; traditional German military command thinking was aware of the necessity to keep wars short by superior operational abilities on account of scanty resources and was an equally important cause of eventual disaster. In fact, the limitation of command thinking to purely military matters and the overemphasis on the operational and tactical aspects of military operations can be traced to Helmuth von Moltke (the Elder). He, unlike Carl von Clausewitz, separated politics from war and wanted no politician to interfere with the generals' responsibility for the conduct of war.[25] Field Marshal Alfred Schlieffen, the chief of the Imperial German General Staff, continued the trend toward further narrowing the theory of war to purely military aspects and developed an almost autonomous, mechanistic war plan, which ignored diplomacy in 1914.[26] The famous Schlieffen Plan was not a comprehensive war plan and did not address the political and economic aspects of a long war. According to Gen. Erich Ludendorff, who epitomized these attitudes, policy had to serve war.[27] This meant the militarization of public and political thought, a common circumstance in Germany in the 1930s. Twice in the history of the German General Staff, in the 1860s and in the 1920s and early 1930s, attempts had been made to widen the horizon of the officers selected for general staff work beyond their ordinarily good broad education and to include a solid education in the natural sciences, technology, politics, and economics.[28] But these attempts had failed, the last time because of Hitler's accelerated rearmament program. Within these limitations were trained the general staff officers of high military competence who were later to become the responsible Luftwaffe commanders and chiefs of staffs. It needs to be added that neither Goering nor Milch had ever seen a general staff from inside; their own military experience had not exceeded the rank of captain in combat units.

Logistics was never prestigious in the German Army. As early as 1848–49, the man who would become Emperor William I considered it the weakest part in the organization of the Prussian Army.[29] Even the famous Moltke[30] and Schlieffen[31] treated it with disdain since it was not directly operational, an attitude that would persist among later officers, especially Rommel.[32] On the other hand, because of Germany's unfavorable geostrategic situation, the priority of the offensive has always been a fundamental element in German military thinking from the era of Clausewitz[33] via Moltke[34] and Schlieffen[35] to Ludendorff[36] and Hans von Seeckt.[37] It was also a fundamental concept in the doctrine of Douhet, whose great influence on all air powers in the 1920s and 1930s already has been noted.[38]

The problem of reserves, like that of logistics, has traditionally been neglected in German military thinking. Douhet,[39] Clausewitz,[40] Moltke,[41] and Schlieffen[42] did not think much of strategic reserves because they thought that the decisive battles took place at the beginning of a war. This was also the conviction of General Jeschonnek,[43] chief of the Luftwaffe General Staff, who had received his general staff training in the Army. This line of thought implied a further limitation on command thinking to military operational aspects and led it into a blind alley because it neglected other possibilities of action and ended in helplessness and improvisation when the initial strike with all available forces failed.

The doctrine of air-ground cooperation did not have such a long pedigree as the other concepts mentioned above, being the consequence of the World War I and Spanish experiences. Seeckt conceived of the Luftwaffe as an auxiliary instrument for land offensives to enhance the power of the attacking armies.[44] Interdiction was to him more remunerative than strategic air bombing. It is significant that the chief of the General Staff of the Army[45] and many a Luftwaffe general originally had come with similar experiences from the Army.

The Socio-Educational Background of Luftwaffe Senior Officers

The German military had a long history of unfamiliarity with things technological. The second-rate treatment of technology and technicians in the German armed forces had deep causes which originated from the social and political changes that took place during the age of industrialization. These brought about what can be called a sociocultural overlap. In addition to being rich, the upper middle-class distinguished itself from other and lower classes by a broad education in the humanities. The ruling aristocracy was still the captive of a preindustrial way of life in which the irrational virtues of man counted more than the rational approach of the technologist. It was a special phenomenon of German social development in the decades before World War I that the educated society endeavored to imitate the ways of thinking and behavior of the doomed, but politically powerful aristocratic class. This so-called feudalization of society and public life tied together politically the educated and wealthy bourgeoisie and the aristocracy. Both social elements stood for the continuation of the monarchical system and for aristocratic and middle-class supremacy in society. Both were opposed to the egalitarian ideas advocated by the lower classes and feared being overthrown by the socialist masses.

The upper class received its education predominantly at the *Humanistische Gymnasium,* the traditional type of high school stressing the humanities, whereas lower class children mostly attended a type of high school known as *Realgymnasium* or *Oberrealschule,* which placed emphasis on modern languages and the sciences. Knowledge of these fields was essential for people who did not possess wealth and had to earn their living. The long controversy about these two types of education, which was so significant for nineteenth-century Germany, was intrinsically sociopolitical. Kindling these quarrels was not only the disdain of the broadly educated member of the upper class toward the more specialized and technically trained member of the lower classes but also the unwillingness of the former to allow the lower classes an increasing influence in society. Although the Army would have preferred officer candidates with a "realistic" education as offered in the *Realgymnasium* or *Oberrealschule,* it expected the graduates of the humanistic high schools to become future officers because they were considered to come from families that stood for throne, altar, and fatherland.[46]

Therefore, in order to understand the Luftwaffe attitude toward technology, it is important to take a look at the social backgrounds and education of Luftwaffe generals. One-half of the 326 Luftwaffe generals whose personal files have evidence about their education (not all the files are so informative) went through the *Humanistische Gymnasium.* Less than one-fourth were educated at a cadet school and about one-sixth at a *Realgymnasium* or *Oberrealschule.* The others either left high school without taking the final graduation examination or had only eight years of elementary school education (*Volksschule*). Thus, it can be said that the majority of these generals received primarily a humanistic education. Only a fraction had high school training in the sciences, and only a quarter had a mixture of scientific and military training. Most of the generals attended high school before World War I, that is, at a time when the philosophies of the irrational and an authoritarian spirit ruled in the schools. The emphasis of those schools was on receptivity and obedience rather than on inquisitiveness and initiative.

As far as the social origin of the Luftwaffe generals is concerned, the Bertram

Collection[47] of short biographical sketches of the 570 Luftwaffe generals reveals some interesting data. Four hundred and ninety-two of the sketches contain information on the professions of the fathers of the generals and show that more than three-quarters of them came from the educated upper middle-class or from officer families. Most came from families of high-ranking civil servants (136), followed by independent landowners, factory owners, merchants, physicians, apothecaries, and lawyers (131). The professions of the remainder included: military officer (91), clergyman (20), and scientist or engineer (11). There was also a large group whose fathers were low-ranking civil servants (69). More than half of the fathers were in academic professions, and seventy-five fathers belonged to the nobility. In some cases, the mothers were of noble descent, and a considerable percentage of the generals were married to women of noble descent. If it is assumed that the factory owners were also technicians or scientists, only a total of 83 among 492 generals' fathers had technical professions. It is well known that before World War I technological impulses rarely originated from the nobility, the army officer corps, the educated upper middle class, or the landowners, but rather from the lower middle class, the source of most of the technicians and engineers. The upper middle class tended to produce scientists. There is a big difference between a scientist and a technician.

These statistics imply that the relatively low development of technological thinking in the Luftwaffe may perhaps also be attributed to the nontechnical origins of its leaders. This is corroborated by their regional origin. Of 547 generals whose places of birth could be identified, 200 came from east of the Elbe River, 64 from Bavaria (especially from northern Bavaria), 64 from lower Saxony/eastern Westphalia/Schleswig-Holstein, and 34 from Austria-Hungary or other regions. Thus, 362 came from regions which had little industrial development before World War I. A minority of 185 generals came from more industrialized central and west Germany, Alsace-Lorraine, and Wurttemberg/Baden. It should be added that only 27 of the 570 Luftwaffe generals had obtained the academic degrees of a diploma engineer or of a doctor of engineering. Five more were civil engineers. Regarding the general staff officers, in March 1940 only 11 of 238 had an academic degree in engineering.

The Unique German Principle of Command

Another cause for inadequate technological understanding was the principle of command called *Auftragstaktik,* which had been developed in the Army, where many Luftwaffe officers had first served. The *Auftragstaktik* allowed the independent execution of orders in accordance with the respective situation, that is, the order did not prescribe how the task was to be carried out but left this to the individual soldier in command. In land operations a quick adaptation to new situations by fresh decisions and orders was generally possible. This was not the case in air operations, however, where technology played a more important role. Decisions in the field of technical development and production of aircraft usually were binding on tactics and command for a longer period. Likewise, air operations, once started, could not be very easily altered. Their conduct required more planning than that of land operations because tactics rested more on technology.

Because of these technological considerations, many a Luftwaffe general, having been trained in the Army, was almost driven to despair. One of them,[48] a former student of a humanistic high school who had been the chief of the general staff of an air fleet and a higher commander for many years, wrote after the war that operational thinking and ability could no longer be regarded as prerequisites for general staff service in themselves. This was true, he claimed, because operations and tactics now depended primarily on the intentions of the Army, conditions of traffic and logistics, or on political, economic, and technological considerations of the high command rather than on such traditional elements as the enemy situation, one's own intention, assessment of the situation, and decision. Command decisions in the Luftwaffe, he continued, were the result of meticulous planning based on technological factors. This process no longer allowed room for imagination and intuitive understanding of one's own and the enemy's situation. The routine of the expert had sufficed; and when the first mission was on its way, there was nothing more to be ordered or led. Along with technology, logistics also enjoyed very little esteem in his eyes because to him it seemed to require "only" organizational abilities and some technical knowledge, but no tactical thought. Important tasks, he concluded, had been transferred from the sphere of "scientific" thinking and artistic planning to the level of plain common sense, which he obviously regarded as ranking below the general staff level. Nothing had been left to the general staff but the pursuit of technology, which had dethroned the general staff in the Luftwaffe. It was not the general staff that placed requirements on technology; rather the general staff had to adapt its requirements to technical reality. Higher command had shifted from operations and tactics to organization and technology, and tactics had neglected to anticipate the technological future. Although younger Luftwaffe officers certainly thought more along the lines of tactics and technology, their lower rank denied them a role in determining Luftwaffe command thinking.

The foregoing evidence about the attitude toward technology makes it clear why, despite the great amount of technology in the young Luftwaffe, its officer corps did not immediately remold the traditional, authoritarian mode of command taken over from the Army. The evidence also shows why the officers did not adapt that mode to technology and shape it into a modern, cooperative style of command which would have paid due recognition to special knowledge and abilities as well as to functionally correct action and discipline.

There are many other reasons for this failure. All technical functions requiring special technical knowledge were left to the Luftwaffe engineer corps, a nonmilitary corps of civil servants in uniform who were only to serve and support the military but who had no independent command authority. On the other hand, the officers were trained for general command functions and often had no technical training or knowledge, and this sometimes led to incorrect decisions with permanent consequences. Also, as a result of the depression in the early 1930s, there was a shortage of technically skilled personnel in the Luftwaffe. During the depression, many young men shied away from technological studies. By the time the Luftwaffe engineer corps was founded in 1935, employment opportunities for technically skilled people had improved remarkably, and many of them were already working in the armament industry. Also, the National Socialist ideology of irrationalism decried the rational approach to life as "Americanism" and glorified preindustrial values like bravery, perseverance, and faithfulness (without which, of course, no society or army could exist). Such an atmosphere was certainly not favorable to the development of a functional discipline and a cooperative style of command within the Luftwaffe which could have complemented the formal discipline and the personal authority of leaders needed in combat.

The Impact of the "Leader Principle"

In the final analysis, the development of a cooperative style of higher command which could accommodate modern technology was impaired by the "leader principle." In the words of Hitler, this required absolute authority from above and unconditional obedience from below. This principle took to extremes the military maxim of order and obedience, which previously had not excluded a certain amount of sober argumentation in the decision-making process nor required the expediency of resignation or transfer in case of disagreement with decisions. Many Luftwaffe commanders, therefore, felt themselves helplessly bound to orders which they considered wrong. This was particularly so with regard to Hitler's increasing influence on the Luftwaffe and his growing habit of bypassing the general staff and giving orders directly to the smallest units. Yet Goering loved the "leader principle" and, as Field Marshal Kesselring wrote after the war,[49] this principle was nursed particularly in the Luftwaffe, where it should have been out of fashion more than elsewhere.

Both Goering and Milch were in essence political leaders. Yet when Milch was in charge of air armament after Udet—at least in the field of technological management—he encouraged discussion and consultation instead of simply issuing orders. On the other hand, the complacent, ambitious, and selfish autocrat Goering imposed the autocratic style of command upon the Luftwaffe. Only his close friends could talk to him freely; he kept other senior commanders with important missions waiting for days, or had them travel after him, or did not receive them at all. Goering, like Hitler, did not think much of general staff work. Having been revolutionaries, both hated the general staff officers for being too aloof and too sober to believe blindly in them. They thought that vesting an officer with sweeping independent powers for a certain task would solve problems. The many special plenipotentiaries—all of them little "leaders"—whom Goering subordinated to himself were one reason why the Luftwaffe High Command gradually disintegrated. Significantly, while the command organizations of the other two armed services always bore the impersonal designation of High Command (*Oberkommando*), the Luftwaffe command organization was known under the personal designation, "The Reich Minister of Aviation and Commander-in-Chief of the Luftwaffe," until the last year of the war.

Finally, Hitler's Fundamental Order No. 1 of 11 January 1940, completely stifled the development of an adequate mode of higher leadership. Every commander had to exercise the utmost secrecy, making discussion, consultation, and coordination of efforts nearly impossible. Not even so much as a handful of Luftwaffe officers were allowed to have a full picture of the overall situation and war effort.

Conclusion

An inadequate and otherwise impaired higher command organization and mode of command thinking, as well as faulty leadership and personal rivalry at the top, caused the Luftwaffe's many problems, errors, and mistakes. It should be pointed out, of course, that other air forces, sometimes for similar, sometimes for other reasons, had these deficiencies, but such shortcomings were not fatal because they had sufficient resources and time to recover. Many problems, especially those of fitting technology into modern command practice, still exist in almost every air force in various disguises. It takes time to become air-minded and thus to develop a mode of command and leadership adequate to war in the air. The German Air Force being only four or, if you prefer, six years old when the war broke out, was still too "young." Clandestine preparations and exercises in the Soviet Union in the 1920s and early 1930s lacked mass and the test of reality. Unlike the three main opposing Allied air forces, the Luftwaffe was never given the time and necessary authority to get settled, to break with the Army and the National Socialist ideological style of command, and to find its own way. The fate of the Luftwaffe proved again that one cannot fight successfully a war for which one is not organizationally and doctrinally prepared.

Notes

1. Luftkriegsakademie Ia, "Abschlussericht über den 3. Kriegslehrgang der Luftkriegsakademie Berlin-Gatow vom 1.9.1943," 12ff., BA-MA (Bundesarchiv-Militärarchiv) RL 5/1031.

2. "Der Chef des Generalstabes der Luftwaffe Nr. 740/42 vom 28.10.1942," BA-MA, Milch Collection, 53:1075-81.

3. Luftkriegsakademie Ia Az., 34 Nr. 12103/44 g vom 15.9.1944, "Richtlinien für den Unterricht im 6. Kriegslehrgang 1944/45," 10, BA-MA RL 5/1032; cf. O. Wien, *Ein Leben und viermal Deutschland* (Düsseldorf: Droste Verlag, 1978), 437-52, and H. J. Rieckhoff, *Trumpf oder Bluff?* (Geneva: Verlag Interavia, 1945), 98.

4. Colonel General Halder (Ret.) in P. Bor, *Gespräche mit Halder* (Wiesbaden: Limes Verlag, 1950), 57.

5. "Vortrag des Generalmajors Wever bei Eröffnung der Luftkriegsakademie und Lufttechnischen Akademie in Berlin-Gatow am 1. November 1935," *Die Luftwaffe. Militärgeschichtliche Aufsatzsammlung.*, ed. Der Reichsminister der Luftfahrt und Oberbefehlshaber der Luftwaffe LA III, 20 Feb. 1936, 1:7.

6. Cf. also R. Suchenwirth, *Historische Wendepunkte im Kriegseinsatz der deutschen Luftwaffe*. USAF Historical Studies. MGFA Lw 35, 26-39.

7. Field Marshal A. Kesselring, *Soldat bis zum letzten Tag.* (Bonn: Athenäum, 1953), S.126; cf. Rieckhoff, 90ff., and "War Diary LtCol von Barsewisch," 25 Sept. 1941; see also Air Division, Control Commission for Germany, British Element, *A Study of the Supply Organization of the German Air Force 1935-1945*, June 1946, 71ff., 78ff., 84f.

8. General H. H. Arnold, *Global Mission* (New York: Harper, 1949), 370.
9. Field Marshal A. Kesselring, "Die deutsche Luftwaffe," in *Bilanz des Zweiten Welt-krieges* (Oldenburg/Hamburg: G. Stalling, 1953), 153ff., and General von Seidel in his speech of 1949, BA-MA Lw 101/3, pt. 2, 51f.
10. MGFA (Militärgeschichtliches Forschungsamt) A-83, p. 24.
11. General von Seidel, 52.
12. Ia/Ausb (II A) vom 30.9.1944, Vorentwurf für eine Studie "Herabsetzung der Flugzeugverluste ohne Feindeinwirkung," BA-MA RL 2 II/181. Cf. *The United States Strategic Bombing Survey*, ed. David MacIsaac (New York and London: Garland, 1976), 3:6.
13. Oberkommando der Luftwaffe—Führungsstab Ia/Ausb Nr. 999/44 g vom 11.4.1944, "Herabsetzung von Flugzeugunfällen"; Oberkommando der Luftwaffe, General der Fliegerbodenorganisation und des Flugbetriebes Gr. IV Az.52bl0/gK 1010/44 vom 13.7.1944, "Die Zahl der Störungen im Flugbetrieb ohne Feindeinwirkung im Monat Mai 1944," BA-MA RL 2 II/181.
14. R. Suchenwirth, "Ernst Udet, Chief of the Luftwaffe Supply and Procurement," in Suchenwirth, *Command and Leadership in the German Air Force* (USAF Historical Division, July 1969), 53-111.
15. Engineer General G. Huebner (Ret.), "Wehrmacht und Technik," BA-MA Lw 103/21, 15.
16. "Besprechung des Reichsmarschalls Goering mit Vertretern der Luftfahrtindustrie über Entwicklungsfragen vom 13.9.1942," BA-MA RL 3/60, 5279-87.
17. K. H. Völker, *Die deutsche Luftwaffe, 1933-1939* (Stuttgart: Deutsche Verlag-Anstalt, 1967), 58f.; E. L. Homze, *Arming the Luftwaffe: The Reich Air Ministry and the German Aircraft Industry, 1919-1939* (Lincoln and London: University of Nebraska Press, 1976), 86f.
18. Homze, 28f., 210ff., 228, 267.
19. Ministerialdirigent Baeumker, chief of the research department in the Reich Air Ministry, to Field Marshal Milch, 10 Jan. 1942, BA-MA RL 3/54, 2430f.
20. General Leutnant A. Nielsen (Ret.), "Die Nachrichtenbeschaffung und auswertung für die deutsche Luftwaffenführung," MGFA Lw 17.
21. Letter Der Staatssekretär der Luftfahrt und Generalinspekteur der Luftwaffe St.713/ 41 gK vom 24.10.1941 to Goering, BA-MA RL 3/50, 455ff.; cf. W. A. Boelcke, *Deutschlands Rüstung im Zweiten Weltkrieg* (Frankfurt am Main: Akademische Verlagsgesellschaft Athenaion, 1969), 23, and R. Wagenfuehr, *Die deutsche Industrie im Kriege 1939-1945* (Berlin: Dunckert Humblot, 1954), 33.
22. O. Groehler, *Geschichte des Luftkrieges 1910 bis* 1970. Militärverlag der Deutschen Demokratischen Republik, Berlin (East) (1975), 135, 148f., 150, 202.
23. Letter LC II Nr. 3201/36 vom 6.5.1936 (Copy), BA-MA Lw 103/50.
24. Homze, 122f., 125, 127f.
25. H. von Moltke, *Militärische Werke*. Hrsg. vom Grossen Generalstab. Berlin 1892-1912, 2:42.
26. J. L. Wallach, *Kriegstheorien. Ihre Entwicklung im 19. und 20.Jahrhundert* (Frankfurt am Main, 1972), 93, 95f., 133.
27. E. Ludendorff, *Kriegführung und Politik* (Berlin: E. S. Mittler, 1922), 23.
28. Cf. D. Bald, *Der deutsche Generalstab 1859-1939, Reform und Restauration in Ausbildung und Bildung. Schriftenreihe Innere Führung, Reihe Ausbildung und Bildung, Heft 28*. Hrsg. vom Bundesministerium der Verteidigung, Führungsstab der Streitkräfte I 15, 1977, 37ff., 87ff., and General H. Speidel (Ret.), "Generalstab und Bildung," in *Militärgeschichte, Militärwissenschaft und Konfliktforschung. Eine Festschrift für Werner Hahlweg...*, hrsg. von Dermot Bradley und Ulrich Marwedel. (Osnabrueck, 1977), 383-96.
29. M. van Creveld, *Supplying War: Logistics from Wallenstein to Patton* (Cambridge, England: Cambridge University Press, 1977), 78.
30. Ibid., 96, 105, 107f.

31. Ibid., 117f., 138.
32. Ibid., 181-201; cf. D. Irving, *The Trail of the Fox: The Life of Field Marshal Erwin Rommel* (London: Futura, 1978).
33. Carl von Clausewitz, *Hinterlassene Werke*, 10 vols. (Berlin: Dümmler, 1858), 5:176 (cited after Wallach, 57, 58).
34. Von Moltke, 4:S.227f.
35. Cf. Wallach, 114.
36. E. Ludendorff, *Der totale Krieg* (München: Ludendorffs Verlag, 1935), 79.
37. Cf. Bald, 69, 84.
38. G. Douhet, *Luftherrschaft*. Deutsch von Rittmeister a.D. Roland E. Strunk, (Berlin: Drei Masken Verlag, 1935), 22f., 69, 82; cf. Wallach, 339f.
39. Douhet, 43, 80, 82.
40. Carl von Clausewitz, *Vom Kriege*. Hrsg. von W. Hahlweg, (Bonn: Dümmler, 1973), 399.
41. Von Moltke, 2, 3:211.
42. Alfred von Schlieffen, *Cannae* (Berlin: E. S. Mittler, 1925), 280, 315.
43. Generalstab 1. Abteilung Nr. 1643/37 gk (Ia) vom 25.10.1937, "Stellungnahme zum Organisationsvorschlag Oberstleutnant Kammhuber," gez.: Jeschonnek, BA-MA RL 2 III/4.
44. Cf. H. von Seeckt, *Gedanken eines Soldaten*, (Berlin: Verlag für Kulturpolitik, 1929), 93-100.
45. Nachlass General Schlemm, file 1, 3. MGFA.
46. Cf. M. Messerschmidt, "Militär und Schule in der wilheiminischen Zeit," in *Militärgeschichtliche Mitteilungen*, 1/1978, 51-76.
47. K. Bertram, Kurbiografien der Generale der deutschen Luftwaffe (not yet published).
48. General Wilhelm Speidel (Ret.), "Gedanken über den deutschen Generalstab." Landsberg/Lech 1949 (unpublished), MGFA MS Nr. P-031a 26, pt. 1:53f., 72, 73, 79; "Der Weg zum Generalstab, Auswahl, Erziehung und Ausbildung des deutschen Generalstabsoffiziers." Landsberg/Lech 1950 (unpublished), MGFA MS Nr. P-031a 26, pt. 2:34, 92f.
49. Field Marshal A. Kesselring, "Der Generalstab des deutschen Heeres. Mit einem Anhang: Berührungspunkte des Heeres-Generalstabes mit dem Luftwaffen-Generalstab," BA-MA Lw 101/11, 182.

Some Observations on Air Power

Ira C. Eaker

Since 1917, I have been an interested and concerned observer of the development and application of air power. Thirty years of that period were spent on active duty, mainly in the Army Air Corps, and now thirty-one years in retired status, but I still maintain an undiminished interest in air power development, weapons, and tactical and strategic employment. My remarks will be in three parts: part one, the development of U.S. air power prior to 1942; part two, air power application and development from Pearl Harbor to the present; and part three, a brief look at air power in the future. Obviously, to cover sixty years of aviation history in less than thirty minutes requires great selectivity.

Air Power Developments to 1918

Since 1978 marks the 75th Anniversary of the Wright Brothers' first powered flight—17 December 1903—it is appropriate that we should review that time span, assess the results of those flights, and catalogue some of the major events which those flights initiated.

Prior to World War I, the principal item in U.S. air power development was the employment of the 1st Aero Squadron, Aviation Section, Signal Corps, U.S. Army, in support of Gen. John J. Pershing's Punitive Expedition against Pancho Villa, the Mexican bandit. That squadron (incidentally the only squadron the Army had), commanded by Capt. Benjamin D. Foulois, had limited success in its observation missions due to the difficulty of maintaining its primitive planes and engines in the sand and wind of the western desert. This poor showing was due to lack of popular interest and congressional support in the fourteen years following 1903. When World War I came in 1914, the United States was thirteenth among nations in military aviation. Even Mexico had more military planes than the United States.

Then came American involvement in World War I. The most significant legacies of World War I to air power developments in the United States included:

(a) Public interest in aviation as a result of the daily news releases on air operations. The lurid stories of the battles in the skies caught the popular fancy; the names of the leading British, French, Canadian, and American aces became household words.

(b) The 200,000 pilots and technicians the United States trained for that conflict became the nucleus for postwar development, civil and military. They were the gypsy pilots who bought war surplus aircraft and carried aviation to every village in the country. They provided the pilots and maintenance crews for the early air mail and the small but enthusiastic air components of the Army, Navy, and Marine Corps.

(c) From the opportunity given American air leaders to meet their European counterparts, exchange ideas, and compare lessons learned came the early evolution of the theories of air power, its organization, tactics, and strategy.

(d) The war demonstrated the need for the best planes, engines, machine guns, bombs, and communications, and provided the essential aviation laboratories and airplane and engine factories for postwar development.

All things considered, World War I was the greatest stimulus to the air world up to that time.

Aviation Developments from 1918 to 1941

Many events and individuals were responsible for the growth of U.S. air power during this period. Some of the most significant events, certainly, must include the First Flight Around the World, the Pan American Goodwill Flight, the Question Mark Flight, the experiments with refueling in air, the first nonstop Flight to Hawaii, and the many successive records set for speed, altitude, and endurance. The Army Air Corps also engaged in semi-annual maneuvers designed to test developing theories of the proper organization, technical improvements, and application of air power.

In May 1927, Charles Lindbergh's nonstop flight from New York to Paris caught the popular fancy as no other aviation event had done. The fantastic response to this feat, here and abroad, gave an unprecedented stimulus to civil and military aviation development. It immediately inspired civil transcontinental airlines and plane and engine factories. The 1931 depression hindered but by no means stopped this development.

The campaign of Gen. Billy Mitchell attracted much publicity and, as controversy always does, made leading headlines and won political partisans pro and con. His court-martial in 1925 stimulated popular, news media, and congressional interest in military aviation. Historical studies of the Mitchell episode, concluding with his trial, demotion, and retirement from the Army, do not always report this event accurately.

Some report, for example, that Gen. Mason M. Patrick, then chief of Air Corps, was antagonistic to Mitchell and his efforts to obtain an independent air force. I occupied an office, as assistant executive, between General Patrick's and General Mitchell's offices from June 1924 until long after the trial, and I can testify that our chief and assistant chief conferred frequently, that each admired the other, and that they were jointly supportive of much the Army Air Corps was doing in those dramatic times.

When the trial began, General Patrick gave me a directive which was to make me, in effect, an assistant counsel in Mitchell's defense. He said, "I want General Mitchell to have any records in our files needed for his defense, but I also want to assure that we get them back. I have told Colonel White, Mitchell's chief military counsel, that you will be responsible for furnishing him any official files he wants." This gave me an opportunity to observe much of this dramatic event.

General Mitchell has also been represented as having been unfairly crucified by a hostile War Department and a prejudiced court-martial. As a matter of fact, Billy Mitchell deliberately goaded President Calvin Coolidge and the War Department into bringing him to trial. He was determined to use the trial as a public relations forum to convince the people and Congress that the Army and Navy were deliberately neglecting their air arms to such an extent that national security was dangerously compromised.

General Patrick has not received the credit he deserves for the part he played in the development of U.S. air power at a critical time. He was a West Point classmate of John J. Pershing. He stood at the top of the class; Pershing was having trouble scholastically, and Patrick was assigned as his tutor. Pershing never forgot his helpful classmate, took him to France, made him chief of the Air Service when Mitchell and Foulois fell out, and also made him the first chief of the postwar Air Service, later Air Corps. Patrick was the leading advocate for many of our cherished Army air ambitions and plans at this time. He was a gifted public speaker and testified before Congress in our favor on many occasions, as well as making many appearances before important groups ably advocating our cause.

During this period, many boards and commissions, beginning with the Lassiter Board and followed by the Baker Board and the Morrow Board, examined the claims of the "young Turks" (some called us Bolsheviks) in the Air Corps and usually approved some part of our demands.

Another aspect, seldom mentioned and never given sufficient credit, was our relationship with the chairmen of the principal committees of Congress—the Military Affairs and Appropriation Committees of the House and Senate. Our successive chiefs of Air Corps provided these congressional leaders air transportation back to their home states or districts, enabling them to make more speaking engagements and public appearances. This gave their especially selected pilots an opportunity to insure that their political passengers became knowledgeable and interested in air matters. Most of them became powerful air power advocates.

In July 1926, Congress passed the Aviation Act authorizing assistant secretaries for Air for the War, Navy, and Commerce Departments. The first appointees to this new task were F. Trubee Davison (War), David Ingalls (Navy), and William McCracken (Commerce). Each became able advocates of aviation in their departments. Secretary Davison brought with him a professional newspaperman whose salary he paid. (In those days public relations men were not authorized in Civil Service ranks.) He taught us many "tricks of the trade" in this important area.

In 1935, as a result of constant publicity and congressional pressure, the War Department authorized the GHQ (General Headquarters) Air Force. This was the first recognition that there might be a mission for military aviation other than tactical air support for armies and navies. General Frank M. Andrews, designated as the commander of GHQ Air Force, was an ideal selection for this critical new assignment.

A skillful and experienced pilot since 1919, with service on the War Department General Staff and a recent graduate of the Army War College, General Andrews immediately established the organization and set about proving that there was a task for air power far beyond the arena of contending armies and navies. Andrews pushed the development and testing of the B-17 Flying Fortress and sent the first of these planes on spectacular missions, such as the six-plane flight to Argentina and the mission to photograph the new Italian liner, *Rex,* four hundred miles at sea. The Air Corps officers who commanded groups and wings in the GHQ Air Force became the commanders and staffs of the sixteen worldwide air forces of the Army Air Forces in World War II.

We must not overlook the major contribution of the Air Corps Tactical School to the development of Air Force doctrine in the years between the wars. Its able, specially selected, instructors became the nucleus of Gen. Henry H. "Hap" Arnold's planning staff in 1941. The war plans they prepared provided for the 2.3 million airmen, the 150,000 planes they manned and supported, plus the great industrial base we established, all of which played a prime role in the conquest of Hitler's Third Reich and the Japanese war lords.

General Arnold recognized that the Spanish Civil War—where Germany and Russia tried out their latest planes, weapons, and tactics—was but a prelude to World War II. He and his staff eagerly reviewed all data reported from the air battles of that conflict. When Adolf Hitler's *Blitzkrieg* bypassed the Maginot Line and conquered France, Arnold sent observers to cover and report every phase of that conflict. When France fell, our observers witnessed Dunkirk, then went to England where they observed the Battle of Britain.

Another great contribution to U.S. preparedness for entry into World War II was a series of maneuvers the U.S. Army and its Air Corps held in Texas, Louisiana, Georgia, and the Carolinas in 1940 and 1941. General Arnold told me that Gen. George C. Marshall had said one of the prime purposes of these maneuvers was to observe, select, and train the ground and air commanders for World War II.

Arnold had a full appreciation of the value of public relations. For example, he invited Hanson Baldwin, military editor of the *New York Times,* Deke Lyman of the *New York Tribune,* Ernie Pyle, and many others to witness the air element of these maneuvers. It was not accidental that Army Air Forces operations in World War II were adequately reported.

Another evidence of Arnold's genius was the relationship he established with Harry Hopkins, the closest adviser to President Franklin D. Roosevelt. This resulted in Roosevelt's dramatic radio speech announcing that the United States would build 50,000 planes in 1940–41. Arnold used this declaration to start a vast program of pilot and technical schools even before Congress had appropriated the money. Arnold also encouraged the British and French to buy U.S. planes for their hard-pressed air forces. The money from these purchases started that great expansion of the aeronautical industry which produced 100,000 planes and 250,000 engines in a single year, 1944.

These were some of the principal events prior to Pearl Harbor which affected the development of U.S. air power. These dynamic events also suggest the names of the political, military, and industrial leaders who played the leading roles in them.

Air Power in World War II

World War II provided the stage for the greatest test of air power thus far. In that war, Allied air leaders were, for the first time, given the resources to validate the theories of earlier air advocates, such as Hugh Trenchard, Giulio Douhet, and Billy Mitchell.

Allied air power in Europe between 1941 and 1945 accomplished three essential missions. First, it destroyed the German Air Force. Second, it supported the Allied armies and navies, enabling them to accomplish their indispensable missions. And third, it successfully demonstrated its strategic capability by destroying much of the weapons-making and war-waging capability of Hitler's Third Reich. These air operations were often interrelated and mutually supportive; combined, they were absolutely essential to victory in Europe.

In April 1942, General Marshall came to the 8th Bomber Command's temporary headquarters during a visit to the British chiefs of staff. After hearing my report on our plans at that early date, he said to me, "Eaker, I do not believe a cross-Channel invasion of Europe will ever be possible until the Luftwaffe is destroyed. Do your plans provide for that?" I answered in the affirmative, assuring him that destruction of the German Air Force was our prime intermediate objective. As our bombers began their attacks on fighter factories, oil production, ball bearing plants, and other key elements of weapons production, Hitler would demand that Hermann Goering's Luftwaffe intervene. The ensuing air battles would result in virtual destruction of the Luftwaffe, I explained.

Again, at the Casablanca Conference in February 1943, General Marshall asked, "Do you still believe that the German Air Force will not seriously interfere with our cross-Channel invasion next year?" I replied, "If the heads of state and their chiefs of staff approve the Combined Bomber plan just presented, and give us the air resources it calls for, I am certain the Luftwaffe will not be a serious factor in that operation."

I was in Russia on the Joint Shuttle Bombing Mission, 6 June 1944, when General Dwight Eisenhower's forces crossed the Channel. I asked Maj. Gen. John Deane, Ambassador Averell Harriman's senior military aide, about the air resistance. He said that early dispatches did not mention it, but he would send a signal to find out. Back came the cryptic reply, "The Luftwaffe did not show today." That moment represented my greatest personal satisfaction in World War II.

The German Air Force did not interfere with our landing, as General Marshall had feared. Neither did it seriously challenge the subsequent land battles as our armies advanced into Germany. Instead, our Allied air forces won and held air superiority throughout, inflicted many thousands of casualties upon the enemy, stopped all road movement by day, and eventually denied the German Army required fuel and weapons, resulting in its unconditional surrender.

I have always assumed that Albert Speer, Hitler's armament minister, was the best authority on the effect of Allied air power on Germany and its war effort. In that belief, Dr. Arthur Metcalf, founder of the U.S. Strategic Institute, and I spent the afternoon of 21 October 1976, with Speer in his ancestral home on a mountain top above Heidelberg. He was fully responsive to our questions.

For example, at one point I said, "Aside from the bombing of German industry, a very high priority with the Allies was the destruction of the Luftwaffe. Since the Luftwaffe did not show on 6 June 1944, when that great invasion armada appeared off the three invasion beaches, we thought we had positive evidence that our Allied air offensive had largely destroyed the German Air Force." Speer answered, "I think your surmise is essentially correct. I was still turning out the required number of fighter planes, but by that time we were out of experienced pilots. We were so short of fuel that the incoming pilots in our flying schools only received three and a half hours of flying training per week. These poorly trained and inexperienced pilots were suffering heavy losses. A pilot only survived for an average of seven missions against your bombers and their accompanying long-range fighter escort."

I recently received a copy of an article Speer wrote on 9 August 1978 which gives further information on the effect of our bombing. From this I quote a few paragraphs. First,

> Your bomber offensive against Germany actually opened a second front long before your invasion of the Continent. Because of the unpredictability of your daily target selection, we had to keep a million men at home to defend against them. These defenses also required 20 thousand anti-tank guns as flak. Other thousands of people were required as fire fighters and to repair damaged factories. Those men and munitions could have provided another 60 divisions for use against Russia or to oppose your invasion in France.

Another significant Speer observation:

> In January 1943 our losses from your bombing of our war-making industry I estimated at 5.4%. In December 1943 it had climbed to 28%. The short fall in critical weapons, according to a memorandum I made at the time, amounted to 36% for tanks, 30% for military aircraft and 42% for trucks.

> Thus the losses inflicted by the American and British air fleets constituted for Germany the greatest lost battle of the war, far exceeding our losses at Stalingrad, in the winter campaign in Russia or during the retreat from France.

> The truly decisive factors were the weakening of the German defensive strength and the immobilizing of German planes and tanks caused by the American and British Air Forces. Even before the encirclement of the Ruhr, the collapse was already final.

Can there any longer be doubt about the conclusion of the Strategic Bombing Survey:

> Allied air power was decisive in the war in Western Europe. Its power and superiority made possible the success of the Normandy invasion. It brought the economy which sustained the enemy's armed forces to virtual collapse.

Prime Minister Winston Churchill told me in January 1944, "The predictions you made to me at Casablanca last February about our combined bomber missions, including around the clock bombing, are now being verified. I no longer have any doubt that they will prove completely valid." That for me was the second great satisfaction of World War II in Europe.

The phenomenal courage and gallant persistence of the combat leaders and crews of the Royal Air Force (RAF) and U.S. Air Force were the main reason for the success of that effort. There were other reasons, such as sound plans, the support of the heads of state, Churchill and Roosevelt, and their chiefs of staff, our overwhelming weapons-making capacity, and the complete support of our civilian population at home. But, if the morale of the bomber crews had ever failed, all the rest would have been to no avail. That is why I shall always give the combat airmen the prime credit for the decisive air victory.

With our victory in Europe assured, Japan remained. General Marshall sent me on a twenty-day trip around the world in May 1945 with a directive to visit each theater, consult the commanders, and report to him how much of our air power in Europe did we need to move to the Pacific Theater, and, of equal importance, how much could they accommodate logistically.

Generals Douglas MacArthur and George Kenney, with their ground and air forces, and Admirals Chester Nimitz, William Halsey, and Raymond Spruance, with their naval forces, had been accomplishing remarkable results, despite limited resources, against the Japanese warlords while the war in Europe was our number one priority. Now that MacArthur and Nimitz were to have available all the needed resources from the European Theater, plus our undivided weapons-making effort—now at maximum capacity—the result was no longer in doubt.

There was still the fear of heavy casualties in the invasion of Japan's home islands. Instead, as we all know, Japan surrendered in three months before any Allied soldier had set foot on Japanese soil. The naval blockade and air power's destruction of their weapons-making industry made it impossible for them to go on. For the first time in history, a powerful nation had surrendered without invasion or occupation. The theories of early air power advocates like Trenchard, Douhet, and Mitchell had been largely vindicated.

What are the prime lessons learned from World War II? Free men make better soldiers, sailors, and airmen than the minions of totalitarian dictators. Sound strategic plans, steadfastly pursued by determined national leaders, are essential to victory in war. The united support of civilian populations on the home front is a decisive factor. The obvious lesson from the final days of the war against Japan is that by the proper employment of sea and air power, a land invasion of enemy territory can be avoided, saving heavy and unnecessary casualties.

This last observation quite naturally raises the question of whether an invasion of Germany would have been required had adequate air power been available. Marshal of the Royal Air Force Sir Arthur Harris, in musing on this possibility, points out that only 17 percent of Allied resources were devoted to air power, 33 percent to the sea war, and 50 percent to the armies. Had these resources been distributed between the three services equally, 33-1/3 percent to each, the RAF could have had 1,000 bombers and the 8th Air Force could have had 1,000 bombers and 1,000 long-range fighters in 1942. In that event, he speculated, a land invasion of the continent might not have been necessary. All the armies would then have had to do was provide an occupation force in Germany as in Japan.

Before leaving World War II and moving on to our subsequent wars in Korea and Vietnam, I propose to pay a deserved tribute to Gen. Carl Spaatz, particularly appropriate before this audience here at the Air Force Academy. He was our senior strategic air power commander, both in Europe and in the Pacific. He was the only general present at the surrender ceremonies both in Berlin and aboard the battleship *Missouri* in Tokyo Bay.

I have often said that he was the only general I knew who never made a mistake. His friendship and influence with General Eisenhower was largely responsible for the generally sound tactical and strategic employment of air power in Europe. His standing with President Harry S. Truman was a prime influence in obtaining a separate Air Force, co-equal with the Army and Navy, and he was the only candidate to be the first chief of staff of the U.S. Air Force.

For the last nineteen years of his life, I was privileged to spend much time with him, from which I have many cherished memories. For example, I was with him when he was inducted into the Aviation Hall of Fame and heard him say, "We must always be prepared to control the air and space above the earth, or join the worms beneath its surface." In 1973, I told him that I had been asked to speak at the Air War College and I would like to take a message from him. He said, "How much time do you have?" When I told him forty-five minutes, he said, with his mischievous chuckle, "Well, you can't do too much damage in that time." He then grew thoughtful and said, "You may tell them, I think we are getting out of the airplane business too fast and not getting into the space weapons business fast enough." Shortly before his death, he said to me, "Partner, I am beginning to worry about you. I think I'll finish my days in freedom, but I'm not sure you will." The memory of that prophecy fills me with foreboding lately.

"Tooey" Spaatz was quiet, reserved, and miserly with words. If he had brought the Ten Commandments down from the Mount, there would not have been ten Commandments, only one –"Always do right." He has not had, to date, half the credit he deserves in the history of U.S. air power.

Air Power in Korea and Vietnam

In our war in Korea, air power was employed very effectively in a tactical role in support of ground operations. If our available strategic air power had been employed properly, China's entry into the war could have been prevented. All that would have been necessary would have been to destroy the Yalu bridges and to lay and maintain a strip of mustard gas five miles wide along the Yalu River.

To those who cringe at the use of poison gas, I ask would that not have been preferable to 33,600 American dead? I also assure you that Russia has learned and profited by that lesson. The Soviets emphasize the employment of poison gas in all their maneuvers and maintain a growing chemical warfare capability. They obviously would employ it to stop a Chinese invasion.

In Vietnam, there was a gross failure to employ our available strategic air power properly. In the first place, our political leaders elected to wage a massive land war in Asia against the advice of all our most experienced military leaders. Then, they took counsel of their fears and failed to employ strategic air power fully, believing, erroneously, that this would bring Russia and China actively into the conflict.

In 1966, President Lyndon Johnson sent for me. (In earlier times we had been close friends. I, having retired in 1947, campaigned for him in 1948, when he first ran for the Senate.) He said, "I understand that in your syndicated columns you have been critical of the way in which I am running this war. All right, wise guy, what would you do?" I replied, "Mr. President, with our preponderant military power and resources, when we cannot persuade Ho Chi Minh to abandon his invasion of South Vietnam, we must be doing something wrong." "Be specific," he demanded. My reply: "What I would do, after issuing an ultimatum, is mine Haiphong Harbor and break the Red River dams, putting North Vietnam's rice production and many of her principal cities under ten feet of water. If this didn't produce the desired result, I would give Hanoi the Berlin treatment, then progressively destroy her war-making potential and all her rail transportation, preventing the transfer of weapons and supplies from Russia and China." Johnson said, "You know our people would not stand for that." I answered, "I remember when President Roosevelt had considerable opposition at first in World War II. Perhaps if you had done as he did, acted boldly, then carried your case to the people and gotten this war over with in a year—which would have been possible with the fuller use of naval and air power—you could have had a united people behind you."

In retrospect, how much better it would have been, if necessary, to destroy North Vietnam than to lose our first war. That would have saved us 50,000 American dead, 250,000 Allied dead, and, subsequently, the greatest genocide in this century. Already the Hanoi butchers have murdered or starved to death more than three million men, women, and children in South Vietnam, Laos, and Cambodia.

Air Power Employment in the Future

A constructive look at U.S. air power in the future must be based on a sound analysis of the history of air power employment in past wars, plus an appreciation of the probable economic conditions and political leadership which will exist.

Obviously, there must be an accurate and continuing study of what nuclear weapons and advancing technology will do to aerospace warfare. The radical change in the time factor, as a result of nuclear weapons, must be realized. Never again will we have years or months in which to build armies, navies, and air forces, or to convert our industrial capacity to its full weapons-making potential. We shall win or lose with military power available when the war starts, and the kind of military power required to prevent war is exactly the kind required to win any war which is forced upon us.

Political as well as economic considerations may be decisive. The divisiveness which prevails among our people and their leadership is a current danger, one which is the result, largely, of the lack of adequate political leadership at this critical time. We need a Churchill with the eloquence to arouse our people and their somnolent, divided Congress and with the courage to "tell it like it is" about declining U.S. power and the ominously increasing Soviet power.

Our national leaders have been telling our people that we must have "rough parity" or "essential equivalence" in order to promote détente in the hope of successful SALT II negotiations. We no longer possess equivalent military power with Russia in either general purpose forces or in strategic nuclear forces. The greatest reason for the current and growing imbalance flows from this decisive fact: Russian leaders are preparing for war, our leaders view nuclear war as "unthinkable."

Warfare today, in a nuclear environment, requires both an offensive and a defensive capability. We have entirely neglected the latter. Russia has put large segments of its defense industry underground; it has provided shelters for its skilled labor force; it has hardened its command and control installations; and all its maneuvers train its military forces to survive and operate in a nuclear environment. We are totally deficient in all these areas.

The Central Intelligence Agency has lately admitted that the national estimates of Soviet strength have been low by as much as 50 percent. Thus, if there were a nuclear exchange between the United States and the USSR today, at least 100 million Americans would die, while less than 20 million Russians would be killed. Seventy-five percent of our industrial capacity would be destroyed, while more than half of theirs would survive. The knowledge and belief in the Kremlin of the accuracy of these estimates destroys the credibility of our current defense posture.

The most likely scenario for our next and last major emergency may be this: One day, over the hot line from Moscow may come this message to our commander-in-chief in the White House: "Mr. President, we order you not to interfere with our operations against Israel. Obviously you will comply for your own Chiefs of Staff will confirm that we have overwhelming military superiority." If present conditions continue much longer, no president of the United States will have any option but to comply with that ultimatum, tantamount to surrender.

Presently, we have, but for only a short time, an alternative. We can begin immediately to regain our military superiority at sea, in the air, and in space. We have ample resources. Only the will and determination of our people and their leaders may be lacking.

The Emergence of the Postwar Strategic Air Force, 1945–1953

John T. Greenwood

President Harry S. Truman's administration marked the transition from World War II to full-blown Cold War. It also encompassed most of the "Air-Atomic Era," a period when the nation's vast superiority in atomic weapons and strategic air power's ability to deliver them apparently dominated American foreign and military policy. Even so, the Air Force's monopoly of the means of delivery in the immediate postwar years did not translate into real military capability or the power to influence the formulation of strategic policy or plans at the national or Joint Chiefs of Staff (JCS) levels. Only after the events of 1948–50 indicated the extent of the threat and opened the Federal purse did the Air Force's strategic capability become more real than illusory.

Origins of the Postwar Strategic Air Force

The leaders of the U.S. Army Air Forces (USAAF) believed that the air campaigns against Germany and Japan conclusively proved the effectiveness of strategic bombing as a decisive weapon of modern warfare. They feared, however, that vested interests in the War and Navy Departments would not welcome an independent postwar Air Force as a competitor for limited defense dollars. On 19 August 1945, Gen. Henry H. "Hap" Arnold, Commanding General, Army Air Forces, wrote to Gen. Carl A. "Tooey" Spaatz, then commanding the U.S. Army Strategic Air Forces (USASTAF) in the Pacific:

> While I am naturally feeling good about peace being effected with Japan, as far as the Army Air Forces are concerned it is, I shall say, unfortunate that we were never able to launch the full power of our bombing attack with the B-29s. The power of those attacks would certainly have convinced any doubting Thomases as to the capabilities of a modern Air Force. I am afraid that from now on there will be certain people who will forget the part we have played. As a matter of fact, I see evidence of it right now in the writings of the columnists—probably inspired by interested parties.[1]

Certainly, the U.S. Navy was one of the most interested parties because it saw the Air Force as its only serious challenger for the strategic mission of being the nation's "first line of defense," a role the Navy considered its own, then and forever. Arnold and other air leaders believed that the Navy's position was sadly outdated in an era of modern air weapons. On 28 May 1945, Arnold wrote to Army Chief of Staff Gen. George C. Marshall about the potential developments in warfare:

> Our Navy, now the strongest in the world, today can protect our shores against attack from any ambitious enemy who might challenge through the sea approaches. However, any Navy, regardless of its strength, would find itself powerless to oppose stratospheric envelopment. . . .[2]

Approaching over the desolate arctic wastes, bombers and, in the future, guided missiles could strike directly at American urban and industrial centers without interference from even the strongest naval forces. Such ideas and similar public statements only exacerbated the increasingly fractious relations between the two services that would mar Truman's presidency and jeopardize national security on more than one occasion.[3]

The wartime advances in air power and the sudden, awesome appearance of the atomic bomb in August 1945 made a great impression on Air Force leaders. They feared the possible consequences of not maintaining an effective air force in constant readiness to defend the United States and to conduct a smashing retaliatory attack against any aggressor. The war had shown that such an operational air force could not be built swiftly, and in the future no time would be available to develop one once a war began. Because of its wartime experiences and basic concepts, the Air Force was the most persistent and vocal advocate of constant readiness, standing most clearly against the old mobilization ideas that permeated the War and Navy Departments.[4] Writing about the postwar Air Force early in September 1945, Maj. Gen. Lauris Norstad, Assistant Chief of the Air Staff (AC/AS), Plans, concluded:

> The day of forming, equipping and training an Army and in particular an Air Force almost overnight is passed. Due to training specialization required and increased production problems of technical equipment, we must have sufficient strength in trained personnel and modern equipment to engage an enemy without being allowed time to build up an Air Force. In the last two wars we have fortunately been afforded up to two years to gear for war. With the character of modern warfare changed so radically in this last war, particularly by new weapons, in the next war we will be in the midst of an all-out war from the start. Our only salvation will be in immediately available modern weapons with sufficient personnel adequately trained in their use.[5]

Neither the aviation industry's production lines nor combat air groups could be readied quickly. Thus, in concert with industry groups and air power advocates in Congress, air leaders pushed strenuously but unsuccessfully for a minimum active Air Force of seventy combat groups to assure national security. General Arnold strongly supported this strength in the JCS discussions on the size of the permanent military establishment in the fall of 1945:

> I am convinced that any careful analysis of future world conditions will clearly show that the contributions which the Army Air Forces must make to the future security of the nation require a peacetime force of approximately 70 groups. . . . In the face of foreseeable world conditions, any greater reduction [below 400,000 men] would be at the expense of national security. The tragic possibilities inherent in long range attacks with weapons as effective as the atomic bomb require us to make plain to the Congress and the President the need for an Air Force mobilized in strength.[6]

Possibly the airmen pushed too hard, boasted too much about air power, threatened too many vested interests, and cried "wolf!" too often to a wary president and war-weary Congress and public to do their cause much good. Being generally younger than their Navy and ground Army colleagues, air leaders were frequently seen as overly pampered, promoted, and pushy airplane drivers in need of some basic military discipline and humility. Actually, the braggadocio concealed a marked inferiority complex and deep-seated fear that their long-sought goal of independence would not be gained now that the war was over. Major General Frederick L. Anderson, AC/AS, Personnel, fully expressed these fears to Carl Spaatz in a letter of 17 August 1945:

> I wish to congratulate you upon proving to the world that a nation can be defeated by air power alone. . . . Regardless of our demonstrated powers, it is now evident that domination of the War Department over the Air Forces is increasing; since V-J Day the vise of control has been closing. If we do not obtain a Department of the Armed Forces, with equal representation by the Air Forces, or a separate Air Force, in the next six months, we will never have it. . . . I also feel it essential that if at all possible you should return to the United States to help us fight this battle. Ninety per cent of the work of Headquarters, Army Air Forces, in the way of planning and implementation of post-war air forces comes to naught in the War Department.[7]

The importance of the Seventy-Group Program and Perry McCoy Smith's distortion of it in his *The Air Force Plans for Peace, 1943–1945* requires a brief digression. The plan originally emerged from the War Department's postwar planning in late August 1945, with the size determined by the War Department's troop basis and the exact composition by the Air Force. The seventy groups remained the centerpiece of all Air Force programs and aspirations until the Korean War expansion to 95 and then 137 wings ended all talk of the Seventy-Group Air Force, except among historians.[8]

A wide variety of postwar plans had bounced back and forth between the War Department and Air Force planners until 27 August 1945. Then, in a meeting with the AAF, the War Department General Staff, the Operations Division, and G-3 directed a peacetime strength of 574,000 men (subsequently reduced to 400,000) and seventy air groups. Although they disagreed, the Air Force representatives could do nothing but accept this verbal directive. Accordingly, after meeting with the headquarters staff, the AC/AS, Plans, decided that with a total of only seventy groups, the retention of the previously planned forty very heavy bomb (VHB) groups of B-29s would produce an "unbalanced" postwar program. Hence, a balanced air force of twenty-five VHB and twenty-five fighter groups was selected to provide the wide range of air power needed to meet postwar requirements. The VHB element was still rather large, twenty-one groups of four B-29 squadrons (eight aircraft each) and four composite bomb groups each with one squadron of twelve SILVERPLATE B-29s modified to carry atomic bombs.[9]

On 10 September, Norstad spelled out the rationale for the seventy groups and the large bomber force. The relatively heavy emphasis on very heavy bombers (720 aircraft) was due to the inability to expand or replace them quickly and the need to have a strong force for the first blow. Readying stored aircraft, opening production lines, training flight and ground crews would all take much longer for bombers than fighters if a war began. Deployments specified in the plan were mandatory, Norstad said, because "only by physically locating VHB's and ancillary air units in the proposed areas will we have operating bases available for immediate operations at all times."[10]

Their plans and worries led Air Force leaders to place heavy emphasis on the strategic force, not an unusual decision given the Air Force's marked preference for bombardment aviation. While tying its quest for independence to strategic air power, the Air Force did not neglect other air missions. The war proved that air superiority was needed for the most effective strategic air operations. This meant fighters for air superiority and bomber escort as well as air defense and support of ground forces. Whatever resources were allocated for tactical aviation, air transport, and other functions, however, the Air Force still saw strategic air operations—both offensive and defensive—as its principal mission and claim to autonomy. Everything else revolved around the strategic mission: doctrine, funds, aircraft, organization, bases, personnel, strategic planning, and technology. Speaking in October 1945, Spaatz succinctly stated the Air Force's basic philosophy: "We have one real defense: A planned and ready air offensive."[11]

The Air Force and the Atomic Bomb

The postwar period showed the crucial impact of technology on the development of the strategic air force. Arnold and other airmen warned endlessly that modern science and technology made the United States vulnerable to devastating air attacks. In his 28 May 1945 memo to General Marshall on future warfare, Arnold outlined the reasons for the Air Force's great concern with technology and also its basic strategic concept for the future:

> It is clear that the only defense against such warfare is the ability to attack. We must, therefore, secure our nation by developing and maintaining those weapons, forces, and techniques required to pose a warning to aggressors in order to deter them from launching a modern devastating war.[12]

To the Air Force, the national security demanded that it has the most modern aerial weapons to deter an attack or else to repel it and then launch a retaliatory counterstrike. Thus, even before the dropping of the atomic bombs and the end of the war, the Air Force had linked national security, technological development, and deterrence of attack.

The atomic bomb presented the Air Force with great prospects but equally great problems. Its ominous implications led to the appointment of the officers' board of Carl Spaatz, Lt. Gen. Hoyt S. Vandenberg (AC/AS, Operations, Commitments, and Requirements), and Maj. Gen. Lauris Norstad to evaluate the bomb's impact on the future of the Air Force. After meeting in September and October 1945, the "Spaatz Board," as it was known, reported to Arnold on 23 October that the atomic bomb did not then call for changing plans for the size, composition, organization, or deployment of the AAF. The report reaffirmed the concept of the strategic air offensive and concluded the atomic bomb provided another, albeit tremendously powerful, weapon for use. The Board recommended a strong research and development program, maintenance of an effective intelligence agency, and the establishment of a special Deputy Chief of the Air Staff for Research and Development (DCAS/R&D) under Maj. Gen. Curtis E. LeMay. While at work, the Board received little support from the Manhattan Project, which zealously guarded the technical details of the bomb and its production. Even the fact that the AAF was the only organization capable of delivering atomic bombs made little difference. One of LeMay's major functions—possibly his primary one—was to open channels to the Air Force for this badly needed atomic information.[13]

Air leaders believed the atomic bomb confirmed their concept of strategic bombardment. The bomb's great destructiveness—roughly four square miles at Hiroshima and Nagasaki—made air power decisive. A concerted air-atomic attack offered the real possibility of knocking out or at least badly crippling an enemy at the start of a war. But the bomb also presented a great many problems. Two of the most critical were the dearth of reliable technical data for planning and the assimilation of the weapon within existing air doctrine and organization without disrupting the basic structure. The Air Force at first embraced the atomic bomb cautiously, even uncertainly, because it lacked sufficient knowledge of the weapon and its production rates to gauge its strategic value or logistical requirements. Due to its current and projected tactical and technical limitations, Air Force leaders did not view the atomic bomb as a panacea for all wartime contingencies and were reticent to scuttle their combat-tested organization and doctrine.[14]

In his third and final report to the secretary of war in November 1945, Arnold concluded with a short discourse on air power and the future. The influence of atomic energy on air power was obvious, he wrote; "It had made Air Power all-important." Foreshadowing later and more elaborate strategic thinking, the AAF commander went on to say:

> . . . it must be recognized that real security against atomic weapons in the visible future will rest on our ability to take immediate offensive action with overwhelming force. It must be apparent to a potential aggressor that an attack on the United States would be immediately followed by an immensely devastating air-atomic attack on him. . . . The atomic weapon thus makes offensive and defensive Air Power in a state of immediate readiness the primary requisite of national survival.[15]

Arnold's remarks contain the different threads of what eventually became strategic nuclear deterrence: strategic air power, the atomic bomb, constant readiness, an air force in-being, and swift, devastating retaliation for aggression.

Despite their uncertainties, air leaders realized that the atomic bomb had to be integrated within the postwar Air Force before their claim was challenged. Following up the Spaatz Board's findings, Hoyt Vandenberg, and especially Brig. Gen. Alfred R. Maxwell and Col. William P. Fisher of his Requirements Division, began pushing early in November for the establishment of an atomic bomb striking force based on the 509th Composite Bomb Group, the only operational atomic unit.[16] Early in December, Lt. Gen. Ira C. Eaker, Deputy Commanding General, AAF, cautioned the Air Staff about designating any special unit for atomic bombing and recommended that the entire strategic force should be atomic: "It seems to me we are very likely to find the attitude of the War Department and of Congress to be that the atomic bombing force is the only strategic air force we will require. If one wing will do the job then one wing will be the size of the strategic air force."[17]

Eaker's advice was adopted. Early in January 1946, the Air Force designated the 58th Bomb Wing and its three bomb groups, including the 509th, as its strategic atomic force, but not in name. The 58th and 509th were to form the nucleus for the conversion of the entire bomber force into the future atomic striking force. The AAF planned to convert its B-29s to carry atomic bombs as well as conventional munitions, thus enhancing their flexibility of employment.[18] Actual modification of aircraft beyond the original forty-six SILVERPLATE B-29s was slow due to security procedures and lack of funding priority. For the next several years, the 509th remained the only unit capable of atomic strikes, and even it was in sad shape. Nevertheless, the bare nucleus of the future strategic atomic air force was in existence by the time the Strategic Air Command (SAC) was established in the functional restructuring of the Air Force in March 1946.

In April 1946, a meeting was held in Headquarters Army Air Forces to determine the future structure and training of the 58th Bomb Wing and 509th Bomb Group, then readying for their roles in the atomic test series in the Pacific that summer, Operation CROSSROADS. In a memo attached to his report, Maj. Gen. Earle E. Partridge (AC/AS, Operations) summarized the basic Air Force and American policy on the employment of atomic weapons:

> Consistent with our national policy it is unlikely that we will attack any nation until we have first been attacked. In such an event. we must have available a unit trained and capable of immediate retaliation against the aggressor nation with our most destructive weapon to effect as much or more destruction than we experienced.[19]

This statement could have been written at the JCS or SAC for decades to come, so little did basic U.S. national strategic policy on the use of atomic weapons change. The only real difference then was that no force existed capable of using atomic weapons if so ordered.

The lack of data on the atomic bomb represented a major obstacle to the combat development of the postwar strategic air force. Clearance and access restrictions limited training of bomb commanders, weaponeers, assembly teams, flight crews, and staffs. Few air officers had more than a rudimentary knowledge of atomic energy and the peculiar requirements and characteristics of the bomb. The lack of technical information and knowledgeable personnel hampered research and development, prevented realistic strategic air war planning, and stifled the healthy evolution of tactics.[20] On 24 September 1946, Col. William H. "Butch" Blanchard, commanding the 509th Bomb Group at Roswell Army Airfield, New Mexico, wrote to Curt LeMay about the practical problems obstructing his progress to combat-ready status. "We have a lot of eager lads here who are chaffing at the bit to progress in the atomic bomb business," he noted, "and I certainly believe we will make a mistake if we have to wait a year or so while someone designs and produces a big bomb for us to practice with." Although he was not overly concerned with his bombardiers' proficiency, Blanchard realized that

> ... there are more people connected with the delivery of an atomic bomb than a bombardier. It can be said that heavy equipment handlers can be trained in a week or two, armament people can be trained in a week or two, and that hoist operators etc. can be trained in a week or two. I will say to you that in two weeks to a month, we can also train pilots to be bombardiers, but we don't do it. . . . While your big thinking is in terms of a year from today, I personally see no more reason for my group to be ready a year from today, than tomorrow, in which case the two weeks necessary to train our ordnance and armament people would not be available.[21]

Three days later, Blanchard again wrote to LeMay about his problems and worries: "Please excuse us for appearing to be a little 'pushy' down here, but we are getting more and more afraid that the Air Force is losing their little toe-hold in the atomic bombing business, and our convictions force us to keep pushing."[22]

A small but major breakthrough for Air Force atomic operational readiness and planning came in the training and operations of the 509th Bomb Group during the CROSSROADS atomic tests at Bikini Atoll in June–July 1946. The group's assignment to drop the ABLE Day test bomb on 1 July provided a wealth of detailed information. Combined with the extensive United States Strategic Bombing Survey reports on the Hiroshima and Nagasaki explosions and the underwater test late in the month, the ABLE Day test gave clearer indications of the vast combat potential of the atomic bomb as well as of its considerable logistical requirements. The JCS Evaluation Board that met to report on the tests provided another opportunity for the AAF to learn more about the weapon that it alone could use. The Air Force hoped that the resulting official CROSSROADS reports would provide the required knowledge of the bomb and thus permit quantitative planning for atomic warfare.[23]

On 5 February 1947, LeMay forwarded to Lt. Gen. Lewis Brereton, chairman of the Military Liaison Committee to the Atomic Energy Commission (AEC), a report prepared by Col. Turner C. Rogers. The full strategic implications of the atomic bomb on warfare, Rogers stressed, could not be estimated without data on the probable supply of bombs. An unlimited stockpile would entirely change the planning, but in the absence of better information Rogers could only assume a limited supply for the next ten to twenty years. Such uncertainties among top Air Force planners accurately reflected the situation in joint and air war planning.[24]

Although organizationally the Air Force opted to integrate the atomic bomb within its existing unit and staff structure, doctrine was a different matter altogether. The wartime concept of strategic bombardment, built on the disruption and destruction of the enemy's war-making capacity and will to fight, relied on precision bombing attacks on carefully selected economic-industrial targets. Before the end of the war in Europe, however, tactical requirements such as massed formations for defense and bomb concentration and technological improvements such as all-weather radar bombing had driven the Air Force from precision bombing to a modified area concept.

Against Japan, precision bombardment gave way almost entirely to urban attacks due to the unusual structure of Japan's industrial economy and the vulnerability of its cities to incendiary raids. Twentieth Air Force struck directly at the enemy's urban population and will to fight and indirectly at its war-sustaining industrial structure. The atomic bomb's tremendous destructive power completed the metamorphosis of strategic bombing from a precision instrument to a bludgeon of mass destruction. The Spaatz Board in October 1945 saw that the atomic bomb was, and would remain, primarily an offensive weapon for use against large urban-industrial targets. Technological and tactical imperatives forced Air Force leaders unwillingly but inevitably toward a doctrine of strategic bombing that emphasized attacking the enemy's most vital and populous urban-industrial centers to gain the maximum effect from the few atomic bombs expected to be available.[25]

Early in 1947, Rogers accurately summarized the extent of this doctrinal evolution:

> Success in a war of the future will depend more than ever before on the industrial capacity and efficiency of the protagonists, therefore destruction of the enemy's industrial capability will contribute most toward reduction of his ability to wage war. This fact coupled with the character of the atomic explosion leads to the conclusion that the most profitable target for the atomic bomb will be large industrial centers.[26]

He then summed up the Air Force's view of American atomic strategy:

> But more important than defensive measures is the prevention of the initial attack. Fear of retaliation has always been the greatest deterrent to any nation contemplating all-out war. Twice in this century our unpreparedness has led a would be world conqueror to believe he could achieve such success. Japan was well aware of our weakness when she struck at Pearl Harbor. The ability to strike back effectively will be our best guard against attack. . . . The possession of a substantial number of atomic weapons and the means of delivering them to any part of the world provides the most potent threat of retaliation known to man.[27]

The theory of strategic deterrence that formed the heart of subsequent Air Force strategic doctrine coalesced in 1945–46 and was well developed by early 1947, far in advance of the war plans, aircraft, or supply of atomic weapons to implement such a concept.

Planning for Strategic Air War, 1945–48

War planning received little serious attention in the Air Force's initial plans except for vague references to readiness and potential aggressors. Some airmen, however, considered possible future enemies even before the war ended. In a January 1945 report to the Theodore von Karman Committee, Air Plans and Intelligence officers singled out the Soviet Union as the only possible enemy in the future and developed a detailed target list of Soviet industrial facilities.[28] Throughout 1945, special air intelligence teams collected information on Germany's secret weapons and all available German intelligence on the Soviet Union and its air force.[29]

In planning their postwar composition, basing, and deployment, Air Force leaders wanted air units sited to fulfill occupation duties in Germany and Japan and to respond quickly to potential aggression. The Pacific and Far East were well covered at first, but leaders were concerned about the Peripheral Basing Program devised for Europe.[30] In September 1945, diplomatic considerations and demobilization pressures forced Gen. Dwight D. Eisenhower, commanding the U.S. Forces, European Theater (USFET), to request a reduction in bomber groups from five to two. Vandenberg informed Eaker that this was unacceptable because it would unbalance the strategic and tactical air elements. "Further," he wrote, "retention of five (5) VHB groups in the European theater is necessary to combat any possible threat from the East."[31] There could be little doubt that he meant the Soviet Union. Thus, as early as the fall of 1945, the Air Force perceived the stationing of B-29s in Europe as a countermove to Soviet forces and intentions.

The Air Force's plans and perceptions, however, were to remain just that. On 27 October 1945, Arnold warned Marshall that rapid demobilization was incapacitating the AAF:

> We amazed the world with the great speed with which we built up our Air Force superiority. Today we are tearing it down even more swiftly—possibly to the even greater amazement of the world and undoubtedly to the comfort and gratification of our potential enemies. . . . Both our Occupational Air Forces and our Strategic Reserve are already weakened to a point where I consider them far below our war standards. Further reductions and further losses of highly skilled personnel will accelerate this loss of effectiveness.[32]

By the following spring, hasty demobilization had undercut the Air Force's ability to act in Europe and elsewhere; its combat-ready units evaporated as trained flight crews and maintenance men returned to civilian life. After an extensive examination, the Peripheral Basing Program was scrapped in May 1946 due to the lack of personnel and suitable VHB bases and to the growing realization that B-29s positioned in occupied Germany would be vulnerable to any Soviet offensive.[33]

Out of necessity and inclination, the Air Force now decided to base strategic units in the United States under SAC's control and rotate them to various theaters for training and orientation. This new policy saved money, guaranteed a strategic air presence in Europe, enhanced readiness while reducing vulnerability, increased morale, promoted training, and acquainted crews and support personnel with all regions, especially Europe and the Arctic, where they might have to operate.[34] Moreover, this change revealed the intimate and delicate interaction of budgets, doctrine, technology, basing, and strategic planning.

This policy was a temporary expedient. The Air Force counted on the development of advanced air weapons not only to facilitate the accomplishment of its missions but also to reduce its overseas commitments. With its range of 10,000 miles, the Consolidated B-36 that was then under development would provide the necessary intercontinental striking capability until long range, jet-powered bombers and guided missiles were ready; it would also remove the need for additional overseas bases. The desire to base all strategic forces in the United States to reduce their vulnerability and to have them constantly ready dominated Air Force strategic thinking and planning in the years to come.

While awaiting the B-36 and the B-52 that would replace it, the Air Force sought other solutions to the problem of range extension of its medium range B-29s, B-50s, and the future all-jet B-47. Although one-way missions were proposed as an emergency measure, aerial refueling was tested and adopted in 1948 as the best practical method of increasing the medium bombers' striking range.[35] The development of some relatively simple technology and flying techniques significantly altered the capabilities and flexibility of the entire strategic force while reducing its vulnerability.[36] Thus, long before Albert Wohlstetter and the RAND Corporation ever thought about the vulnerability of the numerous overseas air bases, the Air Force realized the threat and knew the solution—new technology.[37] It was but a short step from this idea to one of the most basic tenets of strategic deterrence thinking—that the credibility of a deterrent was directly proportional to its invulnerability.

In June–July 1946, Spaatz traveled to England and Europe to discuss possible revised basing arrangements with Air Chief Marshal Lord Tedder, Chief of the Air Staff, Royal Air Force (RAF), and with American theater commanders. As a result of his talks with Tedder, Spaatz received RAF agreement to provide former American bomber bases for B-29s in case of a war emergency in Europe. He also obtained RAF cooperation modifying certain bases to support atomic operations.[38] In August 1946, Col. E. E. Kirkpatrick of the Manhattan District went to England to supervise the construction of the assembly buildings, aprons, and loading pits, and the installation of the required equipment.[39] By early 1947, atomic bomb assembly and loading facilities were in existence in the United Kingdom and the Marianas to support possible atomic bombing operations in Europe or the Far East. These actions set the pattern for future Air Force and SAC war planning because the bases in the United Kingdom would remain the core of all strategic air offensive operational plans through the late 1950s. All that was needed now was an emergency war plan to specify the Air Force's exact role in any future conflict, and the first postwar joint and air war plans were already in preparation in mid-1946.

In February and March 1946, the services and the JCS began serious planning for industrial mobilization and strategic warfare. Joint war planning started with the PINCHER series of studies on various strategic problems and geographic areas. Early emergency war plan studies adopted the strategic defensive and relied heavily on a strategic air offensive to destroy the enemy's war-making capacity through atomic and conventional bombing attacks.[40] Air planners, however, thought that these studies reflected typical World War II thinking that subordinated air power to ground and sea power. Colonel Alvin R. Luedecke, one of the key Air Force planners, questioned the evolving PINCHER studies on 6 September 1946, when he wrote to Brig. Gen. Frank F. Everest, the Air Force member of the Joint War Plans Committee

(JWPC): "Now it seems that the same old thinking of World War II is coming up again with the result that Air Power is treated as an adjunct to Ground and Sea Power."[41]

Possibly one of the things that disturbed Everest and other airmen was the early joint planning for the atomic bomb. In June 1946, Adm. Chester Nimitz, Chief of Naval Operations (CNO), questioned the thinking that called for early atomic strikes to offset Soviet offensives. He thought reference to atomic bombs and planning for their use should be avoided because the bomb might be outlawed or not employed.[42] Spaatz disagreed; he saw the atomic bomb as the decisive weapon. To him, planning to use the greatest American advantage was imperative due to the funds already invested and to Soviet numerical superiority.[43] Nevertheless, studies in December 1946, scratched plans for atomic bombing, assuming the bomb would not be used "for political reasons."[44] These early joint plans still stressed initiation of an early strategic air offensive against Soviet petroleum-oil-lubricants (POL) facilities and urban-industrial targets but without the atomic bomb.[45] Joint planning staggered through the following year without producing an acceptable joint outline emergency war plan or even an agreed concept for one. This failure was at least partly a result of the festering roles and missions controversy.

In joint strategic planning, the Navy worked tirelessly to keep JCS plans focused on Middle East oil and the Mediterranean lines of communication so it could tie its carriers to missions in these areas. The Air Force argued that American carriers could no more operate in the Middle Sea against Soviet land-based air power than the Royal Navy had in 1940–42 against the *Luftwaffe*. In October 1946, Maj. Gen. George C. McDonald, AC/AS, Intelligence, warned Spaatz of an impending Navy push on the Mediterranean to drum up support for a "Big Navy" to keep the Soviets in check. The only problem was that the Navy, even with carrier aircraft, could not reach any significant portion of the Soviet heartland. McDonald surmised that "the real military worth of the carriers may come in trading carriers sunk for time, measured in weeks, while land based air power can arrive on station." Allocation of the land-based air units necessary to allow the carriers to operate with any chance of survival would only reduce the ability to strike the Soviet Union.[46]

Not content with its Mediterranean ploy, the Navy also frustrated Air Force attempts to plan a strategic air offensive against the Soviet Union. Air Force files on strategic planning and roles and missions leave the distinct impression that at every turn the Navy tried to prevent implementation of Air Force concepts of strategic air warfare. Air leaders had struggled in the 1930s to develop their doctrine of strategic bombing and then had used it with devastating consequences against Germany and Japan. Now, admirals with no such experience tried to dictate the bases, phasing, targets, weapons, and strength of the air offensive. These intrusions deeply disturbed the airmen, particularly because they had the primary service responsibility for strategic air warfare. The roles and missions infighting slopped over into strategic plans because functions assigned in them could be used to justify forces and each service's share of the defense budget.[47] Apparently, the Navy sought to buy time until it developed the large carriers and atomic bomb-carrying aircraft to give it a firm claim to a strategic air-atomic mission. The fear of losing this claim became a reality in the spring of 1949 when Secretary of Defense Louis Johnson canceled the *U.S.S. United States* (CVA-58), leading to the Navy's attack on the B-36 program. The Air Force–Navy conflict of the late 1940s culminated but did not end in the B-36/Supercarrier controversy and the "Unification and Strategy" hearings of 1949.[48]

Paralleling joint efforts at planning, the Air Force and War Department prepared their own plans for the early months of hostilities. In the spring of 1946, Air Force war planning began in earnest with a rough outline plan that specified exactly how the available air units would be employed in an emergency. Air Force headquarters soon realized that a basic war plan was required to govern mobilization, deployment, and operations should war break out with the Soviet Union. It is unclear whether these actions resulted from War Department moves to prepare industrial mobilization plans, which had to be based on some realistic estimates of wartime requirements and thus plans, or from the initiation of the PINCHER studies. What is clear is that the AAF decided a detailed, carefully prepared emergency plan showing the commitment of all air units during the first months of hostilities was needed at all times. The Strategy Branch of Plans was to develop this plan, named MAKEFAST.[49]

The real impetus for the formulation of the first detailed postwar strategic air war plan came on 10 September 1946, when Brig. Gen. George A. Lincoln, Plans and Policy Group, Directorate of Plans and Operations, War Department General Staff, directed the AAF to draw up plans for the immediate initiation of strategic air operations against the Soviet Union. The planning assumptions, based heavily on early PINCHER studies (especially JWPC 432/7), included the early deployment and use of the 58th Bomb Wing, as yet the Air Force's lone atomic unit.[50] Air Force planners, however, still lacked adequate technical information on stockpiles and production rates upon which to base any accurate estimate of atomic operations. Consequently, Plan MAKEFAST, submitted in October 1946, was an entirely conventional bombing campaign concentrating on the Soviet petroleum industry. MAKEFAST was essentially the strategic bomber offensive of World War II with much smaller forces, refined by wartime experiences, and directed against the Soviet Union. When MAKEFAST was presented to Spaatz and other Air Force leaders during the December 1946 commanders' conference, he directed that an atomic (special weapons) annex be developed and the plan revised quarterly using the latest information.[51]

Even if an atomic plan had existed in late 1946, the nascent strategic force would have found its execution difficult, if not impossible. The 509th Bomb Group had ten of twenty-three modified aircraft in commission, twenty trained crews, and few atomic bomb shapes-known as Fat Man "pumpkins"—for loading and bombing practice. Only sixteen of forty-six SILVERPLATE B-29s modified during the war were available to operational units, while another eighteen were stripped of equipment and in storage. Six of sixteen VHB groups were activated with aircraft; three had no aircraft; and the other seven were not activated.[52] The Air Force was far short of its goal of an all-atomic strategic air arm to carry out its doctrines of atomic warfare and would have had serious trouble conducting even conventional operations.

Plan EARSHOT, the initial revision of MAKEFAST, appeared in March 1947. It was the first true atomic air war plan but lacked the detailed logistical considerations to support its own execution. While more refined than MAKEFAST, EARSHOT was really little different. It still stressed conventional operations against the Soviet urban-industrial and oil target systems from bases in the United Kingdom and Middle East, plus supporting strikes from Japan, the Ryukyus, and Alaska. Because of the uncertainty about the condition or usability of the Middle East airfields in the Cairo-Suez and Palestine areas, the B-29 units and personnel would primarily deploy to the United Kingdom for strategic air operations in conjunction with the RAF Bomber Command. Although more closely attuned to ongoing JCS studies, EARSHOT was still just a War Department and AAF plan that specified the Air Force's requirements rather than detailed its capabilities.[53]

In forwarding this short-range emergency plan to his major commanders, Spaatz emphasized that EARSHOT left much to be desired. He hoped that their detailed comments would permit a more reliable revision. While specific deployments, operations, and deficiencies in personnel and logistical support were not considered, Spaatz pointed out that EARSHOT nonetheless was:

> ... world wide and of necessity portrays, either directly or by implication, a large portion of the accepted strategic thinking of the Army Air Forces, the War Department and the JCS; however, it is not unlikely that the general scheme of action, and the strategic thought implied in the plan, may be adhered to in whole or in part for many years.[54]

Based on it and its summertime revision, EARSHOT JUNIOR, Headquarters SAC initiated its first detailed planning for the actual conduct of strategic air operations, including the target analyses, mobility plans, combat mission folders, and operations plans and orders that formed the nitty-gritty of the deterrence business. SAC Operation Plan (OPLAN) 14-47 which appeared later that year was SAC's first for a postwar air offensive.[55]

MAKEFAST and EARSHOT covered only the opening months of a war with the Soviet Union, but they showed that planners envisaged a long struggle fought like the recent war. The strategic air offensive was seen as the only weapon with which to strike back at the Soviets, whose offensive powers were considered almost supernatural. Why planners assumed the Soviet Union could launch simultaneous offensive thrusts in Scandinavia, Western Europe, South and Southeastern Europe, the Middle East, India, and the Far East is hard to fathom given the tremendous wartime damage and losses sustained by the Soviets. Whatever the reasoning, the basic war plans were formulated using a mobilization base concept and a classical land strategy similar to that of the 1942–45 European campaign.[56]

Air power was still seen, as Frank Everest had noted, as an adjunct to land and sea power. As in the last war, air power would weaken the enemy, prepare the way for invasion and reconquest of lost territory, and support the ground forces in the seizure and occupation of the Soviet Union. Such planning frustrated and disheartened many airmen. The lessons of air power had not been learned and were not being applied even in War Department planning. In its purest form, a properly planned and executed air power strategy sought to neutralize and disarm an enemy and to destroy his ability and will to wage war and thus harm the United States. By striking at the Soviet heartland with atomic and conventional weapons, a true air strategy aimed to cut casualties and costs to the nation and its allies who would be outnumbered in any contest with the Soviet Union. Matched against the traditional land and sea power strategies and their powerful supporters, this concept had few advocates outside a small coterie of air power enthusiasts.

During 1946–47, the continuing lack of knowledge of the atomic bomb and the grave deficiencies in training, equipment, and priorities both in atomic and conventional bomb groups severely restricted the Air Force's ability not only to plan for strategic air war but also to conduct it if necessary. The inability to get accurate information from the Manhattan Project and then the Atomic Energy Commission made it impossible to plan atomic operations on anything but sheer guesswork. The Atomic Energy Act of 1946 that established the AEC on 1 January 1947 imposed even stricter limits on the dissemination of information on the bomb and its related equipment. Under the new "Restricted Data" category of classification, equipment associated with the handling, loading, and dropping of the bomb was classified, and the crews who were to fly the missions were not yet cleared to see or use the equipment. This situation continued throughout 1947, but by the year's end the Air Force was seeking some relief from the sillier clauses of the Act.[57] Otherwise, training and thus operational capability of the atomic units would continue to be handicapped. For instance, because of the "Restricted Data" equipment in its B-29s, the 509th Bomb Group had not been outside the continental United States for training missions to bases it would have to use if war came.[58] Even without these shortcomings, SAC units would have found it almost impossible to execute current war plans because adequate operational maps and target charts did not exist as of November 1947 to support strategic air operations on a global scale.[59]

During the closing months of 1947, the absence of a viable Air Force atomic program and of an agreed joint war plan still incapacitated Air Force planners. SAC's Eighth Air Force, formerly the 58th Bomb Wing, was maintained in constant readiness as the atomic striking force, but the 509th was still the only operational atomic unit. Faced with numerous problems and obstructions, the Air Force was slow to develop a program equal to the significance of atomic warfare. New priorities in administration, personnel, funding, training, and material were urgently needed. Although sensitive to these requirements, top Air Force leaders were hard-pressed now that independence was finally a reality. Awareness of the problems was one thing, the money for the remedies was another.

The situation in joint war planning was little better. In his September briefing to the president's Air Policy Commission, chaired by Thomas K. Finletter, Maj. Gen. O. P. Weyland (AC/AS, Plans) had stated:

Any realistic estimate of the peacetime requirements of the Armed Forces must be based on joint war plans which, *in turn* [Weyland's emphasis], must be based on an estimate of the *capabilities* of our potential enemies. Until we know what kind of *wartime* structure is needed to fight a particular enemy, we cannot accurately estimate our peacetime military establishment. . . . Although we have an agreed joint strategic concept of how a future war must be fought, we do not have an agreed joint mobilization plan to establish a phased expansion of air, ground and naval forces to the sizes and in the priorities necessary to *win* a war.[60]

After summing up the status of joint war and mobilization planning two months later, Vice Chief of Staff Hoyt Vandenberg told Secretary of the Air Force W. Stuart Symington that "it is believed obvious . . . that *we do not have* a joint war plan. . . ."[61] The basic problem remained that roles and missions and budgets were so closely tied to joint war and mobilization plans that no service was willing to agree to anything that might give another additional functions or force requirements and thus budget claims.[62]

On 16 December 1947, Secretary Symington wrote to Secretary of Defense James Forrestal and Director of the Bureau of the Budget James E. Webb protesting the cut in the Air Force's Fiscal Year 1949 (FY 1949) budget request. The JCS had confirmed the Seventy-Group Air Force as necessary for national defense and approved $5.2 billion for FY 1949 for the Air Force. Although the Air Force asked for only $4.421 billion of this, the Bureau of the Budget slashed the request to $2.9 billion. Symington protested that, with the Air Force already pared to fifty-five groups, this reduction would permit maintenance of only forty fully operational groups and a mobilization structure while severely restricting development and procurement.[63] He told Forrestal that in view of the increasingly tense situation in Europe,

> . . . and any common sense strategic concept as how to get at Russia, we are more shocked at this decision than at anything that has happened since we came into Government, especially as the Bureau of the Budget further limits the relative small percentage of what is considered necessary through specifying in detail how a great deal of our administration and organization should be handled.[64]

Symington's vigorous objections availed him little against an economy-minded president and Congress.

1948: The Year of Change

Budget cuts were but one example of the problems faced by the Air Force during these years. Unfortunately for the service, national leaders remained unconvinced of the central importance of strategic air power. Despite his ostensibly tough foreign policy toward the Soviet Union, Truman had not given the Air Force or SAC any special priorities and had done little to resolve the interservice strife that hobbled planning. The Air Force worked hard to develop and maintain the atomic striking force, but lacked the resources and urgency to make it the number one priority. The strategic force barely held its own in 1946–47 due to demobilization and budgetary limitations. In 1947, further defense cuts reduced the Air Force to an interim fifty-five groups (thirteen VHB and three other very long range groups for reconnaissance, mapping, and weather) for FY 1948, and additional reductions threatened to cut that to forty operational groups.⁶⁵ In this situation, the Air Force strove to retain a strong strategic element without upsetting its balance. While the airmen struggled to keep some strategic capability, joint planning was so disjointed that two different emergency war plans, BROILER and FROLIC, were considered for implementation on 1 July 1948.⁶⁶

Several events early in 1948 then helped to change the course of American foreign policy and to break the logjam in joint planning. The Communist coup in Czechoslovakia in February was the first of the multiple crises of that year that hardened American policy and attitudes toward the Soviet Union and its satellites. In this atmosphere, Forrestal met with the joint chiefs and service secretaries at Key West, Florida, in March to resolve the basic differences over roles and missions.

Although the origins of the interservice squabbling are veiled in the mists of antiquity, the immediate origins of the Key West Conference were in the original Functions paper (Executive Order 9877), hastily signed by President Truman on 26 July 1947, which gave the Air Force the primary mission of prompt and sustained offensive and defensive air operations.⁶⁷ The question left unresolved was the nature and the extent of the Navy's secondary responsibilities in this area. The Air Force clearly believed that the Navy's plans for carrier operations, especially as elucidated in Nimitz's January 1948 retirement statement, infringed upon the responsibilities assigned by the president to the Air Force, the only service with battle-proven experience in all phases of strategic air operations.⁶⁸ This and other vagaries in EO 9877 resulted in the establishment of an Ad Hoc Committee of Lt. Gens. Albert C. Wedemeyer (Army) and Lauris Norstad (Air Force), and Vice Adm. William Styer to iron out the differences. Although it cleared up some problems, the Committee deadlocked over the question of primary and collateral responsibilities in roles and missions, especially between the Air Force and Navy over strategic air war and between the Army and Navy over the role of the Marine Corps. A split paper (SM-9735) was submitted on 4 March, and Forrestal called for the joint chiefs to meet with him at Key West on 11–14 March to settle the issues.⁶⁹

Carl Spaatz submitted the Air Force's view of roles and missions on 8 March, stating, as the Air Force and Army always had, that the service charged with primary responsibility for a mission should determine the nature and extent of collateral participation to insure effectiveness and economy. Spaatz vigorously opposed as "unnecessary, wasteful, and confusing the unilateral establishment by any service of requirements for forces and equipment which are designed to accomplish this or any other primary function of another service."⁷⁰

The net result of the Key West Conference was an apparent settlement of these major irritants. The Air Force retained primary responsibility for strategic air warfare, air defense, and other basic air missions, with collateral functions in sea interdiction, anti-submarine warfare (ASW), and aerial minelaying. The Navy received collateral functions in land interdiction, close air support, and direct participation in the overall air effort as directed by the JCS. At Spaatz's urging, Forrestal added the stipulation that the Navy would not use the collateral strategic air war function to justify additional forces and to develop its carriers into a strategic air force. The Navy agreed to this qualification.[71]

On 21 April, Forrestal circulated a proposed new Functions statement outlining the agreements reached at Key West. The very next day, Adm. Louis E. Denfeld, CNO, sent Forrestal a memo seeking to "clarify" the proposed memo and annex of functions. He contended that the Air Force might be responsible for strategic air warfare but that did not include target selection, which should be a joint responsibility. Denfeld acknowledged that strategic air war was an Air Force function, but said "the Navy shall attack any targets, inland or otherwise, necessary to the accomplishment of its mission." The Navy would participate in the overall air effort as directed by the JCS and, Denfeld continued, "intended that the capabilities of naval aviation will be utilized to the maximum in the air offensive against vital strategic targets." He then stated that joint war plans would soon recognize and exploit the ability of carrier aircraft "in the near future" to deliver atomic bombing attacks.[72] Clearly, then, the Navy had little intention of abiding by the Key West agreements which in reality had provided the wedge it desired.

Denfeld's memo elicited immediate hostile comment from Truman's Chief of Staff Admiral Leahy, Army Chief of Staff Gen. Omar N. Bradley, and Spaatz. In separate memos to Forrestal, they contradicted Denfeld's interpretation of Key West and castigated his memo as an attempt to negate the agreements reached there, which Spaatz felt to be his primary purpose. All three agreed that Key West limited the Navy to air units necessary to support its missions—carrier aviation—and that the CNO had accepted the agreement that no separate strategic air force would be developed using requirements for carriers as a basis for its development. Targets for naval aviation were to be for prosecution of the naval campaign and only as directed by the JCS for the overall air effort.[73] Forrestal approved the majority opinion in a new Functions memo on 1 July, but the situation was clearly not yet resolved.

However tentative they proved to be, the Key West agreements cleared the way for strategic planning. Neither BROILER nor FROLIC were accepted, but a hybrid, HALFMOON, was approved for planning purposes in May 1948. At long last an agreed concept guided service and JCS theater staffs in their war planning. The Air Force now had to redraft its 1948 air war plan, HARROW, which was keyed to the unapproved Plan FROLIC.[74]

The Berlin Blockade soon overshadowed the continuing dispute among the services and made strategic war planning much more critical. The prospect of war with the Soviet Union brought the sense of urgency needed to strengthen the strategic air force. Fortunately, programs initiated earlier were producing results. Operation SANDSTONE in the spring promised a plentiful supply of atomic weapons.[75] More atomic B-29s were emerging from modification centers; B-50s and air refueling squadrons of KB-29s appeared that summer; and the B-36 was approaching service.[76] The Air Force's move to gain control of the Armed Forces Special Weapons Project (AFSWP) to smooth the transfer of atomic weapons from the AEC to SAC in case of an emergency, and the unceasing interservice feuding, led Forrestal to call another special meeting, this time at Newport, Rhode Island. At Newport, Hoyt Vandenberg, now the Air Force Chief of Staff, compromised on most issues to reach accord with the Navy. The Air Force emerged as a limited JCS executive agent for AFSWP until the Navy developed an atomic capability, while the Navy gained a strategic air-atomic role for its carrier aircraft. In the long run, Newport settled some issues, but soon the fighting over the FY 1950 defense budget reopened old wounds and inflicted new ones.[77]

President Truman limited the FY 1950 budget to $14.4 billion, not enough to support all the forces that the three services required to conduct their strategic responsibilities under the recently approved joint war plan HALFMOON. Forrestal established a special Budget Advisory Committee under Gen. Joseph McNarney, then Commanding General, Air Materiel Command and formerly Deputy Chief of Staff under Marshall (1942–44) and theater commander in Europe (1945–47). In a meeting of the service secretaries and joint chiefs with Forrestal, McNarney's committee outlined its findings based on a tentative concept of minimum operations required in case of war, relying heavily on Air Force atomic operations.

From the start of the meeting, the Navy leaders, primarily Admiral Denfeld, CNO, and Vice Adm. Robert Carney, Deputy CNO, were bitter and hostile. In his opening statement, Denfeld attacked the Air Force and questioned current plans for atomic operations from the United Kingdom and Iceland:

> Even if we dismiss the foregoing considerations as unlikely, the unpleasant fact remains that the Navy has honest and sincere misgivings as to the ability of the Air Force successfully to deliver the weapon by means of unescorted missions flown by present-day bombers, deep into enemy territory in the face of strong Soviet air defenses, and to drop it on targets whose locations are not accurately known. For this reason alone, it appears rash to fail to provide some measure of insurance against the chance that the effort may not be effective.[78]

He then criticized the Army and Air Force plans in McNarney's presentation as lacking comprehension of Navy tasks and abilities: "This is not surprising, since no Service can be expected to be expert in any other's business."[79]

Vandenberg and Symington reacted strongly to Denfeld's statement, which reflected the bad feelings between the services on their respective roles in the strategic air war. The Air Force Chief of Staff said:

> I have one comment, as one Service Chief, that I'd like to make at this time. I regret the lack of confidence on the part of the Navy, but I'd like to call attention to the fact that in your own paper you stated, "no Service can be expected to be an expert in any other's business."[80]

Symington was just as straightforward but carried his displeasure one step further. "It seems to me," he said, "and this remark I perhaps should not make, but being very frank, I will—the idea is to substitute a large Navy for the atomic bomb."[81]

The briefings continued throughout the morning, but feelings improved very little. Finally, McNarney objected to Carney's repeated statement that the Budget

Advisory Committee's plan was the Army-Air Force plan for the Navy. Then Gen. Omar Bradley could stand no more:

> Mr. Secretary, I'd like to make the remark that that is the fourth time this morning that there's been side-remarks about the Air Force's and the Army's ideas of what the Navy should be. Of course, they know more about it than we do, but I haven't seen any hesitancy on the part of the Navy to question even how the Air Force is going to carry out their mission. They've been questioning and criticizing. I don't see why we can't express our ideas as well.[82]

From this point on, the fight degenerated into the sad spectacle of the B36-Super Carrier and Unification Hearings of 1949 and the subsequent dismissal of Denfeld and his replacement by Forrest P. Sherman.

Because of the steady deterioration in Soviet-American relations during 1948, Chief of Staff Vandenberg put greater emphasis on the strategic forces. SAC's readiness and operational capabilities were improved, and its mobility and operations plans tested in the deployments of conventional B-29 groups to Germany and the United Kingdom in July and August. What was needed was aggressive and knowledgeable leadership that Gen. George C. Kenney, the present SAC commander, could not provide. Vandenberg considered many possible replacements before finally deciding on the man he would place in charge of SAC if war broke out—Curt LeMay.[83]

When LeMay assumed command of SAC in October 1948, its haphazard development ended. With LeMay came his new team—Tommy Power, Walter "Cam" Sweeney, Emmett "Rosie" O'Donnell, August "Auggie" Kissner, and others already with SAC, Roger Ramey, William "Butch" Blanchard, John D. "Jack" Ryan, Jack Catton, and Clarence "Bill" Irvine, rejoined their old boss. With a firm background in atomic energy from his research and development tour, an unrivaled command of strategic air operations, and his proven staff, LeMay quickly shook the bugs out of SAC and began its transformation into a honed weapon of strategic warfare. SAC's primary mission was the strategic air offensive, and LeMay fought any and all deviations from that assignment. His objective was to train SAC crews and ground personnel into a team that was always ready to execute its primary mission so that it would never have to. This remained SAC's basic philosophy from 1948 on. Although SAC and the strategic mission were granted the Air Force's first priority at the 1948 commanders' conference, not until the Korean War expansion provided the men, money, and materiel for all Air Force missions was the claim honored.[84]

The Korean War was the real turning point for SAC. In December 1949, SAC had 72,000 men, fourteen bomb groups and 610 strategic aircraft, two strategic fighter groups, and six air refueling squadrons. Four years later, LeMay had 171,000 men, thirty-seven bomb wings and over 1,000 strategic aircraft (mostly B-36s and all-jet B-47s), six fighter wings, and twenty-eight air refueling squadrons.[85] This remarkable growth in size, composition, and capabilities was just the beginning. In the 1950s, LeMay built SAC into an air force within the Air Force. He fought to get the best for his command and usually did, but not without making a good many enemies along the way. When he left SAC in Tommy Power's hands in 1957, LeMay had created for the United States a strategic air force vastly superior to any in the world—a deterrent to nuclear war and a guarantor of the nation's security.

The End of an Era

Many authors have concluded that American strategic air power offset the vast Soviet ground superiority in Europe in this period. However, few of them realize the enormous chasm that separated the apparent atomic monopoly of the United States from the actual situation through at least 1949. Truman's foreign policy and his primary strategic trump cards, the atomic bomb and strategic air power, were not in harmony before 1949–50. Many writers, for instance, mention Truman's unspoken message to Stalin in the summer of 1948 when he sent atomic B-29s to England, posed to strike should the Soviet leader make a misstep during the Berlin Blockade.[86] In fact, not a single aircraft capable of carrying the atomic bomb was deployed to the United Kingdom during the Berlin crisis.[87] The possession of strategic bombers, bases, and a stockpile of weapons did not mean that those planes could place those bombs on assigned targets, or for that matter, even carry them. Many people apparently believed it did, but they were not in the Air Force.

The question one must ask, but cannot answer, is how much Truman knew of the actual condition of the strategic force that supposedly backed his policy toward the Soviet Union. If he knew and did nothing, he played a dangerous game of bluff, even for the most audacious poker player. If he did not know, it was not because the Air Force did not tell him or his secretaries of defense. In the early 1950s, this disparity disappeared forever. The Soviet atomic explosion of August 1949, the Communist seizure of power in China in October, NSC-68 in April 1950, and then the invasion of South Korea the following June made national leaders acutely aware of the danger of conflict with the Communist powers. This led to the development of military forces and capabilities commensurate with American defense responsibilities.[88]

Between 1952 and 1954, a number of factors combined to signal the end of the "Air-Atomic Era": Dwight D. Eisenhower's election, his supposedly "New Look" defense policy, the end of the Korean War, and others. Primarily, the strategic situation was altered by the MIKE Shot of Operation IVY on 31 October 1952 that proved the feasibility of thermonuclear weapons (hydrogen bombs).[89] The Soviet hydrogen bomb test the next August merely confirmed this momentous change. For an idea of the revolutionary impact of this new weapon on strategic planning, compare these yields in equivalent tons of TNT: Hiroshima and Nagasaki, under 20,000 (20 kilotons); the YOKE Test in Operation SANDSTONE, 49,000 (49 kilotons); and the MIKE Shot, 10,400,000 (10,400 kilotons or 10.4 megatons).[90] The increase in destructiveness was absolutely staggering, and its implications were easily understood. This "thermonuclear breakthrough" promised production-line weapons with yields of 1 to 2 million tons that weighed under 3,000 pounds, small enough for tactical aircraft and ballistic missile warheads.

This breakthrough supplied the critical technological advance that allowed the Air Force in March 1954 to step up the intercontinental ballistic missile (ICBM) development program for Atlas and later to add Titan and Minuteman and the intermediate range ballistic missile (IRBM), Thor. Once operational, ICBMs could hit Soviet or Chinese targets with excellent accuracy within thirty minutes of launching from hardened U.S. bases.[91] The development of similar Soviet capabilities completed the awesome cycle. The instantaneous retaliation of ICBMS, when teamed with the intercontinental reach and operational flexibility of the B-52/KC-135 force, completely recast the strategic air force and the nature of strategic planning and warfare. The "Era of the Unthinkable" replaced the tentative, oftentimes chaotic, and much less deadly "Air-Atomic Era."

If we speak of maturity as the achievement of full or natural development, then the period from 1945 through 1953 was indeed a search for maturity, more accurately described as adolescence. For the whole Air Force and its strategic force, this trying adolescence produced the great strength and maturity attained during the later 1950s and maintained ever since.

Notes

1. Arnold to Spaatz, 19 Aug. 1945, box 21, Diary Personal August 1945, Carl A. Spaatz Collection (hereafter cited as Spaatz Collection), Manuscript Division, Library of Congress, Washington, D.C.

2. Memo for Arnold, "Potentialities of New Developments in Warfare," 28 May 1945, box 191, 385 Manners and Methods of Conducting Warfare, 1945 Air Adjutant General Central Decimal File, RG 18, Records of Headquarters U.S. Army Air Forces, National Archives (NA), Washington, D.C. (hereafter cited as NA RG 18, 1945 AAG).

3. U.S. Congress. Senate. Committee on Military Affairs, *Hearings on Department of Armed Forces, Department of Military Security,* 79th Cong., 1st sess. See especially testimony of Generals Arnold and Jimmy Doolittle.

4. Perry McCoy Smith, *The Air Force Plans for Peace, 1943–1945* (Baltimore, Md.: Johns Hopkins Press, 1970); Michael Sherry, *Preparing for the Next War: American Plans for Postwar Defense, 1941–45* (New Haven, Conn.: Yale University Press, 1977).

5. Norstad to George, "Arguments for Justification of 70 Group Post-War Air Force," 10 Sept. 1945, box 129A, Operations Division (OPD) 320.2 (4 Apr. 1944) Top Secret Supplement, Deputy Chief of Staff, Operations, Top Secret Plans and Operations File, RG 341, Records of Headquarters U.S. Air Force, NA (hereafter cited as NA RG 341, DCS/O, TS P&O).

6. JCS 1478/4, CG/AAF, "Interim Plan for the Permanent Military Establishment of the Army of the United States," 2 Oct. 1945, JCS Central Decimal File, CCS 370 (8-19-45), sec. 1, RG 218, Records of the Joint Chiefs of Staff, NA.

7. Anderson to Spaatz, 17 Aug. 1945, box 21, Diary Personal August 1945, Spaatz Collection.

8. Perry McCoy Smith, chap. 5, esp. 71–74; Sherry, chap. 7; R. Frank Futrell, "Preplanning the USAF, Dogmatic or Pragmatic?" *Air University Review* 22 (Jan.–Feb. 1971); idem, *Ideas, Concepts, Doctrine: A History of Basic Thinking in the United States Air Force, 1907–1964* (Maxwell AFB, Ala.: Air University, 1974), chaps. 5 and 6.

9. Perry McCoy Smith, 54–74; Ladd to Vandenberg, "Period II Troop Basis," 27 Aug. 1945, in NA RG 18, 1945 AAG 320.3 Troop Basis, vol. 3, box 96; Todd to Air Staff, "Deployment of Strategic Air Force," 18 Aug. 1945, with Norstad to Arnold, "Deployment of Strategic Air

Force," 17 Aug. 1945, in NA RG 341, DCS/O, Top Secret Air Adjutant General Files (hereafter cited as TS AAG), file 21, box 7; see especially JCS 1478 and 1530 series documents, CCS 370 (8-19-45), "Postwar Requirements for Military Forces," in NA RG 218.

10. Ladd to Vandenberg, 27 Aug. 1945; Norstad to George, 10 Sept. 1945.

11. Futrell, *Ideas, Concepts, Doctrine*, 63–94; "Trans-Arctic Air Offensive," October 1945, Misc Folder no. 8: A. R. Maxwell, Subject File, box 329, Spaatz Collection.

12. Arnold to Chief of Staff, 28 May 1945.

13. Report to Arnold, Spaatz, Vandenberg, and Norstad, 23 Oct. 1945; Eaker to Spaatz, Vandenberg, and Norstad, "Orders," 14 Sept. 1945, in NA RG 341, DCS/O, TS P&O, OPD 384.3 Atomic (17 Aug. 1945), sec. 1, box 448.

14. Ibid.

15. General H. H. Arnold, CG/AAF, *Third Report of the Commanding General of the Army Air Forces to Secretary of War*, 12 Nov. 1945, 67, 68.

16. Routing & Record Sheet (R&RS), Fisher to Ch, Air Staff. "The Establishment of Atomic Bomb Striking Force," 30 Oct. 1945, w/incls, in Alfred F. Simpson Historical Research Center (AFSHRC), Maxwell AFB, Ala., Microfilm Roll K1167, in Office of Air Force History, Washington, D.C.; Vandenberg to AC/AS-5 (Plans), "Revisions of the Seventy (70) Group Program," 6 Nov. 1945; Vandenberg to AC/AS-5, "Atomic Bomb Striking Force," 16 Nov. 1945; and Norstad to Chief of Air Staff, "Atomic Bomb Striking Force," in NA RG 18, 1945 AAG 370.22 Campaigns and Expeditions, box 178.

17. R&RS Cmt 1, Eaker to Air Staff, "Atomic Bomb Striking Force," 3 Dec. 1945, in NA RG 18, 1945 AAG 370.22 Campaigns and Expeditions, box 178.

18. R&RS Cmt 2, Vandenberg to Dep CG/AAF, "Atomic Bomb Striking Force," 17 Dec. 1945, in NA RG 341, DCS/O, Files of Assistant for Atomic Energy (hereafter cited as A/AE), 1945 Top Secret 322 Atomic Bomb Striking Force, box 1; R&RS Cmt 4, AC/AS, OC&R, to LeMay, "Atomic Bomb Striking Force," 4 Jan. 1945, and Maxwell to Vandenberg, "The Establishment of an Atomic Bomb Striking Force," 5 Dec. 1945, in NA RG 18, 1945 AAG 370.22 Campaign and Expeditions, box 178; Vandenberg to Eaker, "The Establishment of a Strategic Striking Force," 2 Jan. 1946, approved by Maj. Gen. C. C. Chauncey, 7 Jan. 1946, AFSHRC, Document No 179.061-34A, Office of Air Force History, Washington, D.C.

19. Partridge to Chief of Air Staff, "Conference on Reorganization of the 58th Bomb Wing," 26 Apr. 1946, w/incl, "Organization and Deployment of the 58th Bombardment Wing," in NA RG 341, DCS/O, A/AE, 1946 Secret 008 Policy, box 2.

20. R&RS, Eaker to Anderson, "Personnel for Manhattan Project," 14 Mar. 1946, in NA RG 341, DCS/O, TS AAG, file 22, box 7; Hq Air Materiel Command to CG/AAF, "Air Forces Relationship to Manhattan District," Crawford to Groves, 8 Nov. 1946, with Craigie to CG/AAF, "Air Forces Relationship to Manhattan District," n.d., in NA RG 18, 1946-47 AAG 321 AAF, vol. 1, box 603; LeMay to AC/AS-3, "Inter-Branch Responsibility for Atomic Bomb Program," w/incl, 14 June 1946, in NA RG 341, DCS/O, A/AE, 1946 Secret 312.1 Manhattan District, box 2; Blanchard to LeMay, 24 Sept. 1946, with Blanchard to Ryan et al., 23 Sept. 1946, and Blanchard to LeMay, 27 Sept. 1946, in NA RG 341, DCS/O, A/AE, 1946 Secret 353 Bomb Commanders and Weaponeers Training, box 3.

21. Blanchard to LeMay, 24 Sept. 1946, and Blanchard to Ryan et al., 23 Sept. 1946, in NA RG 341, DCS/O, A/AE, 1946 Secret 353 Bomb Commanders and Weaponeers Training, box 3.

22. Blanchard to LeMay, 27 Sept. 1946, ibid.

23. Partridge to Chief of Air Staff, 26 Apr. 1946; Ramey to LeMay, 29 Mar. 1946, and LeMay to Ramey, 4 Apr. 1946, in NA RG 341, DCS/O, A/AE, 1946 Secret 471.6 Atomic Bombs: Blanchard to LeMay, 24 and 27 Sept. 1946: LeMay to AC/AS-3, 14 June 1946.

24. LeMay to Brereton, 8 Feb. 1947, with Colonel T. C. Rogers, "Strategic Implications of the Atomic Bomb on Warfare," 3 Feb. 1947, in NA RG 341, DCS/O, A/AE, 1947 Top Secret 360.2 Outline of Planning Factors for Atomic Bomb, box 4.

25. Ibid.; Report to Arnold, Spaatz, Vandenberg, and Norstad, 23 Oct. 1945; Power to Eaker, Power, AC/AS-3, "Army Air Forces Presentation for Joint CROSSROADS Scientific Symposium," 1 Feb. 1947, with "Presentation of General Power: Effect of the Atomic Bomb on AAF Tactical and Strategic Doctrine," n.d., in NA RG 341, DCS/O, TS AA.G, file 26, box 9; Maxwell to AC/AS-3, "Army Air Forces' Concept of Strategic Bombing," 1 May 1946, and Partridge to DCAS/R&D, "Army Air Forces' Concept of Strategic Bombing," 7 June 1946, in NA RG 18, 1946–47 AAG 353.41 Bombing, box 629; Maxwell to Air Def and Guided Missile Div and Opns Div, AC/AS-3, and AC/AS-2, "Strategic Warfare at Extended Radii of Action," 15 Apr. 1947, w/ incls, in NA RG 18, 1946–47 AAG 385 Warfare Misc, box 644; Power to CG/AU, "Preparation of AAF Concept and Outline Strategy for War," 11 Apr. 1947, in NA RG 18, 1946–47 AAG 381 War Plans Misc National Defense, vol. 3, box 642; Wesley F. Craven and James L. Cate, eds., "Strategic Bombardment from Pacific Bases," in *The Army Air Forces in World War II*, vol. 5, *The Pacific: Matterhorn to Nagasaki, June 1944 to August 1945* (Chicago: University of Chicago Press, 1953), 507–756.

26. Rogers, "Strategic Implications," 3 Feb. 1947, attached to LeMay to Brereton, 5 Feb. 1947.

27. Ibid.

28. Kuter to Loutzenheiser, 13 Dec. 1944; Loutzenheiser to Kuter, "Character of Possible War in 1965," 28 Dec. 1944; Burgess to AC/AS-5, "Character of Possible War in 1965," 20 Jan. 1945; Guyser to Kuter, "Character of Possible War in 1965," n.d., w/tabs, in NA RG 18, Records of Operational Plans Division, Strategic Planning Records.

29. Spaatz to Arnold, 19 Mar. 1945; Giles to Spaatz, n.d.; Spaatz to CG/AAF, "Exploitation of Air Technical Intelligence Objectives," 8 Apr. 1945; Kuter to CG/USSTAF, "Exploitation of Air Intelligence Objectives," 23 Apr. 1945, in NA RG 341, DCS/O, TS AAG, file 51 Europe, sec. 6, box 13; McDonald to Hodges, 4 June 1945; Quesada to McDonald, 15 June 1945; McDonald to Quesada, 26 June 1945; Quesada to McDonald, 2 July 1945, in NA RG 18, 1945 AAG 386.3 Captured Property, box 191.

30. Norstad to Arnold, 17 Aug. 1945, attached to Todd to Air Staff, 18 Aug. 1945; Norstad to Chief of Air Staff, "Proposed Organization and Deployment of Post-War Air Force (70 Gp Program)," 31 Aug. 1945, with memo for Hinton, "Post-War Air Force Organization and Deployment in Areas other than where major Ground Force Units are Located," 31 Aug. 1945, in NA RG 18, 1945 AAG Interim, Postwar and Peacetime Air Forces, vol. 2, box 99.

31. Vandenberg memo for Dep CG/AAF, "Occupational Air Force Troop Basis," n.d. (ca. 10–13 Sept. 1945), in NA RG 341, DCS/O, TS AAG, file 21, box 7.

32. Arnold to Chief of Staff, "Determination of Permanent Army and Army Air Forces Establishment," 27 Oct. 1945, in folder 114: Letters to General Marshall, box 44, General Henry H. Arnold Collection, Library of Congress.

33. Report, USAFE, "Location of VHB Units in ETO," n.d., and Brief, "USAFE Study on Location of VHB Units in ETO," n.d., in LC/Spaatz, Chief of Staff Papers, "Briefing Materials for European Trip," 21 June 1946, box 264; Norstad to CG/AAF, "Plans for Overseas Deployment of A. A. F. Units," 10 Apr. 1946, in NA RG 18, 1946–47 AAG 370 Deployment, etc, Misc, vol. 1, box 632.

34. See note above; Everest to Spaatz, "VHB Air Base Construction in the United Kingdom," n.d., in LC/Spaatz, Chief of Staff Papers, "Briefing Materials for European Trip," 21 June 1946, box 264; Streett to CG/AAF, "Operational Training and Strategic Employment of Units of Strategic Air Command," 25 July 1946, w/incl, "Operational Training and Strategic Employment of Units of Strategic Air Command," 18 July 1946, w/tabs, attached to Power to CG/SAC, 9 Oct. 1945; Mustoe to AC/AS-3 and Air Staff, "Operational Training and Strategic Employment of Units of Strategic Air Command," 29 Aug. 1945, w/cmts attached, in NA RG 18, 1946–67 AAG 322 Organization and Training of Units, vol. 1, box 605; Weyland to CG/AAF, "Rotation of VHB Groups to ETO," 6 Aug. 1946, and Matheny to AC/AS-4 and DCAS, 20 Aug. 1946, in NA RG 341, DCS/O, TS AAG, file 23, box 7.

35. R&RS, Maxwell to Air Def & Guided Missile Div and Opns Div, AC/AS-3, and AC/AS-2, "Strategic Warfare at Extended Radii of Action," 15 Apr. 1947.
36. *The Development of the Strategic Air Command, 1946–1976* (Offutt AFB, Nebr.: Office of the Command Historian, Hq SAC, 21 Mar. 1976).
37. Albert Wohlstetter, F. S. Hoffman, R. J. Lutz, and H. S. Rowen, *Selection and Use of Strategic Air Bases* (RAND Report R-266) (Santa Monica, Calif.: RAND Corporation, 1954); for a fuller, RAND-ish account, see Bruce L. R. Smith, *The RAND Corporation* (Cambridge, Mass.: Harvard University Press, 1966), 195–240.
38. Eaker to Air Staff, "Decisions Reached in London between General Spaatz and Air Ministry—June 1946," with Bissell memos, 28 June and 6 July 1946, in NA RG 341, DCS/O, TS AAG, file 23, box 7; LC/Spaatz, Chief of Staff Papers, "Briefing Materials for European Trip," 21 June 1946, box 264.
39. Spaatz to Bissell, 8 Aug. 1946, in NA RG 341, DCS/O, TS AAG, file 23, box 7.
40. JCS 1630 series in CCS 381 (2-19-46), RG 218; JPS 789, JSP, "Concepts of Operations for 'PINCHER,'" 2 Mar. 1946; JPS 789/1 PC 432/6, 10 June 1946, in NA RG 341, DCS/O, TS P&O, PD 381 Russia (PINCHER) (2 Mar. 1946), box 949, and CCS 381 (3-2-46), sec. 1–3, in NA RG 218, JCS Geographic File, 1948–50, box 37; Dr. Kenneth Condit, *History of the Joint Chiefs of Staff, The Joint Chiefs of Staff and National Policy*, vol. 2: *1947–49*, in NA RG 218.
41. Caldara to Everest, n.d.; memo for Everest, 6 Sept. 1946; and Everest to Luedecke, 6 Sept. 1946, in NA RG 341, DCS/O, TS P&O, PD 381 Russian (PINCHER) (2 Mar. 1946), box 949.
42. JCS 1630/3, Admiral C. Nimitz, CNO, 13 June 1946, in NA RG 341, DCS/O, TS P&O, OPD 381 Strategic Guidance (19 Feb. 1946), sec. 1, box 382.
43. JCS 1630/4, Spaatz to JCS, 1 July 1946, in NA RG 341, DCS/O, TS P&O, OPD 381 Strategic Guidance (19 Feb. 1946), sec. 1, box 382.
44. JWPC 4861/1, JSPG, "Strategic Guidance for Mobilization Planning," 18 Dec. 1946, in NA RG 341, DCS/O TS P&O, OPD 381 Strategic Guidance (19 Feb. 1946), sec. 1, box 382; JCS 1630 series, in CCS 381 (2-19-46), RG 218.
45. See note above.
46. McDonald to Spaatz, 17 and 18 Oct. 1946, in LC/Spaatz, Chief of Staff Papers, folder: Navy 1, box 262.
47. Condit, *History of JCS*, RG 218.
48. Futrell, *Ideas, Concepts, Doctrine*, 121–34; Paul Y. Hammond, "Super-Carriers and B-36 Bombers: Appropriations, Strategy, and Politics," in Harold Stein, ed., *American Civil-Military Decisions* (Tuscaloosa: University of Alabama Press, 1963), 465–567; U.S. Congress, House of Representatives, Committee on Armed Services, 81st Cong., 1st sess., Investigation of the B-36 Program (Washington, D.C.: GPO, 1949) and *The National Defense Program—Unification and Strategy* (Washington, D.C.: GPO, 1950).
49. Vandenberg to secretary of the Air Force, "Status of Current Joint War and Mobilization Planning," 5 Nov. 1947, in NA RG 341, DCS/O, TS P&O, PO 381 (5 Nov. 1947), box 355; Weyland to Hood, "Formation of Air War Plans Committee," 1 July 1946, in NA RG 341, DCS/O, TS AAG, file 25, box 8.
50. Lincoln to Everest, "Plan for Immediate Initiation of Strategic Air Operations," 10 Sept. 1946, attached to Air Plan for "MAKEFAST," in NA RG 341, DCS/O, TS P&O, PO 381 (10 Sept. 1946), box 380.
51. Vandenberg to SECAF, 5 Nov. 1947; Air Plan MAKEFAST, October 1946 in NA RG 341, DCS/O, TS P&O, PO 381 (10 Sept. 1946), box 380.
52. Status Report (As of 31/2359Z Dec 46), Off of Comptroller, Statistical Control Div, 8 Jan. 1947, in NA RG 18, 1946–47 AAG 332 Organization and Training Units, vol. 2, box 605; Blanchard to LeMay, 24 Sept. 1946; Power to CG/SAC, "Range Bombing Program, Bomb Practice M-107," 15 Oct. 1946, w/incl, in NA RG 18, 1946–47 AAG 353.41 Bombing, box 629; *Development of SAC*.

53. Short Range Emergency War Plan (SREP), Strategy Br, War Plans Div, AC/AS, Plans, "Outline Air Plan 'EARSHOT,'" 15 Mar. 1947, in NA RG 341, DCS/O, TS P&O, PO 381 (10 Sept. 1946), box 380; Vandenberg to SECAF, 5 Nov. 1947; Ritchie to Spaatz, "Cable from General Bissell, London, Dated 27 April 1947," in LC/Spaatz, Diary Jan.–Sept. 1947, box 28.

54. Spaatz to Kenney, n.d., w/incl, SREP EARSHOT, 15 Mar. 1947, in NA RG 341, DCS/O, TS P&O, PO 381 (10 Sept. 1946), box 380.

55. File, PO 381 (10 Sept. 1946), box 380.

56. Air Plans MAKEFAST and EARSHOT, in NA RG 341, DCS/O, TS P&O, PO 381 (10 Sept. 1946), box 380.

57. Spaatz to Whitehead, 6 Oct. 1947, in NA RG 18, 1946–47 AAG 312.1 Operations Letters, vol. 2, box 58; Richard G. Hewlett and Francis Duncan, *Atomic Shield* (Washington, D.C.: AEC, 1972), *passim*.

58. LeMay to CG/SAC, 22 Sept. 1947, in NA RG 341, DCS/O, A/AE, 1947 Secret, box 8.

59. McDonald to DCS/O, "Plan for the Production of Air Objective Folders for Global Air Operations," 25 Nov. 1947, in NA RG 18, 1946–47 AAG 381 War Plans Misc National Defense, vol. 1, box 642.

60. Major General O. P. Weyland, "Presentation by General Weyland to the President's Air Policy Commission, 16 September 1947," in NA RG 341, DCS/O, TS AAG, file 31, box 10.

61. Vandenberg to Symington, "Status of Current Joint War and Mobilization Planning," 6 Nov. 1947, in NA RG 340, Records of Office of Secretary of the Air Force, Off of Admin Asst, Subj File, 1946–50, 1j (2), box 3.

62. Condit, *History of JCS*.

63. Symington to Forrestal, 16 Dec. 1947, with Symington to Webb, 16 Dec. 1947, in LC/Spaatz, Chief of Staff Papers, folder: Secretary of Air Force 1, box 263.

64. Ibid.

65. Ibid.; Darrow memo for the record, "Deployment of a 55 Group Air Force," 15 Sept. 1947, in LC/Spaatz, Chief of Staff Papers, folder: Organization 1, box 262; Futrell, *Ideas, Concepts, Doctrine*, 113.

66. Condit, *History of JCS*, 284.

67. EO 9877, 29 July 1947.

68. Condit, *History of JCS*, 168.

69. Ibid., 173–79; SM-9735, Report of Ad Hoc Committee on Roles and Missions, 4 Mar. 1948, in NA RG 341, TS P&O, PD 660.2 (18 July 1945), sec. 6, box 650.

70. Ibid.

71. Condit, *History of JCS*, 179.

72. Ibid., Denfeld to SECDEF, 22 Apr. 1948, in NA RG 341, TS P&O, PD 660.2 (18 July 1945), sec. 8, box 651.

73. Spaatz to Forrestal, n.d.; memos, Leahy, Bradley, and Spaatz, 29 Apr. 1948, in NA RG 341, TS P&O, PD 660.2 (18 July 1945), sec. 8, box 651; Condit, *History of JCS*.

74. Condit, *History of JCS*, 283–88; for complete details, see CCS 381 USSR (3-2-46), JCS Geographic File 1948–50, boxes 37, 37A, 37B, 37C, 38; Ritchie to Air Staff, "First Draft, Plan HARROW," 12 May 1948, in NA RG 341, TS P&O, PO 385 HARROW, box 472C.

75. Hewlett and Duncan, 163–64, 175–76.

76. *Development of SAC*.

77. Agreed Final Version of Newport Meetings, 20–22 Aug. 1948, in NA RG 341, TS P&O, PD 660.2 (18 July 1945), sec. 8, box 651; cited in Hewlett and Duncan, 170–71.

78. For a detailed account of the FY 1950 budget battle, see Warner R. Schilling, "The Politics of National Defense: Fiscal 1950," in Warner R. Schilling et al., *Strategy, Politics, and Defense Budgets* (New York: Columbia University Press, 1960), 29–266; Condit, *History of JCS*; Minutes, Secretary of Defense Budget Meeting with the Three Secretaries and the Joint Chiefs of Staff, 4 Oct. 1948, vol. 1, in RG 218, CCS 370 (8-19-45), Bulky Package, pt. 11, box 129.

79. Minutes, 4 Oct. 1948 Meeting.
80. Ibid.
81. Ibid.
82. Ibid.
83. General John B. Montgomery, USAF (Ret.), interview with Dr. Murray Green, 7 Aug. 1971, Los Angeles.
84. Gen. C. E. LeMay, USAF (Ret.), interview with John T. Bohn, Command Historian, SAC, in Off of Air Force History, 9 Mar. 1971, Washington, D.C.; C. E. LeMay with MacKinlay Kantor, *Mission with LeMay* (Garden City, N.Y.: Doubleday, 1965), 429–500; Futrell, *Ideas, Concepts, Doctrine,* 145–71; NA RG 341. TS P&O, OPD 337 (6 Aug. 1945); Lt. Gen. C. E. LeMay, "Notes for Discussion with General Vandenberg," 4 Nov. 1948, in LC/General Curtis E. LeMay Papers, Gen LeMay's Diary, folder 1, box B64.
85. *Development of SAC,* 15, 38.
86. See, for example, Louis J. Halle, *The Cold War As History* (New York: Harper & Row, 1975), 166; Joyce and Gabriel Kolko, *The Limits of Power: The World and United States Foreign Policy, 1945–1954* (New York: Harper & Row, 1972), 493.
87. Robert J. Donovan, *Conflict and Crisis: The Presidency of Harry, S. Truman, 1945–1948* (New York: Norton, 1977), 368.
88. Futrell, *Ideas, Concepts, Doctrine,* 135–81; Paul Y. Hammond, "NSC-68: Prologue to Rearmament," in Schilling et al., 271–378; Samuel P. Huntington, *The Common Defense. Strategic Programs in National Politics* (New York: Columbia University Press, 1961), 25–64.
89. Hewlett and Duncan, 590–93.
90. Ibid., app. 4: Announced Nuclear Tests, 672–73.
91. John T. Greenwood. "The Air Force Ballistic Missile and Space Program, 1954–74," *Aerospace Historian* 22 (December 1974): 191–93.

American Postwar Air Doctrine and Organization: The Navy Experience

David Alan Rosenberg

It may be possible for a large military establishment such as the United States Air Force or Navy to find organizational "maturity" through a careful, systematic search of available options. Maturity, however, is generally not discovered as the result of a concerted effort, but is grown into so gradually that it is difficult to identify the precise moment at which it is achieved. A military organization may be said to have reached maturity when its high command has thoroughly analyzed the challenges posed by existing conditions—including the threat of potential enemies, the nature of the domestic political situation, and the state of military technology—and has determined how to manipulate technology and organization so as to operate effectively in pursuit of its chosen objectives.

With respect to naval "air doctrine"—by which is meant, in this paper, basic tenets of strategic rather than tactical doctrine—the Navy had reached maturity by the end of World War II. The service had achieved its wartime goals by skillfully adapting tactics and weaponry, including the most advanced air technology available, to meet the challenges posed by global war. Changing circumstances, however, require continued growth. The development of the atomic bomb, the emergence of the Soviet Union as a major military power, and the postwar struggles at home over budgets and reorganization of the armed services posed a new set of challenges which demanded major adjustments in naval air doctrine.

Between 1945 and 1949, confidence and pride in wartime accomplishments were overshadowed by conflict and confusion as the Navy struggled to redefine its role to meet the needs of a vastly changed national security environment. By 1949, the broad outlines of postwar naval air doctrine were evident, but further refinement and adjustments were required in response to continuing changes in technology, domestic politics, and definitions of national policy goals. By the mid-1950s, the Navy had developed most of the technological and organizational tools necessary to implement the doctrine it had evolved. In this sense, it had once again achieved maturity, at a new level, appropriate for the postwar world. The process by which this level of maturity was reached is the subject of this essay.[1]

The Interwar and Early Postwar Years

A study of the emergence of postwar naval air doctrine must begin with a review of the interwar era. During the 1920s and 1930s, the Navy developed a flexible and generally effective procedure for adapting to strategic and technological innovation. Specialized training was introduced, but special branches were not created. Every officer entering a specialty like aviation or submarines was reminded that his first duty was to the Navy as a whole and that eventually he would be required to assume the traditional responsibilities of command at sea. As a result, new technology and tactics were tested, not within the protective walls of a separate branch, but within the body as a whole during annual fleet problems in which virtually the entire Navy participated. Innovations introduced in this way initially faced greater resistance, but, if successful, could be more fully accepted.[2]

Sea-based aviation was developed partly in response to the naval limitation treaties of the 1920s and 1930s, which forced the Navy to seek alternatives to the large battle fleets in planning for the defense of U.S. Pacific possessions. The aircraft carrier was from its inception an integral, although initially minor, part of naval operations and strategic planning. A generation of naval officers trained during this period witnessed the testing of sea-based aviation, observed its implementation in World War II, and came to accept—sometimes begrudgingly—its proven effectiveness. It was this generation, experienced in the application of naval aviation even if they were not aviators, which assumed control of Navy strategic planning after the war.[3]

The Navy's approach to innovation in the interwar era may thus be described as both conservative and highly flexible. Its approach to doctrine shared the same characteristics. The Navy relied very little on written doctrine as a training tool, preferring that naval officers absorb the fundamentals of naval theory through the practice of their craft. The result was that the Navy's officer corps shared a cohesive, almost mystical, understanding of the principles of sea power based on common experiences and carefully preserved traditions. Although this type of unwritten dogma served the Navy very well before and during the war, it was difficult to define and even more difficult to communicate.[4]

In addition, it tended to hinder rapid adjustment to fundamental change. As Bernard Brodie noted in January 1946, the Navy's "indubitably superb accomplishment in the greatest of all naval wars . . . [will] not facilitate it taking the lead in reevaluating its own place in the national security."[5] The early postwar studies of naval air power focused almost exclusively on past accomplishments. These included an abortive and controversial study for the United States Strategic Bombing Survey called "The Carrier Air Effort Against Japan" and a similar study called "U.S. Naval Aviation in the Pacific, A Critical Review."[6]

The Navy's orientation in the immediate postwar era is probably best reflected by Fleet Adm. Ernest J. King's triumphant third report to the secretary of the navy summarizing the Navy's contribution in World War II. In this report, King extolled the virtues of the "balanced fleet" and pointed out how the aircraft carrier had become an "integral and primary component of the fleet," capable of carrying out a multiplicity of missions, including destruction of hostile air and naval forces, support of amphibious operations, reconnaissance over the sea, and the defeat of "hostile land-based planes over positions held in force by the enemy."[7] Although the theme of flexibility and the emphasis on tactical missions were to constantly reemerge in postwar naval planning, King's presentation gave little indication of how naval air doctrine would be adapted to postwar challenges.

In a pattern that would become all too familiar, at least in the early postwar years, the Navy appears to have first turned its attention to the question of postwar naval doctrine not as a result of its own initiative but in response to outside stimuli. In February 1946, a Joint Strategic Survey Committee report outlined the future missions of the United States' land, sea, and air forces. For the next four months the Joint Chiefs of Staff (JCS) debated the fundamental differences in service philosophy raised in this report. The Army and Army Air Forces, led by Gen. Dwight Eisenhower, army chief of staff, and Army Air Forces Commanding Gen. Carl Spaatz, argued that service missions should be defined in terms of the medium in which each service operated, that is, land, sea, and air, and that duplication in weapons systems should be eliminated. The immediate question at issue was not the combat role of the aircraft carrier—this would come later—but whether the Navy should be allowed to maintain and operate land-based aircraft for such purposes as reconnaissance, antisubmarine warfare, and air transport in support of amphibious landings.[8]

The Navy responded that function, not weapons systems, should determine the role and composition of each service, and that each service should have forces large, varied, and flexible enough to accomplish its mission in the face of any contingency. With interservice tension over antisubmarine operations in the Atlantic still fresh in the Navy's memory, Chief of Naval Operations (CNO) Fleet Adm. Chester Nimitz and Deputy CNO for Air Vice Adm. Arthur Radford angrily resisted the suggestion that they turn over control of vital support operations to the Army Air Forces. Not only would this impair the efficiency and effectiveness of naval operations, but, they feared, it might seriously hamper naval air research and development activity, as had happened to the British Royal Navy when its fleet air arm was controlled by the Royal Air Force during the interwar period.

Radford was careful to point out that in demanding the right to control its own land-based air operations, the Navy had no intention of encroaching on the legitimate functions of the Army Air Forces. He stated:

> The Navy does not contemplate the creation of a land-based strategic bombing command; developing a land-based fighter force for the defense of the United States or of major outlying bases; building a tactical air force for land campaigns, or maintaining a competitive transport service. These are not nor have they ever been the intentions of the Navy. As is well-known, however, a most important part of the Navy is its air arm, complete and adequate, to fulfill naval missions. It includes aircraft based on ships, tenders, seadromes, or fields, with any type of landing gear-floats, wheels, or skis; powered by any type of engine—reciprocating, turbine, or jets; carrying any type of useful weapon—gun, rocket, torpedo, bomb, mine, or atomic explosive. We intend to take full advantage of scientific research and development applicable to air warfare including guided missiles and pilotless aircraft. We will continue to coordinate our enterprises with those of the Army in anticipation that each service will benefit by the progress of the other; unwarranted duplication will be avoided but no promising field of aeronautical science or tactics will remain

unexplored. Our aircraft will continue to be manned by pilots, aircrewmen and technicians who will be unexcelled by any others in the world.[9]

This and similar statements provided the first general definition of what the Navy believed its future to be in the field of aviation. Such a definition, however, was by no means complete. Although Radford described Air Force missions with some care, he failed to describe in a comparable manner precisely what the Navy's missions might be. This led the Air Force to charge that the Navy, despite its protestations, actually hoped to compete for control of strategic air operations.

The dispute was apparently settled in the Navy's favor by the National Security Act of 1947, which created a Department of the Air Force, but also guaranteed to the Department of the Navy control over naval aviation, including any land-based aviation "organic thereto."[10]

This provision, however, proved to be a much more temporary safeguard than the naval officers who had aggressively lobbied for it on Capitol Hill had hoped it would be. The question of the Navy's role in aviation, particularly with regard to the strategic air offensive, was to be a repeated source of conflict over the next several years.

The Atomic Bomb and Naval Aviation

Central to this controversy was the question of the atomic bomb. In the immediate aftermath of Hiroshima and Nagasaki, atomic technology was a matter of critical importance to the Navy because of the threat it appeared to pose to surface fleet operations. If an entire carrier task force could be destroyed by a single atomic bomb, the Navy's future would be in serious jeopardy. In response to public charges that this was indeed the case, the Navy prepared a test fleet to be sacrificed on the nuclear altar at Bikini Atoll in July 1946. The tests carried out during Operation CROSSROADS on 1 and 25 July 1946, however, largely relieved the Navy's fears. The results demonstrated, at least to the Navy, that a surface task force could survive an atomic attack if minor modifications in fleet routine were instituted—in particular, a more widely dispersed steaming formation and washdown techniques for radiological defense.[11]

The question of how the Navy might make use of the atomic weapon received much less immediate attention than the question of whether the Navy could survive any such use by an enemy power. Although Secretary of the Navy James Forrestal and Assistant Secretary Artemus Gates both stated in 1945 that U.S. aircraft carriers would someday be capable of launching an atomic attack, there is no evidence that the Navy was organizing to achieve a carrier-based strategic bombing capability in the immediate postwar years.[12] The position and office of Deputy CNO for Special Weapons (OP-06) had been created in November 1945 when the Office of the Chief of Naval Operations was formally established. The chief function of this position was to organize and implement Operation CROSSROADS. Once this assignment was nearly completed, however, officers within that organization apparently turned their attention to the question of how the Navy could employ atomic weapons. On 24 July 1946, for example, a letter originated by Cdr. Doyen Klein of OP-06 was sent to President Harry S. Truman under the signature of Acting Secretary of the Navy John L. Sullivan proposing the modification of a number of aircraft carriers so that they could handle atomic bombs.[13]

In November 1946, however, OP-06 was disestablished and its functions were reassigned. The DCNO (Operations), who had control of strategic plans, was now charged with overseeing atomic energy development; the DCNO (Logistics) was placed in charge of modification of carriers to handle atomic weapons; and the DCNO (Air), who administered the development of aircraft, aviation ships, and tactical concepts of aerial warfare, was also given responsibility over guided missile development.[14] As a result, dedicated younger officers such as John Hayward, Frederick Ashworth, and Joseph Murphy were forced to attempt to develop a nuclear strike capability for the Navy outside its formal structure. They received aid from Vice Adm. Forrest Sherman, the DCNO (Operations), and Vice Adm. Arthur Radford, the DCNO (Air), but no central office directed their efforts.[15]

One explanation for this lack of attention to the possibilities of atomic technology lies in the fact that in 1946 the Navy was focusing on an apparently more urgent problem. The German development of advanced conventional type XXI and XXVI U-boats toward the end of World War II, and the fact that the Soviet Union had captured a number of these submarines, had led to fears that the Russians could have as many as twenty of these fast submarines by 1948 and several hundred by 1951, thus seriously jeopardizing Allied control of the seas in war.[16] In June 1946, Chief of Naval Operations Chester Nimitz initiated Project GIRDER, a major research and development initiative in antisubmarine warfare. From this time until at least the spring of 1950, submarine technology and antisubmarine warfare development were the Navy's top research and development priorities.[17] This is reflected in the Fiscal Year 1948 shipbuilding program developed between May and July 1946. Although the need for a new aircraft carrier capable of handling heavier airplanes was recognized, the majority of ships proposed were submarines and antisubmarine types.[18]

The Navy was further discouraged from attempting to develop an atomic weapons capability by the confusion that then existed in planning circles as to whether atomic weapons would be available for use in war. During the spring of 1946, initial war plans for conflict with the Soviet Union, code-named PINCHER, were prepared by the Joint War Plans Committee and the Joint Staff Planners of the JCS. These plans called for taking maximum advantage of the atomic bomb.[19] Concurrently, however, President Truman was preparing and presenting a national policy calling for international control of atomic energy and the banning of all nuclear weapons. In response to this policy, Admiral Nimitz urged the Joint Chiefs "to avoid any specific affirmation at this time of any intention to use the atomic bomb."[20] In addition, it was unclear through the summer of 1947 what the limits and capabilities of the atomic bomb might be and how many might be available for use even if international control were not achieved. In the face of such practical and philosophical uncertainties, the

Navy was inclined to confine its attention to the kind of analysis that had been carried out in Operation CROSSROADS. Instead of planning for offensive operations, assessments were made of what damage might result if American ports and the Panama Canal were subjected to atomic attack.[21]

Strategic Air and the Navy

No general statement of postwar naval doctrine was promulgated until early 1947, when U.S. Fleet Publication Number One, *Principles and Applications of Naval Warfare*, was released. In that paper, the Navy identified the "destruction of the opposing will to resist" as "the fundamental objective of the armed forces in war." It noted that this objective could be achieved by attacking the enemy's means of resistance including industrial potential, naval and air forces, and transportation networks—until the enemy was forced to consider further prosecution of the war to be "unprofitable." It pointed out that air attacks would not achieve an early, easy victory, and that "the outcome of the war is dependent finally on ability to isolate, to occupy, or otherwise to control the territory of the enemy."[22]

Because of its mention of industrial targets, this statement suggested to the Air Force and subsequent analysts that the Navy was interested in competing for a role in the strategic air offensive. This interpretation, however, does not do justice to the complexity of the Navy's position. In May 1947, Bernard Brodie prepared a statement for the Library of Congress Legislative Reference Service describing the naval high command's view on atomic warfare that, although unofficial, was described as having been approved by Admiral Nimitz. That statement argued that bombing raids could be performed more effectively by carrier-based jet aircraft and guided missiles than by heavy long-range bombers. Not only were the Navy's planes faster and more maneuverable, but they would be operating from mobile bases over relatively short range.

Although the Navy's leadership did not specify the type of targets which would be destroyed in such attacks, it is clear that they were intended to be relatively limited. They believed that the kind of blanket strategic bombing that had been carried out over Germany and Japan by subsonic propeller-driven bombers would no longer be possible in the postwar world. The technology of aerial defense had so outstripped offensive developments that unacceptably large numbers of aircraft would have to be sacrificed to achieve the kind of massive impact which had been sought by the Army Air Forces during World War II.[23] This view was reinforced by the Navy's belief that relatively few atomic weapons would be available to the United States over the coming decade. If only a small number of bombs could be deployed, it would be necessary to ensure that a high percentage of them would reach their intended targets and that destruction of those targets would have maximum impact.

The Army Air Forces disagreed with this analysis at two critical points. They argued that adequate atomic weapons would soon be available to carry out the type of strategic bombing used during World War II, and, that, despite improvements in aerial defense the long-range heavy bomber was a proven weapon fully capable of carrying out the missions for which it had been designed.[24]

Bernard Brodie's report is interesting because it provides the first statement of a general Navy position on atomic air strategy in the postwar period. It did not, however, accurately reflect official Navy planning. In January 1947, Vice Adm. Forrest Sherman, the flag officer charged with directing naval strategic planning, gave a presentation to President Truman in which he did not mention naval use of atomic weapons, but stressed instead the conventional tactical role of carrier forces. This apparent internal conflict reflects a persistent split between long-term technologically oriented projection and near-term war planning which plagued the Navy as well as the other services after World War II.[25] Whereas actual war planning necessarily focused on currently available weaponry, the kind of forecasting which Brodie was describing was freed from such constraints.

It was not until after the final report of the JCS Bikini Evaluation Committee in July 1947 that the Navy initiated any serious consideration of how it could most effectively make use of the atomic bomb. The Bikini Evaluation Report, presented in detail to the assembled leadership of the nation's armed services on 29 July 1947, argued for a significantly upgraded evaluation of the weapon's potential power. It stressed the need for an effective atomic bomb striking force in being at all times as a deterrent to attack. It also stressed the scarcity of fissionable material and concluded that the bomb would have to be used primarily against urban targets rather than against troops or naval vessels.[26] Concerned about the nation's low level of conventional readiness, and impressed by the results of the Bikini tests, many naval officers—especially naval aviators—concluded that the Navy should attempt in earnest to develop a nuclear weapons capability.

The Navy's shift in attitude toward the atomic bomb in 1947 was indicated by a change in its justification of the construction of its new aircraft carrier, the CVA-58. Design work on this ship had begun in April 1945 in response to recommendations for future carriers developed from World War II combat experience. The new ship was the logical extension of existing carrier technology: it was bigger, had more powerful catapults and arresting gear, and, because it had no "island" superstructure, had a significantly greater flight deck area for parking and operating the heavy jet fighters and multi-engine attack planes which represented the next generation of naval aircraft. The CVA-58 was designed as a multipurpose ship that could handle a wide variety of weapons, conventional and atomic. By the fall of 1947, however, when the new ship was included in the Fiscal Year 1949 shipbuilding program, the Navy was referring to it as an "atomic carrier" both to the public and to Congress. This emphasis reflected not only the Navy's desire to develop its nuclear capability in the wake of the Bikini tests, but also its apparent belief that the carrier would stand a better chance of getting funded if it were defined in terms of the new technology.[27]

Further indications of a growing interest in atomic weapons is to be found in the force projections the Navy developed in November 1947 in support of the first Joint Outline of the Long Range War Plan, CHARIOTEER, being prepared by the Joint

Strategic Plans Group. CHARIOTEER was intended to define what forces, particularly air forces, the United States would need in the event of war with the Soviet Union in 1955. These projections were not solely for JCS use, but were also intended for submission to the president's Air Policy Commission, which was then preparing its public report on the nation's future aviation requirements.[28] The Navy report on its needs projected that it would require four four-carrier task groups by 1955. Each task group would contain one CVA-58 class carrier; all the carriers in each group were to be equipped with long-range multi-engine bombers.[29] The Aviation Plans Division of the Office of the DCNO (Air) provided its own forecasts of the types and capabilities of naval aircraft and guided missiles which would presumably be developed by 1955. At the top of the list were the North American AJ-1 multi-engine prop-jet attack plane, which was to be the Navy's first nuclear-capable aircraft, and a multi-engine jet design, the ADR-42, which, it was believed, could only operate from one of the CVA-58 type ships.[30]

These studies were complemented by a series of additional papers which went beyond technology and force levels to discuss the Navy's philosophy of its future role in warfare. The two most significant public papers were Adm. Chester Nimitz's valedictory statement upon retiring as Chief of Naval Operations, "The Future Employment of Naval Forces," issued on 6 January 1948, and Rear Adm. Daniel V. Gallery's famous memo of 17 December 1947, which was leaked to the public by Drew Pearson in March 1948. Nimitz argued that the Navy had developed carrier technology and tactics to such a point that it could create offshore bases of superior capability and relatively low vulnerability virtually anywhere in the world. He pointed out that since a feasible intercontinental bombing force was not likely to be achieved for several years to come, the Navy was the service best prepared to project power against the enemy in the initial phases of a war. The Navy, therefore, should be assigned continuing responsibility for supplementing Air Force bombing operations.[31]

Gallery, the Assistant CNO for Guided Missiles, went even further. His memo suggested that the Navy would be quite capable of handling most offensive air operations. In fact, he proposed, the entire Navy should be restructured to pursue this objective. "For the past two years," he argued,

> our defense of the Navy has been based mainly on old familiar arguments about exercising control of the seas. Much has been said about anti-submarine warfare. naval reconnaissance, protection of shipping and amphibious operations. It has been assumed, at least implicitly. that the next war will not be much different from the last one. This assumption is basically wrong, and if we stick to it, the Navy will soon be obsolete. The next war will be a lot different from any previous one. It seems obvious that the next time our Sunday punch will be an atom bomb aimed at the enemy capitols and industrial centers and that the outcome of the war will be determined by strategic bombing. The war will he won by whichever side is able to deliver the atom bomb to the enemy. and at the same time protect its own territory against similar delivery. I think the time is right now for the Navy to start an aggressive campaign aimed at proving that the Navy can deliver the atom bomb more effectively than the Air Force can.[32]

If this campaign proved successful, he went on, the Navy's primary mission could be the delivery of atomic attacks, while the Air Force would have the air defense of the United States as its prime responsibility. Gallery recommended as an interim measure the immediate development of a special carrier bomber based on the P2V Neptune until the development of a so-called "atomic" carrier capable of quickly launching a jet-propelled multiple plane strike force. This extreme position met with general skepticism on the part of senior officers of DCNO (Air) and the outright disapproval of the new Chief of Naval Operations, Adm. Louis Denfeld, and Secretary of the Navy John L. Sullivan.[33]

A more moderate position which received wider acceptance was circulated by a veteran naval aviator, Rear Adm. Ralph Ofstie, a Navy member of the joint Military Liaison Committee to the Atomic Energy Commission and a former member of the JCS Bikini Evaluation Board. In a paper entitled "The Composition of the U.S. Military Establishment," Ofstie followed the line taken by Admiral Nimitz, noting that the carrier striking force was peculiarly well suited for offensive air operations against the USSR. Ofstie, however, went beyond previous Navy statements in tentatively identifying the targets for this striking force. In his view, the first wave of attacks should focus on political control centers and urban and industrial concentrations in order to disrupt national organization and command structures. Other tactical targets could then be attacked, using conventional as well as atomic weapons. Ofstie believed "that the day of the great strategic bombing force suited only to aerial bombing is finished." He proposed that emphasis should be placed on developing high performance, high mobility aircraft with improved accuracy, rather than on trying to produce large numbers of "super bombers."[34]

The Nimitz, Gallery, and Ofstie memoranda demonstrate the lack of consensus that existed within the Navy with regard to the basic functions and missions of naval air power. Nimitz had raised the possibility that the Navy might participate in the air offensive; Gallery had taken the extreme position that strategic bombing should be the primary mission of the Navy. Ofstie, probably best representing the views of a majority of naval aviators, proposed a kind of middle course: the Navy should undertake atomic missions, but it should reject strategic bombing as practiced by the Air Force and develop an alternative model based on flexibility and selectivity.

Central to Ofstie's statement, and to the Navy's way of thinking at this time, was the desire to keep as many options open as possible. A series of operations analysis studies, completed in February and March 1948, indicated, as Nimitz had predicted, that sea-based aviation would be equally or even more effective than land-based aviation in delivering long-range air attacks against the Soviet Union.[35] The Navy remained unconvinced, however, that strategic bombing would win a war, despite Gallery's arguments, and certainly was not committed to trying to gain control of the air offensive. Unfortunately, the Navy's insistence on keeping its nuclear options open was interpreted by the general public, and more importantly by the Air Force, as indicating just such a desire. The CVA-58, which was included in the Navy's Fiscal Year 1949 budget, became the public symbol of a presumed Navy campaign to undermine Air Force prerogatives.[36]

The Roles and Missions Debate

Concerned for its new existence as an independent organization, the Air Force was undoubtedly inclined to exaggerate the threat posed by the Navy. In the winter of 1948, the Air Force, with about thirty-five nuclear-capable B-29 bombers, was the only service capable of delivering an atomic attack of any kind. The Navy had modified its three *Midway* class carriers to handle atomic weapons, but the first dedicated nuclear-capable naval aircraft, the AJ-1 Savage, was at least twenty-one months from delivery, and an interim aircraft, the P2V-3C Neptune, had not yet been carrier-tested prior to being modified to carry atomic bombs.[37] In addition, it became increasingly clear that the full resources of both the Air Force and the Navy would have to be used to counter the threat posed by Soviet conventional forces in the event of war. The war plans drawn up in 1948 projected that the situation in Western Europe in case of a Soviet invasion would be so desperate that maximum use of all available forces would be required to meet the emergency.[38] Nevertheless, projections did little to ease growing interservice tension over allocation of responsibility for atomic weapons and bombing operations.

It was against the backdrop of the Nimitz and Gallery statements and the proposed construction of the CVA-58 that the roles and missions disputes of 1948 were carried out. For the most part, these arguments appeared to deal with relatively minor points, such as the question of which service would set schedules for the development of nuclear-capable aircraft,[39] and who would serve as the executive agent for the JCS to the Armed Forces Special Weapons Project.[40] However, the issue of whether the Navy would be allowed to determine what air operations were necessary to accomplish its wartime missions would receive a great deal of attention and provoke extensive controversy. The JCS position paper on armed forces functions was reviewed and revised at the Key West Conference in March 1948 in hopes of resolving the Navy–Air Force split over this issue. The statement issued as a result of that conference, however, as well as a subsequent clarification by Secretary of Defense James Forrestal and a second conference held at Newport in August 1948, failed to satisfy either service. Misunderstanding and conflict between the services was worse by the fall of 1948 than it had been before passage of the National Security Act a year before.[41]

The General Board Sets the Navy's Course

During the winter of 1948, in the absence of any consensus within the national military establishment regarding service roles and missions, and without clear guidance on national security policy from the president or the National Security Council (NSC), the Navy initiated a broad survey of its role in postwar national defense. The need for such a study was abundantly clear by this time. The uses and limitations of nuclear weapons were better understood, and the possibility of conflict with the

Soviet Union seemed increasingly imminent. The General Board was the logical office within the Navy Department to undertake such a study. The Board had been established in March 1900 for the purpose of advising the secretary of the navy on questions of high policy. It did not represent any special interest group within the Navy; rather, it was composed of some of the brightest senior-line officers in the Navy and Marine Corps. Because of this broad representation, the Board was in a better position than the naval aviation community to evaluate the role of air power within the context of the Navy's overall contribution to national defense.

The driving force behind the General Board study of "National Security and Navy Contributions Thereto for the Next Ten Years" was Capt. Arleigh Burke, a surface officer with special training in ordnance explosives who had served as chief of staff to Vice Adm. Marc Mitscher, commander of Fast Carrier Task Force 58 in World War II. It was Burke who recommended that an overall review of national security requirements should be undertaken before attempting to identify the Navy's place in national defense.[42] His 200-page final report, which was released on 25 June 1948, was based on written statements collected from several hundred senior naval officers and distinguished civilians.

The report's conclusions regarding strategic air warfare were in distinct contrast with the views developed in the naval aviation community during the previous two years. The General Board did not offer carrier-based aviation as a major alternative to land-based forces in carrying out the strategic air offensive, and it expressed reservations about the offensive itself as currently envisioned. The report conceded that the air offensive would be vital to the United States in the event of a war with the Soviet Union and that atomic weapons would have to be used to achieve the desired results within a reasonable time. It argued, however, that control of the seas, selective initiation of ground offensives, and other conventional operations must also be considered necessary elements in U.S. strategy and that "sole reliance on the complete success of violent and irretrievable departures from established concepts and techniques of war"—apparently meaning strategic nuclear bombing—would be highly inadvisable.[43] The General Board further expressed the view that the most vulnerable targets in the Soviet Union were not its industrial cities, as attacks on most of these would require too great an expenditure of scarce resources, but the large southern oil fields and the principal naval and submarine bases.

Taking a page from the concerns of Project GIRDER, the report identified antisubmarine warfare as the first mission of the carrier task forces. In an effort that was expected to absorb the greater part of their energy, those forces would destroy and blockade "enemy submarine bases by atomic, radiological, conventional bombing or mining attacks." Additional carrier missions, in order of importance, were support for amphibious assaults to seize advanced bases, air cover for surface forces and convoys in sea lines of communications, and, finally, contributing to the air offensive by attacking targets which could not easily be reached by any other means. In an accurate appraisal of the Navy's current capabilities, which had unfortunately been lacking in earlier studies, the General Board concluded rather pessimistically that:

the Navy's initial tasks of control of the seas, occupation or seizure of advanced bases, attacks on Russian bases and denial of advanced bases to Russia, combined with the enormous logistic supporting effort for the other services and our allies, will place so many demands upon the Navy for immediate operations in widely separated parts of the world that fulfillment of all demands may well be beyond the capacity of the Navy in being.[44]

The General Board study, which was circulated as widely within the Navy as its top secret classification would allow, was a turning point in the Navy's efforts to chart its course in the postwar environment. It provided for the first time a realistic assessment of the demands which would be placed on the Navy in war and the Navy's ability (or inability) to meet those demands. Unlike the Gallery memorandum, it did not endorse the concept that strategic bombing could win the war, and it emphasized the continuing importance of traditional naval tasks. Using the best estimates of the enemy threat then available, the General Board pointed out how impossible it would be for the Navy to attempt to assume a major role in the strategic air offensive while carrying out its own vital missions.

More on the Roles and Missions Debate

From the fall of 1947 on, American joint war plans had called for both an atomic air offensive against the Soviet Union and conventional operations, including naval operations in the Eastern Mediterranean and the maintenance of a substantial foothold in Western Europe. A major reassessment of such war plans became necessary, however, as a result of the $14.4 billion ceiling which President Truman imposed on the Fiscal Year 1950 defense budget and refused to negotiate. A series of heated debates within the JCS in October 1948 produced a plan in line with this austerity budget which called for abandonment of Western Europe and a reduction of the Navy's missions to defense of the sea lines of communication. An atomic air offensive launched from Great Britain, the Suez area, and Okinawa would be the primary U.S. war effort.[45]

This proposed reduction in conventional operations and increased reliance on strategic bombing produced a two-fold reaction within Navy planning circles. First, it led to a concerted Navy campaign to defend the aircraft carrier as a multipurpose weapon which could provide much needed operational flexibility. In presenting this argument, the Navy was not only fighting for its own organizational survival, but was concerned that the strategy being proposed was rigid, one-dimensional, and inherently unsound. Second, the Navy began to voice serious doubts about the proposed strategic air offensive. This concern had been slowly developing during the previous year and a half, beginning with the argument that strategic bombing as it had been practiced during World War II would no longer be possible in an era of sophisticated air defense. By the fall of 1948, as we shall see, the Navy was questioning not only whether such attacks were feasible, but whether they would promote U.S. war aims even if they could be successfully delivered.

The effort to defend the Navy's existing carrier forces in the face of the budget ceiling was only successful in the sense that it averted complete disaster. In the October 1948 JCS debates over the Fiscal Year 1950 budget, the Navy argued that a nine-carrier force level, a cut of two from Fiscal Year 1949, was the absolute minimum that it could accept. Chief of Staff of the Air Force Gen. Hoyt S. Vandenberg recommended that the Navy should have only four carriers, while Chief of Staff of the Army Gen. Omar Bradley recommended six. Secretary of Defense Forrestal on 9 November approved an eight-carrier force level, provided that it could be maintained out of the $4.6 billion allocated to the Navy.[46]

Six months later, on 23 April 1949, after consultation with President Truman, Louis Johnson, who had replaced Forrestal as secretary of defense, canceled the construction of CVA-58, only recently named USS *United States*. Johnson's decision was primarily an economy move, but he justified it on the grounds that the carrier would be an unwarranted duplication of effort since its primary mission was apparently atomic warfare.[47] The eighteen months of Navy propaganda focusing on the nuclear capability of the ship thus proved to be its undoing. While the cancellation of the CVA-58 did not affect the eight-carrier force level established for Fiscal Year 1950, it was an ominous indication of what was to come. On 5 July 1949, Secretary Johnson set a tentative carrier force level for Fiscal Year 1951 of only four ships. Although this was raised two months later to six, the continuing decline of carrier air power appeared inevitable.[48]

The Navy's opposition to continued reliance on the strategic air offensive fared little better than its effort to retain an adequate conventional carrier force. During the winter and spring of 1949, a number of naval officers prepared statements questioning the efficacy and appropriateness of the planned atomic attack on industrial concentrations in the enemy's cities. Those developed by the Strategic Plans Division for Admiral Denfeld's use in the JCS defined the Navy's position in interservice debates. Others, including papers by Ralph Ofstie, Arleigh Burke, and, surprisingly, Dan Gallery, were circulated only within the Navy Department. Gallery argued eloquently that the planned atomic air offensive would be unlikely to win a war:

> ... this kind of war is not as simple as the prophets of the ten day atomic blitz seem to think. Some authorities estimate that the damage done by strategic bombing of Germany was equivalent to 500 Atomic Bombs. But Germany did not surrender until her armies were defeated. This damage is costing the U.S. huge sums of money now. In addition, leveling large cities has a tendency to alienate the affections of the inhabitants and does not create an atmosphere of international good will after the war.

In a total reversal of the position he had developed a year before, he proposed that "we should abandon the idea of destroying enemy cities one after another until he gives up and find some better way of gaining our objective."[49]

In May 1949, a joint ad-hoc committee headed by Air Force Lt. Gen. Hubert R. Harmon completed a study requested by Secretary of Defense Forrestal in October 1948. The purpose of the study was to analyze the probable impact of the planned air offensive under the most favorable conditions. It concluded that, even if all the bombs reached their assigned targets, the air offensive alone would not destroy the Soviet Union's will or capability to make war or prevent a Soviet takeover of Western Europe. Although Air Force Chief of Staff Vandenberg objected to these conclusions and attempted to have them deleted from the final report, an impassioned defense of the study by Admiral Denfeld was effective in preventing major modifications. Since the report was never circulated, however, its impact on strategic planning was limited.[50]

The fear that the Navy would be reduced to little more than a convoy and escort force while the atomic air offensive continued to dominate U.S. strategy finally brought forth within the Navy a clear statement regarding its philosophy of naval aviation. In a postmortem on the cancellation of the USS *United States*, Ralph Ofstie concluded that the Navy itself was to blame for the defeat, because it had "simply remained on the defensive and failed to make its position clear."[51] Ofstie attempted to remedy this failing. On 29 April 1949, he presented what was to be the simplest and yet the most comprehensive statement to date of naval air doctrine:

> Strategic air warfare (SAW) may be defined as the sustained mass attack (when using conventional bombs) or attack with weapons of mass destruction (atomic bombs) against the war making capacity of an enemy. It is essentially based on the wholesale destruction of urban and industrial areas and the civil populace of the enemy rather than direct attack on his active military machine.
>
> ... The Navy does not concur in any view that readiness to conduct SAW should be a major factor in the peacetime air power of the United States. However, the Navy would naturally be ready. within its capabilities to assist in SAW, as a purely secondary function. if directed by appropriate authority.
>
> ... Naval air, representing the mobile air power of the United States, is primarily directed to the delivery of the maximum air strength wherever that mobile air force can be employed against targets of direct and immediate military importance. It considers these targets to be military forces (land, sea, and air) military installations (land, sea, and air bases), and lines of communication (ocean and inland shipping, rail and road transport, and the fuel, therefore—oil).[52]

This statement succinctly spelled out the basic elements of an emerging consensus within the Navy regarding the place of naval air power in national strategy; as such, it provided a basis for future refinements of air doctrine.

These principles also provided a framework for the Navy's continuing efforts to defend its prerogatives within the national military establishment. A renewed movement for additional unification of the armed services in the fall of 1948 had produced a series of proposals for further reducing naval autonomy, including the proposal that the secretary of the navy be demoted from cabinet status. The ensuing debate over revision of the National Security Act, which continued until the approved amendments became law in June 1949, served to keep alive the Navy's fear that it would be forced to accept an increasingly subordinate place in national defense. This round of interservice competition came to a head in the fall of 1949 in the congressional hearings over the B-36, unification, and national strategy.

The Navy's reaction to attack on its position was to strike back publicly through the press and appeals to Congress. The situation appeared desperate, and desperation tactics were employed, including the use of innuendo and falsehoods, which only weakened the Navy's position. The House Armed Services Committee's B-36 Investigation of August 1949 proved especially embarrassing to the Navy because of the use of such tactics by certain zealous individuals.[53] The shadow cast by these proceedings, in turn, made it difficult for the Navy to present its position convincingly during the much more substantive Unification and Strategy Hearings before the same committee in October. During those hearings, a pantheon of high-ranking naval officers, led by Admiral Radford, but including Admirals King, Nimitz, Denfeld, Blandy, Ofstie, and Captain Burke, made a massive, technically oriented presentation that attempted simultaneously to prove the need for a Navy, extol the virtues of the aircraft carrier, criticize the weakening of the Navy's place in the defense organization, attack the technical capabilities of the B-36 bomber, and question the effectiveness of the atomic air offensive in achieving the goals of American air strategy.[54] The emphasis placed by the Navy on the virtues of aircraft carriers while attacking the B-36, the general inarticulateness of most Navy spokesmen, and their inability to document their critique of the air offensive because of the sensitive nature of supporting documents, all combined to leave the cumulative impression on the public that the Navy was condemning the direction of current U.S. strategy primarily because it had not been allowed to dominate the air offensive with carrier-based aircraft.[55]

The new consensus regarding naval air power, which had provided the basis for the Navy's presentation during the Unification and Strategy Hearings, in some ways marked a return to prewar and postwar concepts. Ralph Ofstie's April 1949 memo is oddly reminiscent of Admiral King's 1945 expression of his views on the World War II role of naval aviation. Aircraft carriers were to serve as the United States' mobile striking force, capable of performing a spectrum of missions. from providing naval presence in time of peace (as carriers had been doing in the Mediterranean since 1946) to delivering the atomic bomb in war. The central concept was flexibility, and the focus was on allowing the Navy to define its own strategy in pursuit of its stated missions.

The OP-55 Study

In August 1949, the Air Warfare Division of the Office of the DCNO (Air), OP-55, which was charged with formulating "long range and short range programs for the most effective employment of naval aviation in air warfare," spelled out the specific applications of this new strategy in a study on "The Future Development of Carrier Aviation." This study used as its starting point the JCS 1948 Key West Conference statement on the functions of the armed forces, which identified the Navy's prime mission to be "control of vital sea areas and protection of vital sea lines of communication." The Navy's willingness to accept this definition did not indicate that it had capitulated to external pressures, but rather that the identification of naval air missions suitable to the postwar environment had given the service the confidence to pursue its own path without fear of restriction and encroachment.[56]

The OP-55 study was the most significant single statement of naval air doctrine yet produced. Its most significant immediate conclusion was that enemy air power, rather than enemy submarines, was the most serious threat confronting the Navy. The paper argued that the Soviet submarine fleet was neither as large nor as sophisticated as had been projected since 1946 and that it could be "effectively throttled early in the war and kept under control by a timely and aggressive anti-submarine campaign employing carrier air strikes, aerial minelaying. and antisubmarine subs as the spearheads, backed by the more conventional measures such as barrier patrols, convoy escorts, and antisubmarine warfare hunter-killer carrier groups."[57]

Soviet tactical air forces, which were judged to be experienced, technologically advanced, and potentially extremely dangerous, posed a much more serious threat over the next ten years to U.S. control of the sea lanes in the eastern Atlantic and the Mediterranean. The problem of Soviet air power had earlier been identified in a January 1949 report by the staff of Adm. Richard Conolly, commander in chief, U.S. Naval Forces, Eastern Atlantic and Mediterranean. The report had proposed using aircraft carriers as mobile air bases for fighters engaged in intercepting enemy air attacks on sea lines of communications and advanced bases in the eastern Mediterranean.[58] OP-55 rejected this strategy, arguing that the Navy should take aggressive, offensive action to destroy the air threat by attacks on enemy air installations:

> Carrier aviation must retain the bulk of its strength in *offensive power* if it is to support a truly offensive Navy rather than a defensive one. Our Navy must carry out numerous functions other than defensive antisubmarine warfare and must possess the self-contained ability to move at will and wage offensive war against the enemy in the air, on the surface and below the surface.

The report pointed out that the Navy would expect to make use of long-range heavy attack aircraft and atomic bombs, but that these planes and weapons would be used for tactical missions rather than in strategic air attacks. "It is not military practice to limit the employment of any one weapon to the fulfillment of any one function," the report concluded, arguing that the Navy should be allowed to use all available weapons in pursuit of its assigned objectives.

The OP-55 report made clear that heavy attack aircraft and nuclear weapons would play only a minor role in naval aviation. It recommended that heavy attack planes be kept to a minimum in designing carrier air groups and that emphasis be placed instead on general purpose fighter aircraft with both offensive and defensive capabilities and on day attack and close air support attack airplanes. It emphatically recommended construction of a flush deck aircraft carrier along the lines of the canceled USS *United States* as a "necessary and logical development" in naval technology, but stated that a somewhat smaller new carrier, the size of the *Midway* class, might be substituted if necessary. Failing both of these, it urged modification of one of the existing *Midway* or even *Essex* class carriers to the configuration of the CVA-58 in order to meet the urgent need for a carrier able to accommodate the increasing size and capability of carrier-based aircraft.

The Air Warfare Division's study initiated a process of refinement and implementation of naval air doctrine that proceeded through the mid-1950s along the lines established in the summer of 1949. Its recommendations regarding ship and aircraft construction and deployment were largely implemented, and its analysis of the threats confronting the U.S. Navy and its basic missions underwent only minor adjustment during this time.

Decisive Developments

The problem of gaining support and adequate funding for its projected building programs had been a major stumbling block for the Navy in trying to define its future in the postwar period. Without the constant distraction of interservice competition for limited budget funds, naval air doctrine might have evolved more smoothly and rapidly. In the year after the release of the OP-55 study, however, three developments occurred which were largely to alleviate this source of frustration: a new chief of naval operations with outstanding leadership abilities took office; the fall of China and the Soviet nuclear explosion triggered a movement aimed at upgrading the U.S. military posture; and the Navy's performance during the early months of the Korean conflict proved its particular value in national defense. Each of these developments deserves at least a brief discussion.

Admiral Louis Denfeld was fired as CNO shortly after the Unification and Strategy Hearings. His replacement was Adm. Forrest Sherman, a brilliant, controversial officer who had been one of the architects of unification. Sherman was the first career naval aviator to serve as CNO. He brought to the JCS considerable experience in strategic planning, a clear grasp of naval strategy, and unusual skill at interservice infighting and bargaining. His impact was felt almost immediately. During his first six months as CNO he successfully blocked reduction of the active carrier force below seven despite the tightness of operating funds and helped to heal the interservice rifts that had developed during the previous three years. Until his untimely death in July 1951, Sherman was a leading advocate of conventional preparedness against emergencies short of global war and of keeping as many options open as possible for facing a worldwide conflict.[59]

The willingness of national decision-makers to fund the kind of military establishment that could implement such a strategy increased rapidly following the fall of China in September 1949 and the discovery of the Soviet atomic explosion that same month. NSC-68 of April 1950 crystallized and symbolized the decision-makers' mood. It recognized that a one-dimensional nuclear-oriented defense posture would be inadequate to handle the demands of national security and recommended—after rejection of such other options as isolation, preventive war, and continuation of current policies—that the United States build up its overall military, economic, and political power to avoid defeat in the Cold War and to defend against a possible Soviet attack in 1954.[60]

The North Korean invasion of South Korea on 25 June 1950 was the final impetus to rapid expansion of the U.S. military and of the Navy in particular. Seventh Fleet aircraft carriers were among the first U.S. combat units to respond in a sustained manner to the emergency, vindicating the Navy's claims about the value of mobile, flexible carrier striking forces. On 11 July, the JCS agreed to postpone a scheduled reduction in carrier force levels, and one day later Defense Secretary Louis Johnson, who had canceled the USS *United States* only fifteen months before, told Sherman, "I will give you another carrier when you want it."[61]

By 8 August, Johnson was discussing a defense budget of $50 billion, as compared to the $13 billion budget for Fiscal Year 1951. In December 1950, the NSC decided that the force levels the JCS had recently set as goals for 1954 should be treated as interim levels to be achieved no later than 30 June 1952. Revised final estimates were prepared by the JCS later that month. These called for a ninety-five group Air Force, an eighteen division Army, and a Navy of twelve large aircraft carriers, fourteen carrier air groups, fifteen light and escort carriers, plus large numbers of additional ships. The buildup and renovation of U.S. nuclear and conventional forces continued through the remainder of the Truman administration and for several years beyond. During this period, a new "super-carrier" was included in each fiscal year defense budget, beginning with USS *Forrestal* in Fiscal Year 1952.[62]

The expansion of naval forces was complemented by the achievement in February 1951 of a rudimentary nuclear attack capability for American aircraft carriers. Since the spring of 1949, the Navy had been training for atomic missions using modified P2V-3C Neptune patrol planes, which could be launched from a carrier but could not land on one. That fall the first deliveries of carrier-based AJ-1 Savages were made to the Navy's sole nuclear attack squadron, VC-5. Until late 1950, however, VC-5 had no bombs available nor any assigned missions in which they would be used.[63]

On 14 June 1950, President Truman, in response to a JCS request via the Atomic Energy Commission, permanently released ninety nonnuclear mechanical assemblies from the Atomic Energy Commission to the military. By September, both the JCS and the NSC had apparently approved storage of some of those components on *Midway* class aircraft carriers.[64] Although nuclear capsules for those components were not released by the president to the military until April 1951, nor put on aircraft carriers until at least 1953, a clumsy but workable system of flying those capsules from the United States to carriers at sea, code-named "Daisy Chain," was developed during 1951–52.[65]

The Navy's ability to make use of the weapons once they were received was still strictly limited. In February 1951, six AJ-1s and three P2V-3Cs flew to Port Lyautey, Morocco, in the first operational deployment of Navy nuclear-capable aircraft. Malfunctions in the planes forced their grounding for four months, however, and through October the AJ-1s operated at sea for only nineteen days. Dan Gallery, who commanded the Sixth Fleet's carrier division at that time, was asked years later whether he had ever considered using the planes in the event of war. "We just didn't even think about it," he said.[66]

The mission developed in October and November 1950 for the Navy's nuclear-capable aircraft was specifically a naval one: to destroy the capabilities of the Soviet surface and submarine fleets in areas within a 600-mile radius of the Mediterranean, Norwegian, and Bering Seas.[67] By September 1951, that mission had been expanded somewhat, as indicated by a statement on carrier forces prepared for the JCS in connection with continuing NSC studies of mobilization:

> These forces represent the major striking power of the Navy and are primarily responsible for neutralizing at the source the enemy's offensive capabilities to threaten control of the seas. These forces will destroy enemy naval forces and shipping, attack naval bases, attack airfields threatening control of the seas, support amphibious forces and support the mining offensive. As additional tasks, the carrier striking forces will defend bases and vital areas against attack through the seas, as required. In addition to the above, these forces will provide naval support essential to the conduct of operations by Supreme Allied Commander, Europe (SACEUR), Commander in Chief, Far East (CINCFE) and other area commanders. For example, the 6th Fleet, now in the Mediterranean, will provide naval support to SACEUR in the accomplishment of his missions.[68]

The assignment of carrier forces to support SACEUR's missions was the result of events that had transpired since January 1951. In that month General of the Army Dwight Eisenhower, the newly appointed Supreme Allied Commander, Europe, had briefed President Truman and the cabinet on his vision of NATO's future. Eisenhower's strategic concept for the defense of Europe was to use a "great combination of sea and air strength" in the Mediterranean and the North Sea in addition to building up ground forces on the central front. Then, he noted, "if the Russians tried to move ahead in the center, I'd hit them awfully hard from both flanks."[69]

Tactical support for ground forces had been a mission of the Sixth Fleet since 1948, and the Navy had argued persistently that atomic weapons as well as conventional ordnance should be used for this purpose. In 1949, after the Harmon Report had demonstrated that strikes on industrial concentrations alone would not prevent the Soviet Union from taking Western Europe, and the NATO treaty had been signed committing the United States to the defense of Western Europe, the Strategic Air Command was tasked with holding back a Soviet advance with nuclear weapons.[70] Because of the difficulty of hitting troop concentrations and the scarcity of nuclear weapons, however, plans for carrying out this mission were three years in the making. Before 1952, the JCS did not feel they could allocate atomic weapons or units specifically for tactical use.

Not only was the actual number of bombs small, but the so-called "doctrine of scarcity" held that "there is a definite, positive, and known limit to the number of atomic bombs which can be produced."[71] In early 1950, however, the prospects offered by the experimental breeder reactor and the discovery of new deposits of low grade uranium ore put an end to such thinking. Significant increases in nuclear weapons production were approved in 1950 and 1952.[72] In addition, advances in bomb technology had led to the development of relatively small atomic bombs which could be delivered by general purpose fighters and day attack planes like the Navy's

F2H-2B Banshee and AD-4B Skyraider. In February 1952, the JCS informed Eisenhower that, for planning purposes, a number of atomic weapons had tentatively been allocated for tactical use in the defense of western Eurasia and that Navy as well as Air Force aircraft could be considered as prospective delivery vehicles.[73]

The increased availability of nuclear weapons led to new clashes between the services over control and coordination of offensive atomic operations, particularly the retardation mission. In early 1952, channels for coordination of nuclear strikes and review of unified commanders' nuclear target annexes were established through a jointly staffed war room in the Pentagon.[74] The Navy, however, continued to insist on its right to maintain flexibility in planning for the use of nuclear weapons. Naval leadership repeatedly argued that they could not predict "exactly what targets the Navy will attack on any particular day. It will depend entirely on the situation existing and the requirement for the delivery of the attack."[75] In addition, the Navy and the Air Force strongly disagreed over whether carrier task forces would be able to survive in high threat areas and launch atomic strikes against their targets. A Weapons Systems Evaluation Group (WSEG) study completed in February 1952 indicated to the Navy that atomic missions against tactical targets in Europe as well as targets of naval interest in the Soviet Union would be well within its capability once it received the new *Forrestal* class carriers. The Air Force, however, questioned the premises, execution, and objectivity of the WSEG study and argued that it had not adequately demonstrated that carriers could operate effectively in any capacity in a war.[76] The Navy and the Air Force also disagreed over what kind of nuclear weapons should be stockpiled for naval missions. The Navy preferred to have available some quantity of gun-type weapons which used more fissionable material than the standard implosion bombs, but had greater penetration capability for striking submarine pens. The Air Force held that scarce fissionable material resources should be used to produce a larger number of implosion bombs for less specialized missions.[77] These three areas of conflict continued for the remainder of the decade to limit Navy–Air Force cooperation with regard to atomic operations.

Despite changes in organization, force levels, and technology, naval air doctrine underwent no major revision between 1949 and 1953. The basic concept that the Navy's carriers would serve as flexible mobile striking forces fulfilling tactical missions remained constant. Although some minor adjustments were necessary, they are difficult to pinpoint since no statement of doctrine for this period comparable to the OP-55 study has been uncovered in classified or unclassified sources.[78] In the fall of 1953, however, a subtle shift in the direction of national defense policy reopened old debates and posed a serious challenge to established naval air strategy.

The New Look

President Dwight D. Eisenhower, who for a number of years had been growing increasingly concerned over rapidly rising government spending, decided in the spring of 1953 to take a "new look" at defense policy and defense spending. A series

Postwar Air Doctrine and Organization 195

of studies were undertaken which climaxed on 30 October 1953 with the approval of NSC 162/2 as the administration's statement of Basic National Security Policy. That statement concluded that "the risk of Soviet aggression will be minimized by maintaining a strong security posture, with emphasis on adequate offensive retaliatory strength and defensive strength" based on "massive atomic capability" as well as conventional readiness. It argues, however, that such a military posture would have to be achieved and maintained "at the lowest feasible cost" so as not to "seriously impair the basic soundness of the U.S. economy by undermining incentives or by inflation."[79]

This doctrine of "massive retaliation," as it became known following Secretary of State John Foster Dulles's address on the subject in January 1954, was hotly debated in the JCS. To the Navy, it seemed to threaten the kind of overreliance on the atomic air offensive which had characterized defense planning under the Truman budget ceilings. For reasons of economy, conventional forces were to be frozen or even cut back, while emphasis was placed on the terrible deterrent and striking power of strategic nuclear attacks.[80] This posed a dilemma for the Navy not unlike the one it had faced in 1948. Although naval planners objected to the strategy itself, they were determined to be in the forefront of the nation's defense in line with whatever policy was adopted. On 7 December 1953, Chief of Naval Operations Adm. Robert B. Carney presented to the JCS the Navy's analysis of massive retaliation. Development of such a capability, he argued, was valuable as a deterrent for the time being, but could result in an atomic stalemate once the Soviet Union had acquired a substantial stockpile of its own. Thus, the United States could not afford to be without highly mobile, combat-ready strategic reserves "if we are to continue over the long term to be able to cope with limited aggression and at the same time be prepared for general war."

In a passage that laid out the future of naval aviation, Carney stated:

There is no question but what we must maintain a strong U.S. air capability, including a capability for inflicting massive damage, but not neglecting our capabilities for tactical air support, control of sea communications and vital sea areas, and defense against air attack. U.S. military air power comprises Air Force, Navy, and Marine Corps air power; all three play vital roles in our military posture and none must be neglected if that posture is to be truly effective. Naval air forces, including carrier aircraft, and Marine aviation, all trained to a high state of combat readiness, have repeatedly proved their effectiveness and value. Our entire politico-military philosophy today is based on the concept of collective security, which comprises overseas alliances, overseas bases, and U.S. military forces deployed overseas. The keystone of this entire structure is the confidence felt by our Allies that we can and will maintain control of sea communications in the face of any threat. U.S. Naval air forces as now constituted are essential to maintain this vital sea control in the face of the well-recognized Soviet surface, submarine, air and mining threat to our worldwide sea communications. From both the military and economic viewpoint then, it is unsound drastically to cut back these forces, already bought and paid for, for the sole purpose of making funds available to enlarge other types of U.S. air power.[81]

After the doctrine of massive retaliation was approved, however, the Navy turned from criticizing it to seeking ways of taking advantage of it. In January 1954, Admiral Carney prepared another memo which argued that the offensive striking power emphasized in the stated policy should be interpreted to include aircraft carriers as well as other weapons in the national arsenal. This interpretation was endorsed by the JCS on 5 February 1954.[82]

The dual position taken by the U.S. Navy with respect to national strategy and the atomic bomb in the winter of 1954 marked a culmination of the maturation process which naval air doctrine had been undergoing since 1945. By 1949, the Navy had developed a clear vision of what the wartime and peacetime missions of naval aviation should be. In 1954, it demonstrated an ability to adapt to a changing environment without losing sight of its own goals and identity. Its response to the doctrinal challenge posed by NSC 162/2 was to present a mature statement of its own doctrine, and, when its position was overridden, to attempt to move ahead toward achieving its goals within the context of the policy it had opposed.

For the next several years, the Navy made every effort to conform to the national security policy adopted in October 1953. Nuclear delivery capabilities for all attack aircraft carriers were upgraded as new planes such as the Douglas A3D Skyraider and A4D Skyhawk entered fleet service. In addition, the Navy publicly advertised its aircraft carriers as being thermonuclear-capable weapons systems that were contributing to the national strategic deterrence mission.[83] At the same time, however, the Navy's basic strategic concepts governing the use of all this hardware were much the same as they had been in 1949. The main planned mission of naval aviation at the start of any war was still to be "offensive action (including atomic), against enemy resources that threaten control of the seas, and in support of the land battle."[84]

A Changing Role for the Carrier

Under the leadership of Adm. Arleigh Burke, the author of the seminal General Board study of 1948, and, as director of the Navy's Strategic Plans Division, the original author of Admiral Carney's memo of 7 December 1953, the Navy continued to move away from emphasis on massive atomic retaliatory striking power and toward emphasis on the greater flexibility required to face limited aggression in remote areas of the globe.[85] By January 1958, when the Navy issued a statement of its general long-range objectives for the era of the 1970s, it was proposing that carrier forces be specifically tailored "for limited war, to be the nation's primary cutting tool for that purpose." Although carriers would be equipped with long-range aircraft and nuclear missiles and could serve an auxiliary function in more global conflicts, the Navy's major contribution to nuclear deterrence and nuclear exchange would be in the form of ballistic missile submarines, not naval air power.[86] Despite some modifications, the basic task of attack carriers into the 1970s closely followed the course laid out in this 1958 statement.

Summary

It is very difficult to draw precise conclusions from the records thus far available on the development of naval air doctrine from 1945 through 1958. When one has sorted out the protagonists, identified and analyzed their positions, and traced the course of their arguments, the most striking finding is a negative one: the Navy produced little in the way of doctrinal innovation with regard to air power in this period. Despite the technological revolutions which transformed air warfare during and immediately following World War II, the Navy's vision of the role of the aircraft carrier remained strikingly static. Naval aviators did flirt in 1947 and 1948 with the idea that the Navy might play a leading, or at least significant, role in the kind of nuclear strategic air offensive which seemed to be the doctrine of the future. But the Navy never formally endorsed the concept that strategic bombing could win a war and never seriously sought a role in the air offensive, except to keep its options for the use of air power open.

The doctrine the Navy adopted and maintained grew out of a careful analysis of its technological capabilities and its own combat experience and reflected the service's general philosophy and orientation. Just as naval aviators had been trained in the interwar period to see themselves as naval officers first and aviators second, so naval aviation was viewed in the postwar era as an integral part of the Navy as a whole, to be used in support of the Navy's primary mission—control of the seas. Furthermore, the Navy's conservative approach to acceptance of technological innovation caused it to respond more slowly and cautiously than the Air Force to the challenge of nuclear technology. Whereas the Air Force seized on the new technology and attempted to channel and control it, naval officers, perhaps manifesting a basic fatalism born of long experience at the mercy of the sea, merely attempted to learn to live with it.[87] Although they considered the nuclear weapon to be more than "just another bomb," they did not believe it was an "absolute weapon" which could be used to decide the course of conflicts through its unlimited destructive power. Throughout this period the Navy's leadership continued to focus on traditional objectives, while insisting that they have access to any weapons or technology that might be of use to them.

Naval determination to keep all options open and to maintain the flexibility and mobility of sea-based aviation in the postwar period led to constant conflict with the other services as well as the Department of Defense. At a time when greater precision was being sought in statements of national strategy, and when increasing coordination was necessary in operational matters, naval air doctrine remained fluid and imprecise; and the Navy stubbornly resisted efforts to integrate its carrier forces into rigid and comprehensive war plans.

Independent, intransigent, and inarticulate, the Navy's major contribution to the development of postwar defense was its resistance to change—its insistence that traditional approaches to strategy should not be lightly abandoned, that strategic bombing could never replace actual control of territory or the seas, and that conventional and tactical alternatives to the strategic air offensive must be maintained. Its one truly innovative contribution to the development of postwar defense policy occurred on the level of technology rather than theory: the Polaris missile was to have a profound impact on the national strategy of deterrence. On the conceptual level, it was the Navy's refusal to abandon its priorities under pressure which had the most lasting influence. Its conceptual conservatism and attachment to its own traditions brought the Navy to focus by degrees on a problem much in the nation's future: that of limited war.

Notes

1. The principal sources for this study are declassified papers from the following repositories: Record Group 218, Papers of the United States Joint Chiefs of Staff (hereafter cited as JCS); Record Group 341, Papers of the Chief of Staff of the Air Force (hereafter cited as CSAF); Record Group 319, Records of the Plans and Operations Division, Papers of the Army Staff (hereafter cited as OPD); and Record Group 428, Papers of the Secretaries of the Navy (hereafter cited as SECNAV), all in the U.S. National Archives; the papers of Op-30, Strategic Plans Division; OP-23, Organizational Research and Policy Division; the post-1946 Command File; and the personal papers of the following U.S. naval officers: Arleigh A. Burke, Ralph A. Ofstie, and Forrest P. Sherman, all in the Operational Archives, Naval Historical Center, Washington, D.C. (hereafter cited as NHA). Additional valuable official and semiofficial sources were Kenneth W. Condit, *The History of the Joint Chiefs of Staff: The Joint Chiefs of Staff and National Policy*, vol. 2, *1947–1949* (Washington, D.C.: Historical Division, JCS, 1976, declassified 1978) (hereafter cited as *JCS History*); George F. Lemmer, *The Air Force and the Concept of Deterrence, 1945–1950* (Washington, D.C.: USAF Historical Division Liaison Office, 1963; sanitized and declassified 1975); and Robert D. Little, *Organizing for Strategic Planning, 1945–1950: The National System and the Air Force* (Washington, D.C.: USAF Historical Division Liaison Office, 1964; sanitized and declassified 1975). For an earlier interpretation of much of the material covered in this paper, see David A. Rosenberg and Floyd D. Kennedy, *History of the Strategic Arms Competition, 1945–1972 Supporting Study: U.S. Aircraft Carriers in the Strategic Role, Part 1—Naval Strategy in a Period of Change: Interservice Rivalry, Strategic Interaction, and the Development of a Nuclear Attack Capability, 1945–1951* (Falls Church, Va.: Lulejian & Associates, Contract N00014-75-C-0237 for Deputy Chief of Naval Operations, Plans and Policy), 1975.

2. For a fuller discussion of this process, see David A. Rosenberg, "Officer Development in the Interwar Navy: Arleigh Burke—The Making of a Naval Professional," *Pacific Historical Review* 44 (November 1975).

3. The best published studies of the aircraft carrier in naval strategy are Charles M. Melhorn, *Two-Block Fox, The Rise of the Aircraft Carrier, 1911–1929* (Annapolis, Md.: Naval Institute Press, 1974); Archibald D. Turnbull and Clifford L. Lord, *History of United States Naval Aviation* (New Haven, Conn.: Yale University Press, 1947); Vice Adm. Sir Arthur Heziet, *Aircraft and Sea Power* (New York: Stein & Day, 1970); Norman Polmar, *Aircraft Carriers: A Graphic History of Carrier Aviation and Its Influence on World Events* (Garden City, N.Y.: Doubleday, 1969); and Clark G. Reynolds, *The Fast Carriers, The Forging of an Air Navy* (New York: McGraw-Hill, 1968).

4. Statements of naval doctrine other than tactical doctrine are rare. Classical studies such as those of Alfred Thayer Mahan, Sir Julian Corbett, and Bradley Fiske exist; none are the equiva-

Postwar Air Doctrine and Organization 199

lent of such writings as those of Giulio Douhet or Gen. William Mitchell, however, as basic influences on doctrine. The best examples of basic naval doctrine for the interwar and World War II periods are the U.S. Naval War College pamphlet, *The Estimate of the Situation and the Order Form*, published from 1911 through the 1930s in various editions, and *Sound Military Decision* (Newport, R.I.: U.S. Naval War College, 1943). All of these papers may be found in the Naval Historical Collection, U.S. Naval War College.

5. Bernard Brodie, "New Tactics in Naval Warfare," *Foreign Affairs* 24 (January 1946).

6. David MacIsaac, *Strategic Bombing in World War II, The Story of the United States Strategic Bombing Survey* (New York: Garland, 1976), 127–29, describes the controversy over the "Carrier Air Effort Against Japan." The final disposition of this manuscript is unclear but may have finally been published, possibly in much abbreviated form, as *U.S. Naval Aviation in the Pacific, A Critical Review* (Washington. D.C.: Office of the Chief of Naval Operations, 1947).

7. Fleet Admiral Ernest J. King, Third Official Report to the Secretary of the Navy, in *The War Reports of General of the Army George C. Marshall, General of the Army Henry H. Arnold, and Fleet Admiral Ernest J. King* (Philadelphia, Pa.: Lippincott, 1947), 656.

8. JCS 1478/8, 20 Feb. 1946, through JCS 1478/18, 17 May 1946, CCS 370 (8-1945), sec. 3., JCS.

9. JCS 1478/12, 29 Mar. 1946, ibid.

10. U.S. Congress, 80th Cong., 1st sess., Public Law 253, Chapter 343, Section 206(b). See also Rosenberg and Kennedy, 64–69.

11. For a complete discussion of the impact of the atomic bomb on naval design and construction, see JCS 1691/10, 29 Dec. 1947, CCS 471.6 (10-16-45), sec. 9, pt. 2, JCS, the text of the Bikini Evaluation Report of "The Atomic Bomb as a Military Weapon," esp. sec. 3, "Effects on Ships"; Vice Adm. W. H. P. Blandy, "Bikini: Guidepost to the Future," *Sea Power* 6 (December 1946), 7–9.

12. U.S. Congress, House of Representatives, Committee on Naval Affairs, *Hearing on House Concurrent Resolution 80, Composition of the Postwar Navy* (Washington, D.C.: Government Printing Office, 1945), 1165.

13. Sullivan to the president, Serial 0014P602, 24 July 1946, folder 98C, Arleigh Burke Papers, NHA. See also copy in "Secretary of the Navy" folder, Subject File, President's Secretary's File, Harry S. Truman Library (hereafter cited as PSF-HSTL).

14. A full discussion of these changes may be found in Appendix 1: OPNAV General Organization for Strategic Warfare, 1945–1972, in Rosenberg and Kennedy, *Naval Strategy in a Period of Change*.

15. See Vincent Davis, *The Politics of Innovation: Patterns in Navy Cases* (Denver, Colo.: University of Denver Monograph Series in World Affairs, 1966), 7–22.

16. Nimitz to Truman, 4 June 1946, Memos to and from the President folder, 1946, William D. Leahy Papers, JCS; Nimitz to secretary of the navy, Serial 0008PO3, 23 July 1946, in A8, Intelligence, folder 1, 1946, Op-30 Papers, NHA.

17. "Presentation of Undersea Warfare for the Secretary of the Navy" by Rear Admiral C. B. Momsen, appended to memo, Ruble to secretary of the navy, A8 Intelligence folder, John L. Sullivan Papers, SECNAV.

18. See the memoranda regarding the Recommended Building Program, Fiscal 1948 Budget, in personal folders on the General Board, Burke Papers, NHA.

19. On the development of the PINCHER plans, see the material in the CCS 381, USSR (3-2-46), secs. 1–7, JCS, and in the ABC 381 USSR, 2 Mar. 1946, secs. IA–IG, OPD. See also Glover to Chief of Naval Operations (CNO), Serial 0005P30, 21 Jan. 1947, A16-3(5) War Plans, 1947 folder, Op-30 Files, NHA; David A. Rosenberg, "Planning for a PINCHER War: Policy Objectives and Military Strategy in American Planning for War with the Soviet Union, 1945–1948," unpublished paper presented at the meeting of the Society for Historians of American Foreign Relations, August 1978.

20. Nimitz to Eisenhower, 13 June 1946, ABC 471.6 Atom, 17 Aug. 1945, sec. 7, OPD.
21. See, for example, Parsons to CNO, Serial 005OP36, 14 July 1947, A16-1 folder, Op-30 Papers, NHA.
22. U.S. Navy, Office of the Chief of Naval Operations, *Principles and Applications of Naval Warfare,* United States Fleet Publication No. 1, 1947, quoted in Desmond P. Wilson, Jr., "Evolution of the Attack Aircraft Carrier: A Case Study in Technology and Strategy," published in U.S. Congress, 91st Cong., 2d sess., Joint Senate-House Armed Services Subcommittee of the Senate and House Armed Services Committees, *CVAN-70 Aircraft Carrier* (Washington, D.C.: Government Printing Office, 1970).
23. "U.S. Navy Thinking on the Atomic Bomb," in Bernard Brodie and Eilene Galloway, *The Atomic Bomb and the Armed Services* (Washington, D.C.: Library of Congress Legislative Reference Service Public Affairs Bulletin No. 55, May 1947).
24. "The Effects of the Atomic Bomb on National Security (An Expression of War Department Thinking)," ibid.; Robert Hotz, "Army-Navy Split on Role of Air Power in Atomic Warfare," *Aviation News,* 21 Apr. 1947.
25. Sherman to the president, 14 Jan. 1947, CNO Chronological File, Post 1946 Command File, NHA. As an example of this split, compare PINCHER plans cited in fn. 19 above and JCS 1630, 19 Feb. 1946, CCS 381 (2-18-46), sec. 1, JCS on early postwar planning guidance for joint agencies about a possible future war.
26. JCS 1961/10, 29 Dec. 1947; Rivero to Lalor, 28 July 1947, CCS 471.6, sec. 15, JCS, describes the 29 July 1947 presentation. For a concise discussion of the evolution of U.S. atomic strategy as it related to the Bikini report, see David A. Rosenberg, "American Atomic Strategy and the Hydrogen Bomb Decision," *Journal of American History* 66 (June 1979).
27. For an example of the way the CVA-58 (or rather Project 6A) was discussed prior to the Bikini report, see the Memoranda regarding Recommended Shipbuilding and Conversion Program—Fiscal 1949, in personal folders on the General Board, Burke Papers, NHA. For the way that discussion changed, see Combs to secretary of the navy, 7 Nov. 1947, Al, Plans, Projects, Policies, 1948–49 folder, Sullivan Papers, SECNAV. A full history of the ship is in Burke to the Judge Advocate General, Serial 067P23, 11 May 1949, A21/1-1/1 Carrier folder, sec. 11, Op-23 Papers, NHA.
28. JSPG 499, 20 Nov. 1947, and JSPG 499/2, 3 Dec. 1947, are the various versions of CHARIOTEER, both in CCS 452, U.S. (8-1-47), secs. 1 and 2, JCS. See also *JCS History,* 284.
29. Boone to English, 27 Jan. 1948, with four enclosures, A16-12, War Plans, 1952, Op-39 Papers, NHA. See also Towers to CNO, 21 Nov. 1947 (General Board No. 420) in personal folders on the General Board, Burke Papers, NHA.
30. Duncan to Op-30, Serial 0004P504, 10 Dec. 1947, A16-3, Warfare Operations, 1947, Op-30 Papers, NHA.
31. "The Future Employment of Naval Forces," Navy Department Pamphlet P-514, 7.
32. There were in fact *three* Gallery memos. The most important, from which this quote is taken, is Gallery to DCNO (Air), Serial 00124P57, 17 Dec. 1947. The only copy of this paper in official form is in a special folder in the miscellaneous papers of Arleigh Burke, NHA. The full memo is readily available in the *Army–Navy–Air Force Register,* 11 Dec. 1954. The two earlier memos that were later combined to form the 17 December paper are Gallery to Clark, 14 and 17 Nov. 1947, both in A16-11 folder, sec. 3, Op-23 files.
33. Interview with Rear Adm. Gallery, Oakton, Va., 9 Apr. 1975.
34. Ofstie to DCNO (Air), 7 Jan. 1948, with paper, "Composition of the National Military Establishment," appended, box 8, Ralph Ofstie Papers, NHA. Attached to this paper is a memo by Gallery to DCNO (Air), Serial (X)13P57, 14 Jan. 1948, noting that "this is an excellent summary of the situation during the interval when we have the A bomb and the Russians do not," but that when both sides have the bomb his 17 December 1947 memo should apply. Notes on Ofstie's 7 January memo indicate that Rear Adms. J. F. Bolger, E. W. Litch, and W. Tomlinson had seen and concurred with the views expressed in it.

Postwar Air Doctrine and Organization 201

35. Operations Evaluation Group Study No. 327, Serial LO, 467-48, 26 Feb. 1948, A2.1-1/1 file, sec. 2, Op-23 Papers, NHA; John P. Coyle (author of O.E.G. 327), interview with the author, 27 Jan. 1975, Washington, D.C.; Wright to Distribution List, with studies attached, Serial LO, 317-48, 12 Mar. 1948, Files of the Aviation History Unit, Deputy Chief of Naval Operations (Air Warfare) (Op-05d5) (hereafter cited as AHU Files).

36. The Air Force engaged in a number of studies that were designed to attack the military value of the CVA-58. See "Aircraft Carrier Operations in the Pacific, World War II," folder 168-7017-15, 1947–1948, Aerospace Studies Institute, Archives Branch, Alfred F. Simpson Historical Research Center, Alabama (hereafter cited as AFSHRC). This document is labeled "basic study which was the foundation of Air Force position on the CVA58, 1947–1948." See foundation of Air Force position on the CVA-58, 1947–1948. See also Air War College, "Employment of Carrier Forces for Strategic Atomic Attacks," prepared for the Air Staff, March 1948, RB 72306, folder 168-7017-16, AFSHRC. Another copy of this paper may be found in box 10, Deputy Chief of Staff, Operations, Executive Office, House Investigation, August–October 1949, CSAF.

37. Information on the U.S. nuclear capability in the winter of 1948 is taken largely from JCS 1745/5, 8 Dec. 1947, CCS 471.6 (8-15-45), sec. 8, JCS; Rosenberg and Kennedy, 159–62.

38. *JCS History,* 283–93. See also JSPG 496/1, 8 Nov. 1947 to JSPG 496/4, 11 Feb. 1948, and JCS 1844/4 to 1844/13, 9 Mar. 1948 to 21 July 1948, all in CCS 381 USSR (3-2-46), secs. 8–18, JCS.

39. JCS 1745/5, 8 Dec. 1947 to JCS 1745/12, 19 Jan. 1948, CCS 471.6 (8-15-45), sec. 8, JCS. See also Sullivan to secretary of the air force, 9 Aug. 1948, A16, War, Preparation for, Conduct of, 1948–1949 folder, Sullivan Papers, SECNAV.

40. JCS 1854, 23 Mar. 1948, to JCS 1854/8, 24 Aug. 1948, CCS 471.6, (8-15-45), secs. 9–12, JCS, and Sullivan to secretary of the air force, 9 Aug. 1948.

41. Rosenberg and Kennedy, 79–89.

42. Burke, Memorandum for Members of the General Board, "Very Rough, Very Tentative Outline for Serial 315," 4 Mar. 1948, General Board folders, Burke Papers, NHA.

43. General Board, "National Security and Navy Contributions Thereto, A Study by the General Board," 25 June 1948, G.B. 425, Serial 315, Enclosure (D), 28, ibid.

44. Ibid. Enclosure (D), 6–7; Covering Letter, 4.

45. *JCS History,* chap. 7, 213–55. See also Rosenberg and Kennedy, chap. 5, 1.

46. Memo for Files by Op-(X)3, Subject: Increase in Naval Forces, July 13, 1951, "Naval and Marine Forces" folder, box 1, Forrest Sherman Papers, NHA.

47. Johnson to Truman, 23 Apr. 1949, "Defense, Secretary of, Miscellaneous" folder 1, Subject File, PSF-HSTL; Memorandum by the Air Force Chief of Staff to the Secretary of Defense on the CVA-58 Project, n.d. (but ca. 22 Apr. 1949), in folder "U," General File, PSF-HSTL. Johnson's memo notes that the underscoring in pencil on the JCS memos on the CVA-58 are his. The underscoring on the Air Force Chief of Staff's memo stressed such phrases as "this carrier is designed for bombardment purposes."

48. Memo by Op-(X)3, 13 July 1951. See also Rosenberg and Kennedy, 122–25.

49. Gallery to DCNO (Air), 17 Jan. 1949, MLC-AEC folder, box 8, Ofstie Papers, NHA.

50. The only declassified version of the Harmon Report is "Report by the Ad-Hoc Committee to the Joint Chiefs of Staff on Evaluation of Effect on Soviet War Effort Resulting from Strategic Air Offensive," 11 May 1949, in Al/EM-3/7, JCS folder, sec. 111, Op-23 Papers, NHA. See also JCS 1953/4, 9 July 1949, JSC 1953/5, 19 July 1949, and Vandenberg to JCS, 23 July 1949, all in CCS 373 (10-23-48), secs. 2 and 3, and Bulky Package, JCS, respectively. For the subsequent fascinating fate of the Harmon Report, see Rosenberg, "American Atomic Strategy."

51. Ofstie to Op-05, DCNO (Air), 26 Apr. 1949, A21/1-1/1 Carrier, sec. 2, Op-23 Papers, NHA.

52. Ofstie to Op-05, 29 Apr. 1949, MLC-AEC folder, box 8, Ofstie Papers, NHA.

53. U.S. Congress, House of Representatives, Committee on Armed Services, *Investigation of*

the *B-36 Bomber Program,* 81st Cong., 1st sess. (Washington, D.C.: Government Printing Office, 1949), contains the full story.

54. U.S. Congress, House of Representatives, Committee on Armed Services, 81st Cong., 1st sess., *The National Defense Program—Unification and Strategy* (Washington, D.C.: Government Printing Office, 1950). The full raw data behind the Navy presentation may be found in B-9, Agenda Manual, sec. 3, Op-23 Papers, NHA.

55. Although his analysis tends to qualify this assessment somewhat, this is the primary impression one gets from reading Paul Y. Hammond's otherwise excellent study, "Super Carriers and B-36 Bombers: Appropriations, Strategy, and Politics," in Harold Stein, ed., *American Civil-Military Decisions* (Tuscaloosa: University of Alabama Press for the Twentieth Century Fund, 1963).

56. Air Warfare Division (Op-55), "Study on Future Development of Carrier Aviation with Respect to Both Aircraft and Aircraft Carriers," 22 Aug. 1949, AHU Files. OP-55's mission is described in *Organization Manual for the Office of the Chief of Naval Operations,* OPNAV P 02-100 (Rev. 8-48) August 1948, NHA.

57. Ibid., 4. This conclusion was subsequently confirmed in a special Study of Undersea Warfare, Serial OOIP(X)3, 22 Apr. 1950, by Vice Adm. F. S. Low. This study was requested by Adm. Forrest Sherman in November 1949, possibly to serve as a check on the conclusions of the Op-55 study. The Low paper may be found in completely declassified form under call number N-210-0651 at the Naval War College Classified Library.

58. "Carrier Task Group Operations in the Mediterranean, a CINCNELM Staff Study," Part 1, Conclusions and Recommendations, Tab A, appended to Conolly to CNO, Serial O(K)I, 3 Jan. 1949, folder 79, Miscellaneous, 1949, Op-30 Papers, NHA. Much of the rest of this study remained classified in 1975, but the conclusions were declassified.

59. On Sherman's leadership, see JCS 18(X)/68, 14 Feb. 1950, CCS 370 (8-19-45), sec. 21; JCS 1844/49, 7 Dec. 1949, CCS 381, U.S.S.R., (3-2-46), sec. 42; JCS 1888/3, 11 Apr. 1950, CCS 370 (5-25-48), sec. 2, all JCS, which discuss the proposed reduction in carrier forces, the status of U.S. war plans, and overall U.S. military posture, respectively. See also Ernest K. Lindley, "The New Navy Line," *Newsweek,* 12 Dec. 1949, 27; Undated Memorandum for the Record, prepared ca. April or May 1951, with no author but obviously by Forrest Sherman, in first unlabelled folder, box 1, Sherman Papers, NHA; and Rosenberg and Kennedy, chap. 7, esp., 166–69, and the sources cited in fn. 3, 7, and 8, on pp. 178–79.

60. NSC-68 is most readily available in U.S. Department of State, *Foreign Relations of the United States, 1950* (Washington, D.C.: Government Printing Office, 1977), 1:234–92. The events leading up to that paper are documented in p. 1ff.

61. Undated memorandum for the record, box 1, Sherman Papers, NHA.

62. Ibid., JCS 1800/115, 13 Sept. 1950 and Decision on, 18 Sept. 1950; JCS 1800/116, 18 Sept. 1950; JCS 1800/133, 7 Dec. 1950; and JSPC 851/37, 11 Dec. 1950, all in CCS 370 (8-19-45), secs. 27–29, JCS. See also NSC 68/3 and NSC 684 of December 1950, with supporting documents, in *Foreign Relations 1950,* 1:425–77. Information on authorization dates for *Forrestal* class carriers is in Polmar, 587–88. One carrier was authorized, each year from Fiscal 1952 through Fiscal 1958.

63. Composite Squadron Five (VC-5) Historical Reports, 31 Dec. 1948, 1 Apr. 1949, 30 Jan. 1950, NHA. Note that on 6 Jan. 1950 a second nuclear-capable squadron VC-6 was commissioned. See VC-6 historical report, 22 Feb. 1951, NHA, for details. See also Rosenberg and Kennedy, 159–62.

64. JCS 2019/4, 29 Mar. 1950, and Decision, 6 Apr. 1950, CCS 471.6 (8-15-45), sec. 18A; Lay to acting chairman, Atomic Energy Commission, 14 June 1950, NSC Atomic File, PSF-HSTL; Lalor to JCS, SM-1632-50, 22 July 1950, citing title of JCS 2019/8 as "Storage of Non-nuclear components of atomic bombs on CVB Class Aircraft Carriers" by Admiral Sherman, CCS 471.6 (8-15-45), sec. 18A; and Withington to CNO, September 1950, in A 16-3, Warfare Operations, War Games, 1950, Op-30 Papers, NHA.

Postwar Air Doctrine and Organization 203

65. The "Daisy Chain" system is described, with code-name, in Richard K. Smith, *Cold War Navy* (Falls Church, Va.: Churchill Press and Lulejian and Associates, 1975), chap. 6; on nuclear production, see Richard G. Hewlett and Francis Duncan, *A History of the United States Atomic Energy Commission* (University Park: Pennsylvania State University Press, 1972), 2:525–29, 547–54, 556–68, 576–78; on release of nuclear weapons to the armed forces, see ibid., 538–39; Lalor to JCS, SM-684-53, 31 Mar. 1953, notes that JCS 2019/58 discusses storage of nuclear components of atomic weapons on aircraft carriers, CCS 471.6 (8-18-45), sec. 37.

66. Rear Adm. Daniel V. Gallery, USN (Ret.), interview with the author, 9 Apr. 1975, Oakton, Va.; VC-5 Historical Reports, 26 July 1951 and 14 Feb. 1952, NHA.

67. Director, Strategic Plans Division to Director of Naval Intelligence, two memoranda, Serials 000966P30 of 8 Nov. 1950, and 0001000P30, 24 Nov. 1950, about "Target Data," both in A16-3, Warfare Operations, War Games, 1950, Op-30 Papers, NHA.

68. JCS 1800/166, 7 Sept. 1951, CCS 370 (8-19-45), sec. 34, JCS.

69. "Meeting of General Eisenhower with the President and the Cabinet, January 31, 1951," George Elsey Papers, HSTL.

70. JCS 1844/46, 8 Nov. 1949 (War Plan OFF-TACKLE), CCS 381, U.S.S.R., (32-46), sec. 41; JCS 2056/7, 12 Aug. 1950, CCS 373.11 (12-14-48), sec. 2, JCS.

71. The "Doctrine of Scarcity" and the anticipated changes in it are described in Thomas to Schlatter, 13 Jan. 1950, on Air Force Capability for Atomic Warfare, 13 Jan. 1950, Tab 3, in OPD A/AE 381 (Atomic Warfare), B, CSAF. The Navy view of this doctrine was described in an interview with Vice Adm. John T. Hayward, USN (Ret.), 15 July 1975, Tyson's Corner, Va.

72. Hewlett and Duncan, 2:525–78; Thomas to Schlatter, 13 Jan. 1950.

73. JCS 2220/4, 31 Jan. 1952, CCS 471.6 (4-18-49), sec. 7, JCS. For a later statement of the same thing, with upwardly revised forces and weapons figures, see JCS 2220/19, 6 May 1953, in sec. 10 of the same file.

74. JCS 2056/24, 29 Feb. 1952, as promulgated in SM-597-52, 3 Mar. 1952, CCS 373.11 (12-14-48), sec. 7. See also the various debates that appear in the following JCS Papers: JCS 2056/27, 5 May 1952; JCS 2056/31, 20 June 1952; JCS 2056/34, 13 Aug. 1952; JCS 2056/35, 9 Sept. 1952; JCS 2056/38, 10 Feb. 1953; and JCS 2056/42, 25 Feb. 1953, all in the same file, secs. 7–10, JCS.

75. JCS 1854.16, 7 Sept. 1951, CCS 471.6 (8-15-45), sec. 23, JCS. Later statements of the same nature appear in the papers just cited above.

76. JCS 2131/1, 28 Feb. 1952, contains the Weapons Systems Evaluation Group Study of the Evaluation of the Offensive and Defensive Capabilities of Fast Carrier Task Forces in 1951, including Appendices B and C which contain Air Force criticisms and the Navy's rejoinders. See also the decision on JCS 2131/1, 2 July 1952, in which the JCS agreed to only "note" the conclusions of the study, rather than approve them: all in CCS 045.92 (228-50), sec. 2, and Bulky Package, JCS.

77. Memorandum for the JCS by the Chief of Staff of the Air Force on Military Requirements for Atomic Weapons, CCS 471.6 (11-3-51), sec. 1, JCS; Vice Adm. John T. Hayward, USN (Ret.), interview with the author, 4 June 1976, Newport, R.I.

78. The only comprehensive statement of naval air doctrine uncovered between 1949 and 1954 is the now-declassified presentation by Rear Adm. James S. Russell on "The Carrier Task Force" to the National War College, 29 Jan. 1954. AHU Files. That presentation must be seen as the logical fulfillment of the 1949 Op-55 study; its discussion of carrier mobility, flexibility, and self-sufficiency; its description of air group composition (three fighter squadrons, one light attack squadron, with a detachment of heavy attack aircraft) and its list of possible carrier air targets (352 Russian or Satellite air fields, 54 shipyards, 44 naval bases, 66 areas suitable for mining) are all in keeping with the broad outlines laid out in 1949.

79. The best published discussions of the "New Look" may be found in Glenn H. Snyder, "The New Look of 1953," in Warner R. Schilling et al., *Strategy, Politics, and Defense Budgets* (New York: Columbia University Press, 1962), 379–524; Samuel P. Huntington, *The Common De-*

fense: Strategic Programs and National Politics (New York: Columbia University Press, 1961), 64–88; and Douglas Kinnard, *President Eisenhower and Strategy Management: A Study in Defense Politics* (Lexington: University of Kentucky Press, 1977), 1–36. For the background of Eisenhower's concerns on defense spending, see the Eisenhower Diaries from 1948 to 1952, especially his comments on defense spending following Truman's release of the Fiscal 1953 budget in January 1952, Dwight D. Eisenhower Library. NSC 162/2, 30 Oct. 1953, from which the quotes were taken, is most readily available in *The Pentagon Papers* (The Senator Gravel Edition; Boston: Beacon, 1971), 1:412–29.

80. On the continuing battles over the place of strategic air power in war and defense planning, see SM-298-53, 26 Mar. 1953, Memorandum for the Joint Strategic Plans Committee by name, with enclosed revision of "Format B" to the Joint Strategic Capabilities Plan, CCS 381 (11-29-49), sec. 5; JCS 1844/152, 19 Oct. 1953, discussing splits in the same plan, in sec. 10 of the same file; and JCS 2101/107, 23 Oct. 1953, and Decision, 27 Oct. 1953, which discuss NSC 162/1, in CCS 381 U.S. (1-31-50), sec. 30, JCS.

81. JCS 2101/112, 7 Dec. 1953, CCS 381 U.S. (1-31-50), sec. 32, JCS.

82. Carney to JCS, Serial 0005P35, 18 Jan. 1954, and JCS Info Memo 922, 10 Feb. 1954, in CCS 381 U.S. (1-31-50), secs. 32 and 35, respectively.

83. For examples of such conformance, see JSPC 851/134, 18 Jan. 1955, in CCS 370 (8-19-45), sec. 49; Polmar, 587ff; and Philip A. Dur, "The Sixth Fleet: A Case Study of Institutionalized Naval Presence" (Unpublished Ph.D. diss., Harvard University, December 1975), 66–68.

84. JSPC 851/134, 18 Jan. 1955. See also such unofficial but highly representative statements of approved doctrine as Cdr. Malcolm W. Cagle, USN, "A Philosophy for Naval Atomic Warfare," *U.S. Naval Institute Proceedings* 83 (March 1957).

85. On Burke's leadership and views on carrier air power, see United States Senate, Committee on Armed Services, 84th Cong., 2d sess., Subcommittee on the Air Force, *Study of Airpower* (Washington, D.C.: Government Printing Office, 1956), pt. 18, 18 June 1956, 1339–86; CNO to secretary of defense, Serial 012P00, 6 Nov. 1956, in originator's File, 1 October–30 Nov. 1956, and Burke to Schindler, 14 May 1958, Personal File, 1 April–30 June 1958, Burke Papers, NHA; and David A. Rosenberg, "Admiral Arleigh A. Burke," in Robert William Love, Jr., ed., *The U.S. Chiefs of Naval Operations* (Annapolis, Md.: Naval Institute Press, 1979).

86. CNO to Distribution List, Serial 04P93, 13 Jan. 1958, with enclosure of "Statement of U.S. Navy Long Range Objectives, 1967–72 (LRO-57)" (also known as the "Navy of the 1970 Era"), Naval War College Classified Library, Call No. NA 50.062.

87. For an early example of such Air Force attempts, see U.S. Air Force Field Office for Atomic Energy, Draft "Doctrine of Atomic Air Warfare," 30 Dec. 1948, OPD A/AE 381 (Doctrine of Atomic Warfare), CSAF.

The Interaction of Technology and Doctrine in the U.S. Air Force

Robert Perry

Successful new military weapons routinely derive from proven technology and virtually never emerge from efforts to contrive, shape, or push immature technology in the name of perceived requirements, however well conceived they may be.[1] Second, the evolution of USAF military doctrine since World War II has been very largely driven by expectations about the rate and direction of future weapons development. Third, many—if not most—of the postwar difficulties of Air Force research and development, and many problems of defining and applying appropriate air doctrine, have developed because the consequences of basing doctrine on unrealistic technical expectations are widely misunderstood or ignored.

Some definition of terms may be in order. "Successful," for something like a military aircraft system, means timely delivery of a weapon that satisfactorily performs its assignments and at a price not much in excess of that anticipated when it was ordered. The aircraft must be able to cope adequately with its opposition. "Immature technology" can be unripe in many respects, including (and perhaps most important) being too costly; incautious efforts to exploit immature technology have routinely brought about the delayed delivery of operationally unready systems in numbers smaller than planned (because unit costs increase more often than program budgets, and quantities are reduced as prices go up). The problems begin when planners or engineers or makers of doctrine or requirements are obliged to decide whether some technology is indeed suitable for application to military needs and then to act on that decision. The tendency has been to insist that such decisions be made very early; indeed, retrospective judgment suggests that many are made prematurely.[2]

Finally, the premise underlying both strategic and tactical doctrine in this country since World War II—that quality has an inevitable advantage over quantity—may well be overdue for revision. There are many reasons for questioning the prevailing judgment about quality versus quantity, but the most pressing is that many of the performance attributes traditionally credited with ensuring qualitative superiority in, for example, aircraft have become irrelevant to combat outcomes. Marginal advantages in speed, range, and altitude head that list. One is entitled to ask, in a similar vein, if a "50-percent better" CEP (circular error probable, or accuracy probability) makes one ballistic or cruise missile "better" than another when both carry nuclear warheads and the "poorer" performer has a CEP of 100 feet or less! Like excellence in close order drill as a measure of soldierly qualities, many widely acclaimed achievements of military enterprise and some assumed advantages of "advanced" technology have outlived their usefulness.

The Application of Technology to Military Ends: Hazards and Limitations

Even the most determined, brilliantly managed, well-funded effort to develop and apply technology to military ends will not in every instance prove successful. First, technology can be stubbornly intractable: in thirty years of trying, no operationally superior aircraft or missile system dependent on ramjet propulsion has emerged. The autogyro and the flying wing were extremely ingenious but flawed conceptions; they were militarily redundant because whatever promise they had was overtaken by alternative, cheaper, and better technology.

Second, the incorporation of marvelous improvements at frequent intervals has not guaranteed the continuing usefulness of some fundamental system that has outlived its time. Adding jet engines to the B-36 and B-50 extended their operational lives by several years and probably did no grave harm to the national interest; but, like adding a swept wing and an all-jet propulsion package to the B-36 fuselage—as was done to create the prototype B-60—such expedients mostly demonstrated that 1940-style airframe concepts ill satisfied the needs of the Strategic Air Command in the 1960s. Nor, in the judgment of most specialists in strategic theory, did the development and production of the B-47 and B-58 contribute greatly to the achievement of strategic superiority. Although both incorporated the very latest in aerodynamics, propulsion, and avionics, they were generally regarded as poor strategic bombers.

Third, not even the most attractive experimental development, however soundly based and well proven, can find operational employment if a matching requirement does not appear: the liquid-air-cycle engine has been languishing "on the shelf" for decades in search of an application since its laboratory demonstration.

Fourth, a capability developed skillfully and effectively, against great odds and at enormous expense, can be wholly negated by the appearance of a superior (or cheaper) means of performing a function: heat-sink and transpiration-cooled reentry bodies became museum curiosities with the development of ablative coating for missile warheads; and the attractive qualities of the Navy's XF-10F, a variable sweep fighter of the early 1950s, became irrelevant once the Royal Navy demonstrated that steam catapults could satisfactorily launch jet fighters from carriers. And there are areas where technological improvement effort is wasted because no need for the improvement ever develops: Wright Field engineers of the 1950s demonstrated that a supersonic turboprop powered fighter aircraft probably could be built, and there was great applause for the development of a "steerable" nose gun for fighters in 1948—but where was the need?[3]

The moral of all this is that marvelously impressive technology that contributes nothing to the solution of a critical national problem is worth no more, and no less, than a persuasively worded requirement that calls for some technological achievement scientists and engineers cannot provide.

Faith in Future Technology

Traditionally, the armed forces of the United States have been committed to the use of large numbers of conservatively designed, highly dependable weapons that could be capably operated by quickly trained soldiers, sailors, and airmen. About 1945, American military planners became excessively enamored of the promise of threshold technology. Why? The question is one to which, sadly, historians of military technology have not yet turned their full attention, though it deserves as much. At the least, it is worth notice that by 1945 soldiers and sailors and (particularly) airmen (and most of all American airmen) had been very much impressed by the effectiveness of such, to them, revolutionary devices as radar, jet propulsion, and rocketry, and, most of all, nuclear energy. Without those and a very few similar military innovations, World War II could have been very much like a quick-march World War I. Note that these innovations had truly revolutionary military implications, although most informed scientists and engineers were surprised that the military had waited so long to exploit their known (but not always fully demonstrated) potential. Also frequently overlooked was the circumstance that where new technology could provide only marginal advances, and in situations where marginal advantage in, for instance, flight performance could have a much less pronounced effect than better training, or superior tactics, or great numerical advantage, the benefits of new technology became uncertifiable. Such wry calculations came less readily to mind than the injunction that "Guided Missiles Could Have Won"[4] or other military memories of the many technical and scientific advances that affected the course of World War II.[5]

Like most of their countrymen and contemporaries, American military leaders of the late 1940s concluded that almost any conceivable weapon could be built if one were willing to make a sufficient investment of ingenuity, resources, and time. To the failure of their late adversaries to put sufficient faith in that credo was widely attributed the outcome of the war.

That is how faith was born. But to understand how it was that technology and doctrine became so interdependent and pending technology such a dominating influence, it is necessary to turn to the perceived lessons of World War II, to review some relevant events of the 1945-57 period (the pre-Sputnik era where American technological supremacy was virtually unquestioned), and then to consider what doctrinal consequences have befallen the Air Force by reason of the introduction of new weapons. To keep that process to reasonable lengths, it will be necessary to discuss a few representative cases and to extrapolate from them.

To many observers looking back from the perspective of 1945, the state of doctrine and weaponry in 1939 must have seemed as hopelessly confused as before Bull Run. Notwithstanding some holdover of older aircraft, the U.S. Army Air Corps had by 1939 mostly completed the conversion to monoplane fighters and bombers with retractable landing gear and fully enclosed cockpits. That transition seems remarkable only if one recalls that of all the world's major air forces only those of Germany and Japan had done as much. The Royal Air Force, the Russian and Italian air forces, the Poles and the French—and the U.S. Navy—still were producing and operating various aircraft with fixed landing gear or open cockpits—or both. The RAF, the Russians, the Italians, and the U.S. Navy still were buying wire-braced, fabric-skinned, wood-framed biplane fighters. In August 1939, when every operational shipboard fighter in the inventory of the U.S. Navy was a biplane, the Germans began testing the world's first turbojet-powered airplane.

Six years later, as World War II ended, three nations had substantial stocks of operational jet fighters, electronic warfare was a reality, cruise missiles and comparatively long-range rocket missiles were commonplace, nuclear weapons had destroyed two cities, and virtually every military aircraft in the world was both new and technologically obsolete.

What were the prospects for the future? On design boards or in test stations, but either untried or available only in imperfect prototypes, were aircraft with swept and delta wings, turbojet engines with two to five times the output and efficiency of their two-year-old predecessors, air-launched radar-guided and heat seeking missiles, rocket engines that if clustered could propel warheads over distances of two or three thousand miles or put satellites in orbit, automatic navigation devices that could accurately direct winged vehicles over courses of several thousand miles, sophisticated airborne radar for both fire control and bombing, and aircraft that could exceed the speed of sound. A bomber capable of flying 8,000 miles with 10,000 pounds of bombs was ready for test; virtually every new bomber in development in the United States in early 1946 was faster than the highest performance propeller-driven fighter of World War II.

It is not a great oversimplification to suggest that the new U.S. Air Force sensibly concluded—*en masse*—that new technology, shaped and applied as rapidly as circumstances (most budgetary) would permit, could dominate military capabilities for the forseeable future. The difficulty was that there seemed to be too many promising opportunities—and they all promised to be costly. In a sense, the first flurry of post-demobilization activity consisted of attempting concurrently to explore an infinity of opportunities: to develop fighters and bombers, and cruise and ballistic and air-to-air missiles; simultaneously, to exploit the promise of unexplored frequency bands and rocket propulsion and turbojet propulsion and rocket-assisted takeoff; indeed, to make everything in the Air Force inventory perform "better."

But how would all this marvelous new technology affect Air Force tactics and strategy? The usual early USAF response to that question was to propose modest variations on the concepts and doctrines developed during World War II. Evolution of doctrine through the gradual application of revolutionary new technology was preferred, but that sometimes created difficulties. For example, without much regard for the probability that air defense missiles would diminish the primacy of strategic bombardment as it had developed by 1946, analysts concluded that the greatest threat to long-range bombers in the 1950s would be—as in 1945—the fast, heavily armed, air defense fighter. Thus, one development sidelight of 1945–55 was an effort to perfect escort fighters for B-36s and B-52s that could extend the role of the P-51s over Germany and Japan in 1944–45. Because designers could not provide jet fighters with cruise ranges of several thousand miles, they turned to such alternatives as parasite aircraft to fit the bomb bays of B-52s and B-36s, and fighters towed behind bombers or coupled to their wing tips. Aerial refueling was seriously considered for escort fighters well into the 1950s.[6]

The menu of promising opportunities seemed endless and irresistible. How was one to know in 1948 if he were overlooking an opportunity that could lead to the 1960s equivalent of BTO (bombing through overcast), or radar gunlaying, or jet propulsion? The legend of the secret weapon that might have won, and the bomb that did, persisted as both threat and goal. The Manhattan Project was proof that if sufficient manpower and brainpower and money were wisely invested in a well-managed undertaking, however difficult, it could be successfully concluded. To ask "How difficult?" was to display an unseemly lack of faith. "How much?" was heard more often, but "this will be worth any price, we can't afford not to have it" was the usual response. If the Manhattan Project seemed too grand an example to be emulated every few years, the Air Force had its own model: the development of the B-29 during the war. If the Hiroshima bomb represented a triumph of scientific will, the B-29 was certainly the most remarkable instance of integrative engineering successfully brought off by any air force and perhaps the best instance of how money and manpower could humble virtually any intransigent technology. Or so it seemed.

Technology and Doctrine: The Case of the Ballistic Missile

It was in that environment, with those perceptions, that the new U.S. Air Force set out in 1948 to develop and put into its squadrons the weapons that would make the nation secure for generations to come. Possessed of a nuclear monopoly that had yet to experience challenge, with its native scientific talent reinforced by the German creators of the V-1, the V-2, the Me-163 and Me-262, the United States seemed capable of nearly anything. And although the principal architects of Air Force doctrine foresaw several decades during which bombers, and then bombers complemented by long-range cruise missiles, would continue to dominate strategic forces, they recognized that the intercontinental ballistic missile (ICBM) had eventually to be reckoned with.

Early U.S. work on ballistic missiles was stimulated by German example, the V-2, but was to some extent constrained by the premise that an intercontinental missile was at least a decade away and by the virtually unchallenged conclusion of American airmen that the long-range bomber was by a considerable margin a more effective weapon. Indeed, until June 1944, all Allied intelligence authorities, if not all analysts, were agreed that no German rocket would ever be fired at England. The British assumed that no warhead smaller than five tons could be militarily effective (which ignored British and American experiences with half-ton and one-ton bombs) and that a missile large enough to deliver such a payload could not be developed in time to affect the course of the war.[7] The second assumption was reasonably sound, although it was dependent on the length of the war. But the first expressed a preference rather than a valid conclusion; as it happened, the Germans had different perceptions. Although the notion of an "invulnerable" delivery system underlay the German decision to build large quantities of V-2s, it was true that to the Germans a ballistic missile carrying one ton of explosive and costing about $45,000 was quite as sensible an investment as a $50,000 bomber that was likely to be shot down—together with several skilled airmen who could be put to better use elsewhere.

In any event, although the Americans were attracted to the concept of a ballistic missile, they felt no particular sense of urgency about it. Initially, they settled for a slightly funded research project that had as its goal the somewhat unrealistic combination of 5,000-mile range and 5,000-pound payload, although all concerned were privately willing to concede that what was immediately sought was a means of exploring and demonstrating various aspects of technology. Actually, the vital ingredients of a medium-range ballistic missile were mostly within reach by 1947, excepting a nuclear warhead and proven means of warhead reentry, but few realized it. In any case, the delayed consequences of the postwar budget cutbacks caused the Air Force to drop all plans for ballistic missile development in July 1947. Of the twenty-eight missile projects established a year earlier, the two chief survivors with nominal application to the strategic mission were what later became Snark and Navaho. Both were cruise missiles also expected to carry 5,000-pound warheads 5,000 miles, although most other specifications were vague. (At various times Snark was to be subsonic or supersonic, air launched or ground launched, and was to rely on any of several different guidance schemes.)

The logic of the decision to drop ballistic missile development has been derided in later years, but mostly for the wrong reasons. American decisionmakers chose what they thought to be the safe assumption that evolutionary development along familiar lines would surely lead to the availability of a strategic missile. What were the "safer" lines? Application of turbine engine improvements, aerodynamic advances, and perfected autopilots to early cruise missiles, with ramjet propulsion and stellar-inertial guidance following along at some later date, was a preferred policy. The appreciation that a highly accurate 5,000-mile ramjet powered cruise missile—and particularly one that was boosted to 70,000 feet by a rocket and thereafter flew at sustained supersonic speeds—was a more ambitious undertaking than the atomic bomb, much less the B-29, was seldom to be heard. Fifteen years and several hundreds of millions of dollars later, that appreciation still was not widespread.

Nor were the scientists on whom the American services relied for advice in the late 1940s notably enthusiastic about the future of a ballistic missile. Indeed, their most vocal spokesman—and the individual whose advice was most eagerly sought by the Air Force openly ridiculed advocates of the long-range rocket. Vannevar Bush told Congress, "I say technically I don't think anybody in the world knows how to do such a thing [make an accurate, nuclear armed ICBM] and I feel confident it will not be done for a long period of time to come."[8]

Technical uncertainty, then, was a prime reason for discounting both the need for and the probability of obtaining ballistic missiles. The chief obstacles were still assumed to be guidance accuracy, rocket engine adequacy, and means of warhead reentry. None of these difficulties and nothing comparably unfamiliar troubled the perceptions of the cruise missile developers. They had no reason to doubt that they could perfect their chosen instruments of strategic warfare. For that matter, bomber program managers had only lately conceded that many of their casually accepted performance, schedule, and cost goals were unachievable.[9]

In 1950, the Atomic Energy Commission demonstrated experimentally what many nuclear scientists had contended earlier, that both lightweight fission weapons and eventual fusion weapons were feasible. The era of nuclear scarcity ended. At the same time, some members of the Air Force began to question whether manned bombers could successfully operate in an environment that included both sophisticated radar and high performance missiles. Finally, continuing work on rocket engines, principally in support of the rocket-boosted Navaho, made the ICBM propulsion problem seem manageable, although most military experts thought that completing development would take another ten years. Striking optimism about guidance technology was also evident, but the implications of stable-platform research and of miniaturized electronic components were not widely recognized.

Early in 1950, Air Force planners began to reassess their long-term missile requirements, and by the end of that year the Air Council had recommended establishment of a relatively slow-paced ballistic missile development project, the Atlas, to extend over nearly fifteen years. Underlying that recommendation was the assumption expressed as strategic doctrine—that manned bombers would remain the backbone of strategic air power until at least 1965 and that increasingly more effective long-range missiles (Snark, followed by Navaho) would gradually be inserted into the inventory in the mid- and late 1950s. The first operational ICBMs were slated for deployment toward the middle or end of the 1960s.

Even though by that time progress had been considerable, there was mounting evidence that expectations of a smooth transition from increasingly proficient manned bombers to "unmanned aircraft" (as cruise missiles were officially called) had been unrealistically optimistic. And any sensible military engineer expected ballistic missiles to be even more difficult to perfect. Moreover, to those concerned with the effectiveness of the strategic forces in the early 1950s, the uncertain accuracy of the ballistic missile made it unattractive as compared to a manned bomber. Consequently, the Air Staff felt no particular urgency about accelerating the slow-tempo program initially approved. Indeed, had it not been for the sudden increase in military appropriations that attended the start of fighting in Korea, it is unlikely that the Atlas program would have obtained even the slight funding needed to get it into preliminary development.

In 1953, after disengaging from the Korean affair, the Eisenhower administration announced a new doctrine of national defense policy, subsequently dubbed "Massive Retaliation." It expressed a pronounced national aversion to any future commitment of American ground troops to large-scale fighting anywhere on the periphery of the Asian Communist world and the intention of using nuclear weapons delivered by the Air Force as the primary instrument of national force.

Almost concurrently, but for different reasons, the Department of Defense began to give serious consideration to accelerating the undernourished ballistic missile program. Although no great technical advances had been demonstrated since 1951, a number of respected experts now believed that an ICBM could be developed in no more than four or five years. The Strategic Missiles Evaluating Committee, established in late 1953 to identify and cancel the less promising of several Air Force missile programs, endorsed an Atlas program acceleration and expressed grave doubts about the probability that several of the more "conservative" aerodynamic missiles would ever see active service.

That judgment certainly was not equally welcome in all areas of the Pentagon. Senior air officers remained convinced that cruise missiles were better prospects and, that if more money were to be spent anywhere, it should be on the new bombers (B-52 and B-58) due for delivery during the next three years. The Strategic Missiles Evaluating Committee, chaired by John von Neumann, had the ear of the secretary of the Air Force, however. It concluded that the existing accuracy requirements for ballistic missiles, expressed in hundreds of feet, were preposterous; in the absence of good targeting information, any missile exchange would involve the Americans in "city busting" whatever their inclinations, so extreme target accuracy was an illogical requirement. The von Neumann group and its supporters agreed on the need to develop a ballistic missile quite different in many respects from those hitherto sponsored by the Air Force. Recognizing the institutional obstacles to rapid development of ballistic missiles, the group also urged that the development effort be separated, in all respects, from those being conducted by the regular air establishment, thus insulating it from prospectively hostile treatment.

By January 1955, a new ballistic missile program had been shaped. Although obstacles were numerous and varied, within a year parallel development of alternative rocket engines, guidance systems, airframes, reentry bodies, launch techniques, and staging methods was in train. By 1956, when Army Chief of Staff Gen. Maxwell Taylor first urged that a strategy of flexible response be substituted for massive retaliation, Atlas, Thor, Titan, and Jupiter were impending; and work had begun on what would become Minuteman. In 1957, Henry Kissinger argued the likelihood of and the prudence of preparing for limited nuclear war rather than for expanded conventional war responses to peripheral threats. Sputnik and the "missile gap" allegations followed, not much ahead of the first salvos in the presidential campaign of 1960.

When the Kennedy administration came to power in January 1961, it enunciated a new set of strategic principles embodying graduated response and a counterforce strategy. All were fundamentally predicated on the use of the Atlas-Titan-Minuteman force plus the nearly ready Polaris missiles. Thor and Jupiter were to be retired from their European emplacements, having outlived their usefulness in only three years. The subsequent decision to abandon the costly emplacements because their vulnerability made them attractive targets represented a further application of the same logic. None of the withdrawn missiles had real second strike capability and their existence represented a temptation to a missile armed opponent to strike first, perhaps touching off the sort of spasm war the graduated response policy was intended to offset. Technology had proceeded so rapidly during the first five years of the ballistic missile era that by 1963 the missiles of 1960 were as obsolete as B-17s, and, unlike old bombers, nuclear-tipped missiles were not suited to fighting brushfire wars.[10]

Perhaps the most striking indication of evolutionary change in technology and weapons concepts over the years from 1945 to 1963 is that by the later date ballistic missiles were the nation's chief instruments of strategic warfare, entrenched so securely in both doctrine and force structure that proposals for alternative or supplemental strategic weapons encountered impressive objections. That was precisely the reverse of the situation before 1957, when the ballistic missile was the handicapped competitor in a contest with cruise missiles and manned bombers, the chosen instruments of those earlier years.

For nearly a decade, objections to American reliance on ballistic missiles began with the contention that they were technologically incapable of doing what was required of them. Whether they could be developed at all was argued into the mid-1950s; their military qualities were widely discounted for another five years. These were not academic exchanges, confined to the isolation of staff offices and briefing rooms. They concerned important issues, both doctrinal and institutional, although the advocates of extensive reliance on ballistic missiles accepted with uncharacteristic indifference the evidence of a perpetuation of the bomber concept enunciated in the mixed force concept of 1957. In the event, the deployment of substantial numbers of missiles permanently affected the composition and structure of the strategic forces, working to the ultimate disadvantage of the manned bomber. Between 1951 and 1962, nearly 3,000 jet-powered strategic bombers entered the inventory; by the end of the latter year, it was plain that relatively few would be replaced as age and wear had their effect. Ballistic missiles had become the dominant weapons of the United States. Yet until the certainty of that event had become apparent, there was relatively little discussion of its consequences.

Technology and Doctrine: Efforts to Ignore the Ballistic Missile

The analysis that underlay the 1954 decision to proceed with ballistic missile development was characterized by technical astuteness and scientific insight far above the ordinary, but doctrinal consideration was narrow and tightly hedged. For the next decade, many of the wide-ranging consequences of that decision were ignored; institutional infighting was frequent, as though the most important national issue was whether the airplane drivers or the missile sitters should rule the Air Force.

Most critics who had misgivings about the eventual effects of heavy reliance on ballistic missiles had a mistaken instinct for the sorts of arguments that best supported their case. They ignored significant factors that might have buttressed their case until missiles had expropriated many of the functions of the bomber. Early suggestions that it would be advisable to create an alternative to massive deterrence—a "graduated response," or "limited nuclear option"—were initially spurned by "the airplane drivers" until they found it expedient to argue for preserving bomber aircraft capabilities on remarkably similar grounds. It was paradoxical that a strategy and doctrine based on the creation and preservation of response options was opposed by its ultimate beneficiaries until the most attractive option had all but lapsed.

Cultural and institutional resistance to the innovation represented by ballistic missiles was only one reason for their relatively slow acceptance by strategic planners. Failure to take appropriate account of the unpredictability of technology was another. The notion that ballistic missiles would eventually come along in the evolutionary wake of increasingly complex cruise missiles dominated strategic planning from 1946 to 1954. Indeed, it was not until 1957, when it became all too obvious that an operational Navaho could not possibly precede an operational Atlas, that the "orderly evolution" misconception finally decayed.

Where strategic doctrine was realistically coupled to capability which meant in the near term only—it remained relevant. The expectation that intercontinental cruise missiles would be developed quickly and at modest cost, however, reinforced the tendency of planners to extrapolate doctrine from existing capabilities; and the restatement of bomber-plus-cruise missiles concepts in long-range planning documents reinforced the tendency to fund cruise missile development first and ballistic missiles least. The consequences of a different sequence of events from that postulated were not much explored.

Were there other instances in which fuzzy doctrinal assumptions interacted with immature technology to lead planners astray? Single cases do not adequately prove broad theses; yet multiple instances are not much more likely to convince when conventional wisdom is being challenged. Indeed, even if all the evidence points to conclusions that contradict ingrained belief, the evidence is likely to be ignored. Nevertheless, there is abundant support for the proposition that planning preferences expressed as requirements have little influence on the course and rate of advancing technology.

Engineering logic suggested, even insisted, that a turboprop variant of turbine propulsion would be the first successful application of jet propulsion to long-range aircraft. Russian designers were so committed to that assumption that they made no serious efforts to develop long-range turbojet bombers or transports during the years when American and British designers were creating the B-47, KC-135, Comet, B-52, and V-series bombers. That was one reason why Soviet progress in the development of long-range aircraft lagged well behind that of several Western states, notwithstanding a considerably larger Soviet investment and a starting point that was not greatly different. In the United States, the turboprop-powered bomber never appeared (the only promising candidate was expeditiously redesigned to become the B-52!); only one American civil transport designed for that power plant was widely used (Lockheed's Electra); and, except for reengined passenger aircraft, the turboprop was mostly applied to a few specialized military transports which had slight commercial attractiveness. (The C-130 was the only important exception.) Notwithstanding all that, until the late 1950s many military planners both persisted in the conviction that turbojet engines would ultimately give way to turboprops in many applications and invested substantial development resources in the effort to bring that about. It never happened.[11]

Similar circumstances surrounded the recurrent assumption that air-to-air missiles would entirely supplant conventional (or slightly unconventional) rapid-firing guns as fighter armament, a thesis that on two occasions in two different decades led to the production of missile-only USAF fighters, and twice had to be abandoned. However compelling the anticipation than an operationally adequate Identification Friend or Foe (IFF) device would appear, and however attractive the doctrine that prospect induced, the "adequate IFF device" has eluded designers and engineers for the three decades since 1944, when R. V. Jones first convinced the British Bomber Command that German night fighters were using IFF as a target indicator. Finally, nuclear propulsion for aircraft remained so attractive that the United States continued to support its development well after its economic and technical infeasibility had been convincingly demonstrated.

The Gap between Invention and Application

These and similar cases share a common trait. It is not the laboratory scale demonstration of some technical capability that proves difficult, but the eventual transition to operational utility. Science is not the obstacle; it is engineering. All of the devices mentioned above demonstrated nominal feasibility; every major feature of Navaho and Snark worked modestly well—in isolation; marvelously ingenious IFF devices were successfully tested; the turboprop was a more efficient user of turbine power than a turbojet; variable-sweep aircraft were safely flown by 1950; flight-capable nuclear heat sources were both designed and constructed, several times; ramjets have powered many flight vehicles, and so on. What was lacking, in all of these cases, was an application that engineers could cope with at reasonable cost and in reasonable time, and a level of operability that did not demand greatly more in precision or reliability or effectiveness than had actually been demonstrated. In the genre of ballistic missiles, the Thor is a marvelously useful example of the successful application of the art of the feasible: it represented the packaging of available propulsion, guidance, and warhead reentry technology in a relatively conventional airframe, and hence proceeded from project definition to initial demonstration in little more than a year and to full-scale operational test in precisely three years. In that same period, the Navaho program schedule slipped one year for each year the missile remained in development. The contrast is instructive.

In the end, ballistic missiles were rather quickly carried through advanced development and, in their employment, significantly altered earlier doctrinal concepts and national policy. Intercontinental cruise missiles became quaint themes for military historians because guidance, propulsion, and reliability elements refused to conform to the expectations of those who prepared strategic doctrine. Bombers, whether manned or unmanned, did not remain the dominant elements of strategic force in the 1960s and 1970s, as planners had anticipated. Indeed, the effort to upgrade the bomber, to make it the technological equivalent of the ballistic missile, and thus to extend its dominance, failed in the 1960s and 1970s, perhaps because it paralleled the course of cruise missile development in the 1950s: costly and demanding efforts to achieve high but not necessarily relevant performance when more direct, less expensive, and politically more attractive courses remained available.

Similar consequences have followed efforts to subordinate technology to doctrine in areas other than strategic warfare. Interdiction and close air support are important responsibilities of the Air Force; their performance, in future battle areas, has been imperiled by effective anti-aircraft weapons against which conventional defense suppression techniques, however enriched by improved technology, have been less than fully effective. Suggestions that air doctrine be altered to conform to such "realities" has not been wholly welcomed.

Quality versus Quantity

The quality-versus-quantity issue is not an obvious candidate for treatment in such terms. But to the extent that qualitative advantage is honored much more highly than quantitative in U.S. air doctrine, there is a connection. Some areas of technology are much more likely to promote rapid changes in operational concepts and applications than others. Thus the introduction of a novel device with great effectiveness—an omnidirectional air-to-air missile, for instance—would presumably have a pronounced effect on the outcome of air battles. It would impart to its possessors a very substantial advantage, perhaps making conventional air-to-air tactics and the fighters designed for them obsolete.

The addition of a marginal improvement that gave one nation's fighter aircraft a 10-percent edge over others (however, that was calculated) would have a much less marked effect. Indeed, if the air force with the "inferior" technology had superior pilots, or better tactics, the difference might well go unnoticed. That was more or less what happened in the Winter War between Finland and the Soviet Union, in initial engagements between F-80s and Mig-15s in Korea, and in some other well-known cases. There appears to be little doubt that superior numbers and roughly equal quality give one a clear advantage: American and British fighter pilots benefited from that circumstance in the closing months of the war with Germany. It is generally conceded that the German introduction of Me-163 rocket fighters in significant numbers could have made Allied bomber losses unbearable, and the Me-262 fighters were completely superior to all operational Allied fighters of 1945. But there is no real evidence that slight performance advantages can be decisive or that they cannot be readily countered by either modest infusions of new technology or intelligent changes in tactics or doctrine. Certainly the experience of the U.S. Air Force in Southeast Asia between 1965 and 1972 would bear out that premise, confirming the relevance of air combat outcomes over Korea nearly two decades earlier![12]

These broad statements are but slightly supported by careful historical analysis. It is an area in which historians have traditionally deferred to statisticians and operations analysts. But there is much to be said for exposing such issues to historical research techniques—which, to the dismay of some and the glee of others, have been appreciably altered by the widespread availability of devices that comfortably can process enormous quantities of data. It is prospectively as risky to trust in the "quality beats quantity" premise as it was to repose faith in the notion that evolutionary development would insure the ultimate appearance of intercontinental cruise missiles, or that the bomber always gets through. The assumption that technology and doctrine will alike change in traditional, evolutionary ways is comfortable, but it is not necessarily true; and, as some of the instances noted above suggest, it may also be an invitation to disaster.

Has all this been an abstract exercise in historical research? Or does a better understanding of the interaction of technology and doctrine and the derivation of requirements have current relevance for the Air Force and for the nation? Part of today's history, and tomorrow's, is a complex of doctrinal and technological issues that includes an aging bomber force, parity in ballistic missiles, a risky basing policy for U.S. strategic missiles, cruise missiles of a new sort, enlargement of an old arms limitation agreement, and a host of issues having to do with modernization of the equipment base for NATO. The century may afford few years in which it still will he possible to bring changing doctrine and emerging technology into concert with one another or to choose among technologies on grounds other than their seeming budgetary advantages.

The history of the thirty years following World War II suggests that Air Force doctrinal planners may again find themselves compelled to alter operational concepts, basing options, and force structure to be consistent with the attributes of a set of technologies favored because of their nominal cost effectiveness, political attractiveness, or apparent availability. In the past, the consequences of such compromises have not often been advantageous to the Air Force—or the nation. There is much to be said for promptly acknowledging and realistically dealing with the influence of technology on the formation of doctrine.

Notes

1. Horst Boog (see his essay in this volume) has observed that the inability of Luftwaffe leaders to appreciate that technology does not obey military commands, however smartly voiced, was a substantial contributor to the 1943–45 decline of that air force; and Kenneth R. Whiting has remarked (in his paper also in this volume) that a delayed but sufficient Soviet appreciation of the relevance of air technology to tactical air doctrine had much to do with the revival of Soviet air power between 1943 and 1945. So many and so blatant are similar examples that the historian of technology is bewildered by the persistent assumptions (of those who compose

"requirements") that the pace and direction of technology can be forced into channels arbitrarily selected by the armed services.

2. A. J. Harmon et al., *A Methodology for Cost Factor Comparison and Prediction* (Rand Corporation, RM-6269-ARPA, August 1970); R. L. Perry et al., *System Acquisition Strategies* (Rand Corporation, R-733-PR/ARPA, June 1971); R. J. Art, "Why We Overspend and Underaccomplish," *Foreign Policy*, no. 6, 1972.

3. See J. Jewkes, D. Sawers, R. Stillerman, *The Sources of Invention* (London: Macmillan, 1958); M. M. Postan, O. Hay, J. D. Scott, *Design and Development of Weapons* (London: HMSO, 1964)—by far the most fascinating discussion of the interaction between innovation and requirements is E. E. Morison's classic, *Men, Machines, and Modern Times* (Cambridge, Mass.: MIT Press, 1966); a more limited but relevant piece is Robert Perry, *Innovation and Military Requirements: A Comparative Study* (Rand Corporation, RM-5182-PR, 1967).

4. J. W. Angel, "Guided Missiles Could Have Won," *Atlantic Monthly*, December 1951.

5. The effects of that remarkable set of years are well represented by, for example, R. V. Jones, *The Wizard War* (New York: Coward, McCann & Geoghegan, 1978) and R. S. Macrae, *Winston Churchill's Toyshop* (New York: Walker, 1971). There are many others in the "gee whiz" category, of course, most notably the accounts of German weapons development and its almost realized (!) potential. See, in particular, the fourteen *Toward New Horizons* reports prepared by the U.S. Army Air Forces Scientific Advisory Board under Theodore von Kármán, Chairman, 1945.

6. See M. S. Knaack, *Encyclopedia of U.S. Air Force Aircraft and Missile Systems* (Washington, D.C.: GPO, 1977), 1:312-13; G. Swanborough and P. M. Bowers, *United States Military Aircraft Since 1908* (London: Putnam, 1963), 466-67.

7. David Irving, *The Mare's Nest* (London: Kimher, 1964), 44-45.

8. Vannevar Bush, who had been chairman of a special committee on new weapons for the JCS, in testimony before the Special Senate Committee on Atomic Energy, December 1945; also see *Hearings Before the Preparedness Investigating Subcommittee of the Committee on Armed Services, Part 1*, November 1957 (85th Cong., 1st sess.).

9. See R. L. Perry, *The Mythography of Military R&D* (Rand Corporation, P-3356, May 1966); and idem et al., *System Acquisition Strategy*.

10. For the early history of ballistic missile development, see E. M. Emme, ed., *The History of Rocket Technology* (Detroit: Wayne University Press, 1964). Other near-contemporary accounts of interest include John L. Chapman, *Atlas: The Story of A Missile* (New York: Harper & Brothers, 1960); and Julian Hart, *Mighty Thor* (New York: Duell, Sloan & Pearce, 1961). The "missile gap" issue, and pre-1957 concepts and doctrine, are addressed in *Hearings Before the Preparedness Investigating Subcommittee . . .* , U.S. Senate, 85th Congress, 1st and 2d sess., November 1957 and January 1958.

11. It sometimes is forgotten that the turboprop variant of the B-47 actually was built for test purposes and that a similar modification was scheduled for B-52s at one point. The curious history of military interest in turboprop propulsion has yet to be completed, although it clamors for attention.

12. See R. F. Tolliver and T. Constable, *Fighter Aces* (New York: Macmillan, 1965); J. N. Merrit and P. M. Sprey, "Quality, Quantity, and Training," in *USAF Fighter Weapons Review* (Summer 1974); R. F. Futrell, *The United States Air Force in Korea, 1950–1953* (New York: Duell, Sloan & Pearce, 1961).

Korea, Vietnam, and NATO: An Airman's Perspective, from First Lieutenant to Four-Star General

Bryce Poe II

In recent years, it has become fashionable to refer to the Korean War as the "Forgotten War." Caught as it was between the enormous scale of World War II and the contentious nation-dividing Vietnam experience, the Korean War and those who participated in it apparently deserved neither recognition nor serious examination. Until recently, with few notable exceptions, this view was shared by politicians, some in the military, and scholars alike. Happily, the situation has been slowly changing. Although much of the renewed interest in the conflict has been associated with the well-deserved acknowledgment of the sacrifices of the veterans and recognition of what they accomplished, too little attention seems to have been given to "lessons learned" for consideration by military professionals.

Generals Henry H. Arnold and Carl Spaatz did not want the experience gained through the expenditure of so much blood and treasure in air warfare to be lost. Just two months before the end of the Korean War, General Spaatz established the Air Force Historical Foundation as a "quasi-official" organization of the then new United States Air Force. The founding members, including names like Ira C. Eaker, Hoyt S. Vandenberg, Nathan F. Twining, Alexander de Seversky, Benjamin D. Foulois, William F. McKee, Earle E. Partridge, Werner Von Braun, Thomas D. White, and Laurence S. Kuter, took very seriously Spaatz's charge "to preserve and perpetuate the history and traditions of the USAF and its predecessor organizations." Added to this was the task of making "lessons learned" available to the active forces.

This essay attempts to provide a personal view of how those lessons were applied, or perhaps ignored, in various circumstances during my own service from just prior to the Korean War through my 1981 retirement and beyond. These experiences and observations are offered in the hope they will spark additional scholarship.

Prewar Situation

To understand the events of the Korean War, it is necessary to examine the state of the U.S. armed forces during the years immediately prior to the conflict. Although at the time I was a relatively junior 1st lieutenant RF-80A reconnaissance pilot, my additional duties in Japan gave me a clear understanding of both on-scene readiness and combat support. There was a wide difference in capabilities within the services.

One problem was with quality of personnel. Although the majority of American servicemen and women were fine people, we had too many serious personnel problems. In eighteen months as trial judge advocate of a military court, I tried, and convicted, over thirty individuals for everything from murder and rape to grand theft and sale of fuel and weapons. As commander of a 5th Air Force Air Force Training Center, along with a cadre of tough noncommissioned officers, I retrained assorted bullies, thieves, and malcontents—plus any poor fellow unfortunate enough to get venereal disease—by means of boot-camp-type discipline and forty miles hiking each week with full packs. The center's tents usually had a population of about ninety, with each man serving thirty to ninety days based on a demerit system.

Throughout Japan of the late 1940s the special courts filled the stockades with discipline problems. Discussions with my counterparts in the Army indicated that this sort of problem was a great deal larger and more difficult for them. Fortunately, we airmen had real and visible missions; work to do on the flight-line and in the shops supporting airstrip alert; and even airborne patrols to keep an eye out for Soviet aircraft. With less demanding occupation duties the Army had less important and useful work to carry out.

Training was another area where we were much better served than our Army friends. The sky offered almost unlimited areas for nearly any kind of flying training and ranges were available for weapons delivery. On the ground, densely populated areas restricted Army ground maneuvers. When working these exercises with the Army, I nearly always found "enemy" columns poorly camouflaged or even bunched up bumper to bumper on roads or in open areas.

World events also impacted readiness of the Far East Air Force (FEAF). The Berlin Blockade was significant in two ways, for example. First, a substantial number of pilots were transferred out to fly in the airlift. Second, the Berlin crisis accented the growing concern about Soviet intentions not just in Europe but worldwide. Along with another lieutenant, I was chosen to conduct carefully controlled, highly classified, reconnaissance flights to determine Soviet air force composition on the mainland of Asia as well as on Sakhalin and up the Kuriles. Only one such prewar sortie was scheduled for Korea. It aborted due to weather and was not rescheduled. After August 1948, the State Department was responsible for Korea. General Douglas MacArthur declared, "from June 1949 to June 1950, constant intelligence reports of increasing urgency were submitted to Washington, advising of a possible North Korean thrust." Despite this, it would seem that the national concern was with the Soviets, not their satellites.

That concern was well founded. Soviet-controlled areas had many military airfields loaded with numerous aircraft. While a few of the bases were covered with Lend Lease P-39s and P-63s as well as older model Yaks, there were plenty of more modern fighters to include some of the late model La-9s and 11s. Although piston-engined, the La-11 had more than enough performance to catch our attention when they attempted interception and we had clumsy, extra long-range tip tanks on our RF-80As.

Nevertheless, in many ways our flight operations were improving, especially with regard to safety. Although the United States Air Force's flying training loss rate had been somewhat reduced, especially since the end of World War II, the rate was still terrible. In 1946 the accident rate of 61 per 100,000 flying hours resulted in 879 fatalities and something had to be done. Air Force leadership went to work, nowhere more seriously than in the FEAF. Although fighter and reconnaissance pilots resented some of the resulting rules and restrictions on flying, accidents became less frequent and flying discipline improved. The 8th Tactical Reconnaissance Squadron (TRS), the only jet reconnaissance unit in FEAF at the start of the Korean War, was fortunate to have two very well-qualified, no-nonsense commanders in succession and set an Air Force–wide record of 10,000 hours without an accident. By 1950, the accident rate in the Air Force as a whole was cut almost in half to 36. As a result of the increase in flying time, however, another 781 airmen died that year. To put rate and casualty figures in perspective, in 1993 the entire USAF accident rate was 1.34 per 100,000 flying hours and 42 fatalities.

Logistics support for the armed forces was sporadic and unpredictable. All services suffered from and resented the stateside longshoreman strikes that reduced the flow of foodstuffs to a trickle. Australian bully beef, Japanese white fish, and withered boxes of wartime rations were staples for the troops and their families. What we now call "Combat Support" was uneven. Some spare parts, ammunition, and weapons were in good supply just because they were left over from World War II, but records were so poor that locating them was difficult. The situation was manageable for older aircraft in the process of being replaced, but support for the new inbound equipment was sparse. Cannibalization of all equipment, ground and air, was common. That, plus the traditional "moonlight requisition," even between units, made on-scene determination of readiness or even planning of flight time problematic. Estimates of item deliveries from stateside for support or modification were grossly inaccurate.

These conditions summarize the situation in Japan in mid-1949. The situation was worsened by President Harry S. Truman's late-1949 cuts in military spending and Secretary of Defense Louis A. Johnson's harsh response to that reduction of funds. The impact on FEAF was immediate and very serious, especially in the case of personnel. As a result of the reductions, a large number of people had to be removed from service very quickly. Deadlines were short, with some commenting they had been given "holiday" presents in the form of demobilization and a boat ticket home. This applied to a surprising number of active aviators.

Worse still was the method of selection. According to official pronouncements at the time, the Air Force "wanted to retain people with the broadest skills and those members with only one skill were marked for release." This demonstrated both poor judgment and lack of knowledge when applied to the flying units in Japan. It very often resulted in the least qualified pilots being retained. Prior to the drawdown, for example, there were too few people trained to assume many of the required additional duties. In order to keep the best pilots in the air, these less than desirable jobs were often given to air crew members who were considered substandard in flying skills, or even unsafe for crew duty. Unfortunately, this procedure technically qualified these poorer air crew members in more than one specialty. Accordingly, they were too often retained. As one frustrated commander put it at the time, "If we had a pilot who always landed wheels up or a navigator who couldn't find his way to the bathroom, we gave him the job of running the Officers' Mess or base housing. Now he has a second 'skill' and I lose a number of highly qualified combat veterans and have to put that clown back in the cockpit!"

In addition, a concurrent cost-saving program took a considerable number of air crew officers off flying status. As a result, some of these men resigned in anger, reducing still further the available combat force. My own 8th TRS was fortunate in that we were nearly all regular officers. Accordingly, we were less vulnerable to the rapid reduction. This circumstance, stemming as it did from the days when we had the only jets and were manned with hand-picked people, applied to very few flying units. In most cases, regulars were in a minority. Despite all these difficulties, hardworking commanders paid increased attention to training and safety. A cadre of seasoned combat airmen kept 5th Air Force units combat capable, but the drawdown of experience and blow to morale took a toll. This was recognized in May 1950 when the word came that the usual operational readiness inspections had been suspended. At 0400 hours on 25 June 1950, the North Koreans crossed the 38th parallel and started the Korean War.

Call to Arms

FEAF's 5th AF had a contingency plan designed "to provide for the safety of American nationals in Korea." The plan had been updated only three months earlier and the focus became evacuation. It was quickly put in motion. General MacArthur had added the requirement to be prepared to attack hostile ground and surface targets in support of the evacuation if issued orders to do so. In accordance with this, FEAF fighters covered evacuation by ship on 26 June and by air on the 27th, that day successfully beginning the air war by shooting down seven enemy aircraft and driving off the remainder of the thirteen sent in to attack the transports. By 29 June, almost 1,600 people had been safely removed, about half by air and half by sea, without a single loss or injury.

Meantime, on the 27th, three of us joined our commander in a flight of RF-80As, followed by a transport with ground crew and photo lab personnel, in a move to Itazuki in Southern Japan. We were told that the Republic of Korea (ROK) Army was doing well but by the time we landed it was clear that the situation was critical. Late that day, General MacArthur published orders ordering FEAF to attack and destroy military targets south of the 38th parallel.

Although there were no contingency plans for these operations, FEAF reacted promptly. The 5th AF launched Douglas B-26 Invaders to conduct night attacks (made ineffective by weather), moved additional B-26s south to Ashiya, Japan, and set up a schedule of strike and ammunition shipment sorties for the 28th. Twentieth Air Force began the move of "all combat ready" B-29s from Guam to Kadena with orders to quick turn on arrival and attack enemy forces south of the 38th parallel. Additional Lockheed F-80s and North American F-82s were to enter the fight, this time to concentrate on attacks on ground targets.

The weather was miserable, and forecast to stay that way on 28 June but, fortunately, when I arrived over the battle area at dawn I found holes in the overcast, got down to the targets and confirmed that attacks could be made on the advancing North Korean forces. What followed saved the Republic of Korea from almost certain and immediate disaster. B-26s went in with bombs and then followed up with rocket and strafing attacks along roads and railroads. These were not easy targets; some aircraft went down and nearly all suffered battle damage. F-80 pilots took real risks in attacking targets at maximum range from their home base during periods of clouds and rain, but it paid off. The burning tanks, artillery, and trucks they struck left fires that could be seen for fifty miles. By afternoon, 19th Bomb Group B-29s were in the fight, bombing roads and railroads.

The rout of the day before had not stopped the North Korean airmen from trying again on 29 June. This time they were more successful, damaging and destroying aircraft at Suwon and getting away with no losses. FEAF was completely involved with blunting the drive of the enemy army. Even some of the F-82s, initially airborne to protect the transports, were involved using napalm—a first for that aircraft type.

Despite the emphasis on enemy ground targets, the North Korean air threat was not forgotten. Over the next three days I photographed airfields deep in North Korea—Wonsan, Kanko, and Pyongyang. My unit continued to cover every known enemy air base within 170 miles of Suwon on a regular basis. Many, such as Hungnam and Hamhung, were so far away that we had to use and jettison the hard-to-come-by special tip tanks. The photography inevitably showed none of these bases had been attacked, even those with many unprotected aircraft. There had been a successful B-26 strike on Pyongyang airfield on the 29th, but nearly all sorties were against ground targets in the forward area. This was understandable, considering the desperate situation of the ground forces, a situation that would continue after U.S. Army forces from Japan were engaged. Enormous effort and great individual courage did not make up for undermanned units, substandard equipment, and poor training. In the early days of the war, air attack was often the only means of stopping the advance of North Korean forces on the ground.

The situation grew increasingly difficult. Enemy airfields based a large number of combat aircraft which were much closer to the front than U.S. aircraft operating from Japan. Quick and flexible action was often necessary to ensure success. I saw this when I flew my second reconnaissance flight to Wonsan. On that flight I discovered that the base had been attacked at some point after my first mission. Unfortunately, despite a single string of bomb craters running across the base, nearly all of the forty-five enemy aircraft I had counted before were gone. In short, the enemy had successfully evacuated the air base. That same day, a fellow reconnaissance pilot subsequently found a small strip and about two dozen fighters at Pyongyang just north of the 38th parallel. After our film was developed and reports sent on up the line to what we thought was the "dead letter office," we walked across the base to the 8th Fighter-Bomber Group and gave the commander a package of the material on Pyongyang. He reacted immediately, leading a force of seven F-80s in destroying or damaging almost all the aircraft as well as their fuel dump and communications facility. As they departed for home a large number of Navy aircraft arrived, to find little left to shoot at. The system had finally worked and two lieutenants were in real trouble but we thought it was worth it.

Two lessons stayed with me as a result of those early days. First, it does no good to provide even the best intelligence information unless it is promptly delivered to those who have the ability to act on it. Second, the U.S. Air Force had been so capable at providing air superiority—even air supremacy—that it has been taken for granted by people who should have known better. Fortunately, it was not taken for granted by the commanders of the FEAF and 5th Air Force. Supported by General MacArthur, they began a campaign, using modest resources, that essentially kept enemy aircraft on the ground until the arrival of the MiG-15 in October.

Air Superiority

The most important thing to recall about air superiority in Korea was not how it was established, but rather that soldiers, sailors, airmen, and people to the top level of government took it for granted. It was widely believed that air superiority could be counted on no matter what the circumstances.

As the ground fighting in Korea stabilized a reconnaissance pilot needed no map to tell the difference between friendly and enemy territory. In the north, targets could only be discerned by multiple and hazardous low-level passes or by agonized scanning of film. In the south, United Nations vehicles moved in tight columns, parked in neat rows, artillery batteries fired from open revetments, and some regimental and division headquarters even had whitewashed stones marking paths and outlining unit mascots. At night, lights across the front line were almost nonexistent, while on the UN's side storage areas, bivouacs, and lines of communication were clearly illuminated. Lack of realism and a sense of ignoring reality could have been very costly.

Back in the United States such dangerous practices also existed and were ignored. During an exercise in Texas called "Longhorn," for example, aggressor aircraft caught forty-four tanks, bumper to bumper in bright sunshine on an open field. Forty-eight P-47s, with two 1,000-pound bombs each, simulated an attack. The Army general officer serving as the exercise's umpire ruled only four tanks destroyed and eleven aircraft shot down! Nor were problems confined to the United States. In NATO, they manifested themselves in another way during a subsequent major northern Europe exercise called "Mainbrace." A naval force, approaching the Norwegian coast from Iceland in terrible weather, came under simulated attack by F-84s for several days. Carrier aircraft could neither reach hostile land bases nor make timely launches to intercept their attackers. Nevertheless, the task force was assessed zero losses.

These exercises demonstrated that the luxury of operating under a friendly air umbrella for so long had blotted out what many should have remembered about the potential threat posed by enemy air activity. In 1964, General Leon Johnson, then chairman of the Net Evaluation Subcommittee of the National Security Council, wrote me, "the U.S. Army has drawn the wrong lessons from history regarding the employment of air power. We had almost complete air superiority in the ground war in Europe and total superiority in Korea as well as in Vietnam."

A few years later, when I returned from Southeast Asia to command a wing in Germany, I found things somewhat improved. There was no question that the Soviet air threat was real, even massive, but measures to meet that threat were distressingly inadequate. Construction of aircraft shelters had finally begun but other important actions were in limbo. Key support activities were neither hardened nor dispersed. With the complexity of new aircraft, a single bomb through the roof of the soft avionics shop could have made most pilots lethal only if handed an M-16 and transferred to the infantry. Many mission essential functions, such as air base defense and movement of bombs from depots to flight lines, were a responsibility of the host nation. Sometimes these functions were not practiced or even specifically assigned. All of this was true in the face of superior numbers of Soviet combat aircraft practicing daily attacks on target ranges that were carbon copies of NATO air bases, down to the last revetment.

When I returned to NATO years later in the position of Vice Commander-in-Chief, United States Air Forces Europe (USAFE), I found that while most aircraft were sheltered only a small percentage of the other key facilities were protected. Cuts in defense spending at home and among allies had been justified by political artificial reductions in the threat. Among these were the notion that "the SU-24 Fencer cannot air refuel; and you can count on at least 15 days warning before the enemy attacks." Additionally, we were told that any major Warsaw Pact offensive was remote because of the potential use of nuclear weapons and even if it did happen, NATO's ability to go nuclear immediately would make training seriously for anything else superfluous.

Regretfully, lack of concern for the conventional air threat was not limited to politicians. Some senior airmen seemed blind as well. Eventually, however, our concerns for the vulnerability of personnel and support facilities led to immediate, inexpensive, and largely home-grown protective measures. These included berms pushed up around buildings, sandbags, fifty-five gallon drums filled with gravel, and sections of pipe set in the ground as personnel shelters.

After seeing real progress at most USAFE airfields, I visited a wing where there was no indication at all of any effort at protection. The commander indicated that what we had directed was not cost effective—berms would eventually be washed down by rain, rock-filled drums would rust out in a couple of years, and sandbags would rot. He went on to suggest that the plan for corrections had been submitted through proper engineer channels. Construction would start in three to four years! Even those unacceptable dates were doubtful in the extreme.

The unrealistic estimates of warning time also resulted in problems. At stations where aircraft shelters were still under construction we demanded dispersal, normally using pierced steel planking (PSP). The commander of one wing advised me that his aircraft were lined up wing-tip to wing-tip because, although the PSP was on hand, it was excess material shipped in from Southeast Asia without the steel pins needed to assemble it. The commander was not worried, however, because, in his words, "the pins were on order." This overly optimistic and unrealistic state of affairs was a far cry from the real world of 1950 in Korea. There, when we first arrived at Itazuki, the first order of business was dispersal and digging foxholes. We took the commander over to his own machine shop and introduced him to a chief master sergeant who assured us that he could produce the PSP pins "quicker than you can pick them up off the floor, general!" I could not help but compare these lackadaisical attitudes with regard to preparedness with that of the pragmatic paratrooper, Lt. Gen. James Gavin, at a meeting in Germany in 1953. A newsman took down an M-1 Garand rifle from the wall behind the general's desk and Gavin cautioned, "Be careful, that's loaded!" When the man expressed surprise Gavin said, "I jumped on a corps headquarters, why couldn't someone jump on mine?" I am also reminded of the comment of my friend, the commander of the Royal Jordanian Air Force, who, when recalling the Israeli attack in the Six-Day War, said, "I lost my whole air force in 90 minutes!" In short, lack of preparation or failure to plan realistically could result in disaster.

After the Korean War the situation was treated more realistically in Asia. On the Korean peninsula U.S. Air Force and Army personnel as well as South Korean military professionals have worked hard to prepare to respond immediately to any threat. Unfortunately, those efforts have all too often been weakened by political actions that either sent dangerous signals to the enemy or actually cut U.S. forces and equipment.

Combat Support

Once war broke out in Korea, the ineffective state of on-scene combat support became all too apparent. In the early days the available equipment was hard to get for two reasons. First, regular supply and communication channels were such that you often had to have a friend on the ground at the air base or depot in Japan. Acting as a broker, this friend listened to your troubles and then searched for what you needed. Normal requisitions simply did not work. Second, you could just about forget surface transportation, even in Korea, because movement was controlled, depending on location, by either Eighth Army, the X Corps, the Japan Logistical Command, or all of these. It was impossible to ignore the bureaucracy when it came to fuel or bombs, but most spare parts came by air.

During the Korean War, my additional duty as assistant maintenance officer involved both test flights and "scrounging" everything from engines to bench stock bits and pieces, even fasteners. Film was hard to come by along with the de-mineralized water necessary to process it. Once we moved to Taegu Air Base (K-2) in Korea we depended on phone calls to friends in Japan to locate spare parts. When new aircraft arrived in Japan, I would fly back, pick them up, scavenge wrecks at a couple of airfield graveyards, and put the spare parts contraband in the camera bay before returning home.

When an aircraft needed depot work we would strip it before turning it in. Aircraft instruments were especially hard to come by. On three occasions I can remember flying back to Japan with only an airspeed indicator and tailpipe temperature gauges operating. If the weather was bad this required finding some aircraft to fly in formation with. After landing I reversed the system of junkyard search before hitching rides on empty medical evacuation flights to get the parts back.

Spare parts support was a continuing problem during the Korean War. Although the system matured, it was complicated by the fact that FEAF began the war with outdated B-26s, F-51s, and F-80s. None of these was in production at the time. Throughout the conflict, the FEAF lost out to both Strategic Air Command and USAFE in priority for modern aircraft replacements.

Fuel shipments that survived bureaucratic delays, movement by ship, truck, and the rare train or pipeline, were generally of poor quality and the fuel itself often came in corroding 55-gallon drums. Flakes of metal in the fuel played havoc with aircraft engines. A pilot's life became exciting when simple throttle movements threatened flameouts in combat. Straining fuel through chamois did not solve the problem; a single jet wing would use some 125,000 gallons per day and the time involved would have been excessive.

Air base construction also suffered from bureaucratic problems. The changeover from the Army Air Forces to the independent United States Air Force had an impact on engineers in both the Army and our new service. Engineering equipment was too often in short supply and poor condition. The working conditions, weather, soil, aircraft damage to PSP, taxiways, and runways made construction conditions extremely difficult. Construction debris caused damage to aircraft engines. Tires did not last long on the rough runways. Subjected to a rate of some 10,000 takeoffs and landings per month, the PSP runway at Taegu went out of service in May 1951. Even so, the 5th Air Force was kept flying by innovative scheduling and hard work by armies of hardworking engineers.

By the summer of 1951, newly trained and equipped personnel arrived from the United States. By then the USAF had begun its own engineer training and no longer had to rely on the Army, but the problems never got easier. In later years it was easy to look back with pride, almost awe, at how very well the brand new USAF did in those extraordinarily difficult days.

Shortly after returning to the United States from Korea, I found myself running a replacement training outfit to update F-51 pilots to the RF-80A. A few weeks later we were essentially grounded due to lack of fuel. This situation had occurred as a result of a civilian labor strike. It cost some of my students three weeks of training they should have had prior to departing for combat. We also raided our own spare parts to provide a ready source of hard-to-get items to the units in Korea. This informal supply line depended on friendly airlift support for delivery.

In the years following the Korean War, I moved through many overseas and stateside jobs before arriving at the most demanding of combat support situations, the 460th Tactical Reconnaissance Wing (TRW) at Tan Son Nhut in South Vietnam in 1968. It was a big wing, consisting of eight squadrons on five bases, and flew 10,000 to 12,000 hours per month. By then, the spares support was well in hand, as were aircraft battle damage repair and modification operations. Engineer Red Horse teams were on top of any number of other jobs. Among their best was supervising the building of aircraft shelters. This was critical, not only because rocket attacks were a danger but also because shelters reduced the high incidence of cockpit system failures due to rain. In sum, support personnel on the base were doing a great job.

Nevertheless, a serious problem resulted from the fact that, unlike the normal wing, we did not own some items absolutely critical to combat operations. Sadly, this situation was predictable because we shared the station with all the major headquarters and could expect low priority for many things. The trouble was primarily that the base commander owned the motor pool. Such items as aircraft tugs, tractors, and cranes were key to our sortie generation and these vehicles were just not available as required. It was neither correct nor prudent to run to the 7th Air Force commander to get a crane, but there were occasions when I was greatly tempted to requisition items at gun point. Things finally came to a head when, during a period of nightly rocket attacks, we could not muster enough in-commission warehouse tractors to push some of the RF-4Cs into revetments. As a result, they had to be flown to Thailand each evening. The hope was to get useful mission sorties en route home the next morning. Eventually, we worked out a way to provide tractor repair on our own, but the problem did not go away until the base commander rotated and a mission oriented replacement arrived.

Across the world in USAFE there were many more examples of support not being tied to the combat mission. Logistics is a national responsibility in NATO. This means that logistics support of air weapons in the theater was the responsibility of the U.S. European Command (USEUCOM), an organization neither manned nor postured to do the job. Despite this, the system worked pretty well, with the wings themselves functioning directly with each other, depots, and other sources, but circumstances were such that plans, obligations, and even doctrine too often originated in the wrong place. This could lead to out-of-channel procedures much more complex and troublesome than those we used to hustle parts for flying units in Korea. One of the most serious problems in NATO resulted from the higher headquarters unilaterally obligating parts, weapons or even aircraft to an ally without prior notification to those who must provide the transferred item and obtain a replacement for it.

Another difficulty in Europe was the expectation of a number of allies that—national responsibility or not—the United States would provide them major and timely logistics support in war. Munitions and fuel were two important examples. Arguably, this was a result of desperate military logisticians or commanders who had been unable to get funding for their own stockpiles. Some NATO countries planned for and financed military budgets based on the assumption that if war came it would be nuclear. Therefore, so the feeling went, expenditures on conventional munitions were senseless. Such "logic" often defied common sense. When asked how he had arrived at a national requirement for only enough artillery rounds for four days of fire and just 100 rounds of .50 caliber ammunition per gun a foreign official said, "We measured the number of rounds on hand and divided by the number of tubes, artillery and machine gun, to arrive at the answer dictated to us by the budget people: no additional ammunition required."

As they were during the Korean War, the availability and movement of fuel and bombs were probably our most serious concern in NATO. Unfortunately, logistic support for these essential items was planned, managed and carried by people and organizations who seemed to plan only on peace rather than war. There were occasional flights of complete fantasy. One example was an earnest plan to move fuel through pipelines that carried raw oil in peacetime and would require weeks to flush and clear. The same NATO planning document counted on movement by river barges that would be unable to navigate waterways because the bridges spanning them would have long since been dropped by enemy action. There was to be neither tracking to locate and divert items in movement nor communications to direct such changes. Bombs from major depots were to be moved by unspecified civil transport with no who, when, how or where, nor any provision for peacetime identification or exercising. These overly optimistic appraisals almost made one wish nostalgically for the 1950 triumvirate of the Japan Logistical Command, Eighth Army and X Corps!

To be fair to NATO's planners, they were operating under circumstances that insisted on unrealistic warning time and, as a result of public law, had already cut much of the "fat" from overseas forces. In the U.S. Army this had meant the disappearance of a very large number of trucks. For the same reason, at one time the number one limitation on USAFE sortie generation in war would have been the availability of refueling vehicles. The good news was, however, that we had drastically improved the reliability and maintainability of aircraft from the end of the war in Southeast Asia.

Intelligence

Early disaster in the Korean War was averted only by great effort, raw courage, and rare good fortune. Had attention been paid to available intelligence, or, if that material had not been considered credible but backup measures taken to confirm or deny it, then those early days might not have been so tragic. In fact, action might have been taken to cause the invasion to be delayed or even canceled. The Korean

War was also a classic case of the intelligence warning being primarily heeded only by people who had neither the responsibility nor the authority to act. The timely provision of intelligence information to people who can do something with it other than to admire and talk about it is a problem that still exists. In addition, holding such data close, whether by classification or "not invented here" assessments, can make impotent the application of the force and violence required in war. Nowhere is this more consequential than in the application of, or the defense against, that most lethal of all principles of war, *surprise!*

During the war in Vietnam our wing took as much as four million feet of film a month, and I was not sanguine about how much of it was put to worthwhile use. Both the fighter wing people and friends in the Army used to needle me about the lack of photography for their purposes. My concern was heightened by some of the requests we got either for odd targets or blowups of "pretty pictures." The same day we printed a special order for a dozen large prints of a truck on the Ho Chi Min trail (identical to nearly every other truck) I was unable to confirm any interest in what I considered a coup in photo interpretation—location of a large enemy headquarters via tracing telephone lines in multilayered jungle. Finally—shades of Pyongyang and the 8th Fighter-Bomber Group—we set up a small photo cell at Phu Cat, home of the F-100 Misty Forward Air Controllers (FAC). We could land our RF-4Cs there, process the film in minutes, review the wet negatives with a FAC crew for lucrative targets, and they would be airborne within the hour. It worked very well. The people who could do something about it got the intelligence material.

In addition to four jet squadrons, the wing had our "Antique Air Force," three squadrons of EC-47s. Although older than many of the air crewmen, these aircraft had an important job and did it well. I flew at least once each month with an EC-47 crew and on my very first such sortie noticed that the analysts and interpreters in the cabin threw covers over their equipment and stopped all activity as I passed en route to the cockpit. Worse yet, they did the same for the aircraft commander and his crew. Not the fault of those technicians, but of their orders to keep their equipment secure from "uncleared personnel."

This was not a minor annoyance but a real problem in mission accomplishment. To pin down an enemy target the aircraft had to be repositioned for several fixes. The aircraft commander took change of course directions from a technician in the rear end who could not see out the window and was not familiar with aircraft performance limitations. Sometimes this even put them all at risk to flak or rough terrain. Fortunately, a note to a classmate in the Pentagon then brought prompt clearance for these EC-47 air crews. This not only improved mission effectiveness but also markedly raised the morale of everyone on the crew. Most of those lieutenants in the right seat had signed up to shoot down MiGs and were not wild about flying the EC-47. Once they really participated, and, as one told me, "the enemy came up on the radio and asked for rations and we delivered him fire from five artillery batteries followed by a B-52 drop," they became more enthusiastic. I took pride in flying with them, although

I became a bit apprehensive when they took me places at 90 knots where I had been uncomfortable the day before at more than 500 in the RF-4C.

In Europe during subsequent years there were wild differences in essential elements of information (EEI) on the same subject, depending on the source. This was a real problem compounded by political bias attached at higher level to whatever finding did not fit current budget or fashionable persuasion. This situation was sufficient to elicit occasional sadness, rage, disbelief, or resignation. The culprits were seldom the intelligence people, or even their in-theater commanders. There were scores of examples. In one, the U.S. forces had excellent intelligence on Soviet airfields but, according to directives, we were never to use national data but rather rely on official order of battle provided by higher echelons in NATO. Not surprisingly, target folders made with this largely out-of-date data could have resulted in attacks on airfields that had long since been abandoned. We forwarded photos to prove this and got no response other than repeated admonitions to use the NATO baseline information!

When Tactical Electronic Reconnaissance (TEREC) was introduced in USAFE it more clearly pinpointed Soviet dispositions and movements. Unlike the larger surveillance platforms, which the other side could anticipate and shut down operations, a TEREC aircraft was just another RF-4C Phantom flying on the periphery of the iron curtain. We were able to locate hundreds of threat emitters on a single sortie. Since this allowed positive confirmation of any buildup or threatened deployment, it was both a real peacetime coup and key to wartime operations.

Sadly, however, the information we collected was taken from the reconnaissance base to Wiesbaden for initial review, then forwarded to Washington, and in about six months we received a beautiful bound volume showing the exact location of the emitters. Of course they were mobile systems, mounted variously on artillery, antiaircraft, armor or other vehicles, so the information was worthless. In short, all of our efforts were of little use to either those who were watching for a possible surprise attack or to the wings who would be charged with destroying the weapons. A parallel problem existed with updating settings on jamming pods. Instead of a straight shot from the airborne aircraft as it located the threat to the wing preparing to launch to destroy or avoid it, that information also had to run a tortuous course all the way to the United States and back before the pod was reset. Both of these problems were subsequently resolved, but clearly signaled the difficulty of getting information from a reliable source to someone who can do something about it.

Gambling on Lives and Readiness

I understand a Soviet general once said, "The better is the enemy of the good." This is a fitting description of too many decisions that put American forces at risk, both personally and in the ability to perform their mission. Our commanding lead in technology was a major factor in winning the Cold War. Pride in that, however, often resulted in "not invented here" budgets or even political decisions to reduce or deny support—in replacement parts or modifications—of on-line weapon systems facing potential enemies. We also bet for too many years that no conflict would occur before the replacement system was fielded in strength.

An early example of this can be found in the history of the RF-80A I flew in Korea. When we first got the aircraft we found it was safe and a delight to fly. It had very few mechanical problems. One dangerous defect, however, cost us several fatalities. The airplane had a six-inch hydraulic accumulator, which was too small to handle the aileron boost requirements. When the landing gear was dropped or even dive brakes extended, the pilot frequently found the ailerons locked until the hydraulic pressure built up again. This cost several lives when pilots pitched up in the tight landing pattern and went into the ground before they were able to regain aileron control. Mid-air collisions occurred when formation commanders called for dive brakes to slow the formation, and then attempted to turn before a wingman's system reset.

In early 1947, a clever crew chief in our squadron salvaged a nine-inch accumulator from the Douglas RB-26 night photo aircraft maintenance and found it a perfect replacement for the six inch culprit in our jet. Thus, as a result of one sergeant's creative problem-solving the danger went away. But when technical specialists at Wright-Patterson Air Force Base were advised of this unofficial field modification, they declared that we took action at our own risk. As a result, it was only through ad hoc meetings on cross-country flights that word of the unofficial but effective remedy was passed to stateside Shooting Star outfits. In 1950, just as the Korean War began, the system finally provided us with a "legal" 7½-inch accumulator that worked. But, during the interim, lives had been put at risk, probably lost, and mission effectiveness reduced, for the three years when a solution was at hand but not accepted.

Another example of curious priorities had to do with our primary mission, photography. One of the first things we learned after our 1946 update from the F-6D (tactical reconnaissance version of the F-51 Mustang) was that the camera equipment that had performed so well on the F-6D and the F-5 (photo reconnaissance version of the P-38 Lightning) cycled far too slowly to enable us to take advantage of the higher speed of the RF-80A. The most useful photo product is stereoscopic. This requires an 80 percent forward lap of film frames. At the high speeds we normally flew at low altitude we could not always count on any overlap. This sometimes resulted in no target coverage and was therefore unacceptable.

We were advised that the problem was being worked at Wright-Patterson but no solution had been found by the time the war began. In spite of our motto, "Alone, Unarmed and Unafraid," we considered slowing down over a defended target too much to ask. Accordingly, there was great anticipation when camera experts from the United States came in with assorted new equipment. Unfortunately, no high-speed camera was included, but they delivered a number of other products for us to flight test. I drew a new color film and a request for "pretty pictures." I found some T-34 tanks and made several passes before returning home to find only half the film had cycled and none of it would develop. Several other pilots had similar lack of success with various systems and the exercise ended when one of my squadron-mates flew a "G-Force Resistant" 35mm movie camera that took a 20mm hit right in the lens that blew off most of the nose of the aircraft. We celebrated his safe return and the departure of the team of experts. A few months later we finally got newly developed cameras that came close to doing the job.

Cameras did not present the only equipment difficulty. The venerable F-51s and F-80s took the brunt of the war longer than they should have. The resulting aircraft losses, in particular those of the Mustangs, were very serious indeed. It was a wonderful airplane, but even in World War II it had been shown to be vulnerable to ground fire. The coolant system was particularly vulnerable. A single hit too often caused engine failure within seconds. With North Korean and Chinese infantrymen heavily armed with automatic weapons, low-level air support was extremely hazardous. After being turned down on a request for more Republic F-84 jets to replace the Mustangs, the FEAF commander asked for the old F-47 Thunderbolts, which were far less vulnerable to ground fire as a result of their rugged air-cooled engines. Sadly, this request was denied because it would have meant two obsolete propeller type aircraft to support in theater.

Despite its relative lack of sophistication, the F-80 was among the best all-round aircraft in the world for its time. It took battle damage better than anything else, was a perfect no-torque gun platform, and was highly maneuverable. Unfortunately, by the outbreak of the Korean War, it was well past its prime, having been flying for over six years. Its obsolescence became very clear with the advent of the Soviet-built MiG-15. These modern jets enjoyed a great speed advantage. As a result of the predictable routes of flight we were forced to fly down the Yalu, the initiative was too often left to the Soviet or Chinese pilots.

Senior policy decisions delayed the replacement of the F-80. The 5th Air Force was initially equipped with older type aircraft because they were the only ones available in quantity. Moreover, many leaders considered the Soviet Union as the real threat, especially in Europe. Finally, there were always overly optimistic estimates about when the Korean War would end. These and other factors combined to delay the arrival of new equipment and limit its quantities. Despite requesting a 50 percent fighter reserve, for example, the FEAF commander was allocated only 10 percent. Similar problems plagued the size of the bomber force. Things were even worse for U.S. ground forces at the start of the Korean War. Many Army divisions began the war with only two battalion infantry regiments instead of the standard three. Similarly, artillery battalions contained only two batteries instead of three. Shortages in equipment, both on the ground and in the air, really constituted huge gambles with the lives of American service personnel.

Sometimes the potential costs of these gambles went unrecognized and the lessons unlearned. Many years later in Europe we were faced with blocking several thousand choke points to blunt any Soviet attack on NATO. The laser pod for bomb delivery was the best answer but when we requested it the response was, "It really isn't good enough, we are going to give you a wonderful system, called Position

Emitter Location System (PELS)!" Despite clear evidence that the laser pod was the only existing system available to meet the threat and that PELS had only begun testing, it took me two trips back to Washington to get five pods. Spare parts for complex systems were also difficult to acquire. When we began having trouble getting spare parts for TEREC-1, we were told not to worry. The TEREC-2 was coming and it seemed foolish to spend money for spare parts on an out-of-date system. Our ability to locate hostile targets suffered for more than a year. Eventually we began to get spare parts to the original system. So far as I know, PELS was never fielded.

Tragedy was often the result of such shortsightedness. One of our most capable helicopter pilots, a Son Tay raider and standardization and evaluation chief for USAFE, had an accident caused by simple mechanical failure. As a result of a minor ground impact, he and his crew of three were killed in an explosion of an external fuel tank. Sadly, the lives of these airmen might have been saved had the Air Force been aware of a procedure already in use. Recognizing the potential for cracks in the tank, the Army had routinely strengthened them so that they were fireproof. All too late, we retrofitted our USAFE aircraft with the Army version of the tank and passed the word. Three years later, as Commander, Air Force Logistics Command (AFLC), I found that the substitution of Army fuel tanks had been halted pending tests on the validity of those items to determine if an Air Force solution might not be better. The Army had not had an accident such as ours. The Air Force tests themselves were stalled. In short, we were still putting crews at risk for no good reason. Even at senior rank it took me another eighteen months to get the job done.

The Future

The lessons of the Korean War were not lost on the Soviets and their allies. Even so, their system was, for the most part, less adaptable to new ideas or changing priorities. Apparently, they were very poor at many intelligence efforts, especially assessing the threat. Russian post–Cold War commentary on the Yalu air battles lacks credibility. Nevertheless, it's clear that the North Koreans will never forget the difficulty of operating under UN air superiority. Their current sheltering of aircraft and airfields, command and control, supply and maintenance, fuel and munitions, bivouac areas, and ground combat units may be the best in the world. This protection is not limited to military installations and the forward area but is nationwide and includes cities, transportation hubs, and industry as well, with many arms production and repair facilities deep underground.

There are many examples of lessons the North Koreans learned from the war. An intriguing one had to do with the potential conduct of any new attack on the South. They know, from their experience, that they will have to move quickly, before the United States can reinforce. Accordingly, it is apparent they plan to support any future attack with a number of unconventional actions. The existence of tunnels under the demilitarized zone is well known but not all are pinpointed. Infiltration of special forces wearing friendly uniforms is to be expected. Even now, in peacetime,

North Korean "skunk boats" insert terrorists and agents. Their use will be expanded to large numbers should war break out again on the Korean peninsula. Another very serious threat to South Korea has its roots in the famous "Bed Check Charlie" PO-2 open cockpit biplane raids during the war. In the early stages of the conflict they were little more than a nuisance, but as time passed, they became a more serious problem. One night in May 1951, two PO-2s attacked Suwon and, according to subsequent reports, "did more damage to the F-86 Sabers than had all air combat with MiGs up to this time."

Contemporary North Korean plans for biplane attacks may be even more lethal. Instead of the tiny PO-2, the North Koreans can now call on the Antonov An-2 Colt. It is supplied in large numbers as the primary transport for their airborne light infantry brigades. Carrying up to thirteen heavily-armed soldiers down valleys and around mountains, it can fly low enough to avoid radar detection. The primary targets would be airfields, command and control facilities, and radars. Their range is such that they can reach any target on the peninsula, inserting the combat team either by airdrop or landing. These aircraft and the infantry they carry constitute a serious threat. It might legitimately be asked whether or not United Nations forces are fully prepared for the night drop of a dozen or more enemy airborne troopers on their flight-lines or command facilities.

Borders and Sanctuaries

During the Korean War, many aviators found themselves operating under orders that seemed diametrically opposed to the importance of what they were charged to do. Severe limitations on combat sorties forced airmen to do things that reduced success and increased risk. Early in the war, General MacArthur passed on to the FEAF commander orders from Washington which reminded everyone that air operations in North Korea were to stay well clear of the frontiers of Manchuria and the Soviet Union. In the following weeks we flew lengthy missions along the Yalu River to photograph bridges. It was important to be sure that the centerline of the reconnaissance pass was at a right angle toward China; a vertical photo across the border was worth a court martial. Aviators were forced to do ridiculous and hazardous things like lifting a wing on the border side and slipping the aircraft at slow speed down "MiG Alley" in hopes of securing an image of Antung airfield. Months later, when the Communists made a serious effort to obtain air superiority, the 5th Air Force commander acknowledged understanding of the international sensitivity but twice asked permission to pursue the enemy across the Yalu and destroy air bases. These requests were denied.

Late in 1950, I was again called back to Japan to take another look at the Soviets. I flew more reconnaissance flights than anticipated since my friend and classmate, who had split the prewar missions with me, had been killed in action by that time. The Soviets responded over time with increased air defense reaction from the mainland. Occasionally, it was necessary for shorter range USAF F-80Cs to come out and meet me about halfway home to discourage any Soviet La-11s that might follow. All these missions out of northern Japan were flown right across the borders and down the target runway. Back in Korea my flights reverted to the more difficult efforts to stay legal and still get useful photographs. Toward the end of my tour my squadron commander took over the clandestine missions.

My RF-4C tour in Southeast Asia brought similar problems years later. Although those pilots flying reconnaissance missions had fewer restrictions than airmen with weapons, the frustration was higher because too many of the latter were shot down by ground-fire directed from sanctuaries. Many others parachuted safely, only to drift across a "neutral" border and be denied air rescue. Scores were killed or captured as a result. The air war over Vietnam was made even more difficult by another factor. This was the insistence on using detailed mission plans prepared at high level in Washington. Very often this meant less effectiveness and was more costly. Reconnaissance F-101s were sent to Hanoi every day, for example, at the same time and altitude. They even were forced to use the same call sign. Not surprisingly, several were lost in succession. It took too long but eventually scheduled mission times and call signs were varied.

It seems clear that air power's most spectacular successes have been obtained when air forces have been allowed to be operated by on-the-scene professionals with minimum interference. On the other hand, serious problems and meager results inevitably result when air operations have been directed, or constrained, by "absentee," and, apparently, unqualified authorities. Nevertheless, despite occasional concerns about civilian control of the military, American airmen have consistently operated with the highest degree of professionalism and patriotism. Can there be a more compelling example in history, than that of pilots, whether over the Yalu or Hanoi, who saw the fireball that was their comrade-in-arms, downed by guns or missiles from sanctuary, had weapons aboard, and did not destroy those targets because their orders forbade it?

The first-rate performance of the United States Air Force during the difficult combat actions in Korea and Vietnam and the successes of the Gulf War have clearly prepared it for its role as the preeminent aerospace force in the world today.

Air Power in the Cold War

R. A. Mason

> Combat is the only effective force in war; its aim is to destroy the enemy's forces as a means to an end that holds good even if no actual fighting takes place because the outcome rests on the assumption that if it comes to fighting, the enemy would be destroyed.
> —Carl von Clausewitz, *On War,* Book One, Chapter Two

The Cold War aligned the United States, its NATO allies and sympathizers on the one hand, against the USSR, its Warsaw Treaty allies and sympathizers on the other, in a confrontation which dominated international relations for forty-five years: more than half the total existence of air power. Between 1945 and 1990, the air forces of both sides experienced combat, but with rare individual exceptions, not against each other. Lessons were learned, and sometimes expensively relearned and revised by the superpowers and their friends in Korea, Vietnam, Afghanistan, the Middle East, and the Falklands. Experience in these conflicts modified and occasionally deflected the evolution of air power doctrine, procurement, and operations in the two alliances but the constant overriding influence on air power throughout the period was the contest for putative superiority between them. This essay suggests that the impact of air power on the establishment, progress, and outcome of confrontation was always dominant, and, in at least two critical periods, decisive.

The Strategic Environment in 1945

The strategic environment in 1945 was overshadowed by the atomic bombs on Hiroshima and Nagasaki. During World War II, air power had already strongly influenced the outcome in almost every theater of operations except in the battles of attrition between German and Soviet armies. With the advent of "the bomb," the forecasts of two generations of air power protagonists appeared to have finally materialized. General Carl Spaatz encapsulated the view that strategic air power was:

> the most powerful instrument of war thus far known, because of its ability to concentrate force from widely dispersed parts on to specific targets, because it could penetrate deeply to destroy vital targets beyond the reach of armies and navies and because it could be economical in the force required to concentrate on a limited number of vital target systems, in sum, strategic bombing was the first war instrument of history capable of stopping the heart mechanism of a great industrialized country.[1]

Nor were other generals or admirals under any illusions about the implications of air power for their own activities. General Omar Bradley observed:

granting the axiomatic and supreme importance of air superiority. . . . A proper conception of the term regards it as securing control of the air in order to ensure the unrestricted use of that element in carrying out defensive operations against the enemy not only in the air but on land and sea.[2]

British Admiral of the Fleet Peter Hill-Norton subsequently wrote, "One lesson [of World War II] was that no naval commander could hope to survive long without air power while his fleet was within range of enemy air power."[3]

Before 1939, a study of international air power would have covered the air and naval forces of several countries, including Britain, France, Germany, Italy, the Soviet Union, Japan, and the United States. By 1945 the vanquished nations, Germany, Japan, and Italy had been deprived of the ability to exercise any air power. France and the countries of Western Europe were emasculated by five years of German occupation. The opportunity to exploit air power's postwar potential was confined to the three victorious nations: the United States, the Soviet Union, and Great Britain. Of the three, only the United States was in a position to take full advantage of the opportunity.

Britain still displayed the trappings of great power—its mid-twentieth century empire and protectorates girdling the globe. In reality, it was economically exhausted, with an immediate impact on its military strength. Within eighteen months the Royal Air Force (RAF) was reduced from a total of 55,000 to a little more than 1,000 frontline aircraft. International leadership in jet fighter aircraft was eroded by constrained resources; reequipment of fighter squadrons was slowed down and the entry into service of a jet bomber was postponed until 1951. No long-term postwar military planning had taken place and Britain had been excluded from participation in the American atomic weapons program. The international prestige and professionalism of the wartime RAF would be carried forward, but there would be little impetus for dynamic exploitation of air power from the descendants of Trenchard.

Eastern Europe was occupied by the armies of the USSR. The economic devastation and dislocation suffered by the Soviet Union between 1941 and 1944 was largely concealed from the West. Conversely, its aircraft industry was perceived to be producing 42,000 aircraft a year before four-fifths of all German aircraft production fell into Soviet hands in 1945.[4] Some 300,000 German missile and aircraft technicians and engineers were believed to have been transported to the Soviet Union, while German surface-to-air, air-to-air, and surface-to-surface missiles, jet aircraft, jet and rocket aircraft engines with their construction lines and development plans were all extracted for installation, development, and production back in the USSR.[5] Nonetheless, it was not the 20,000 aircraft believed to be in the Soviet Air Force in 1945 but rather the 3 million men in 100 or more divisions, many still deployed in Europe, who projected Soviet power after 1945. The Soviet Union would make gigantic strides to catch up with Western air power but with one important exception, in the development of surface-to-air missiles, it would not determine the direction of future air power evolution.

The United States, on the other hand, was uniquely placed to exploit the opportunities of the World War II air power legacy. In 1944, it had produced 93,623 aircraft in an aircraft industry which was becoming politically and economically influential.[6] In 1945, it had more than 31,000 combat aircraft but whereas those of the USSR were largely single-engined and obsolescent in airframe design and avionics, American aircraft were almost entirely modern, with a large long-range bomber and escort fighter element. Within twelve months, however, U.S. forces in Europe had reduced from 3,100,000 to 391,000[7] and U.S. Army Air Forces (USAAF) demobilization was proceeding at such a pace that General Henry H. Arnold and USAAF colleagues were concerned at the residual war-fighting capacity of the service.[8]

But above all, there was considerable awareness in the United States of air power's potential. The most serious debates were no longer over air power's effectiveness, but about whether it should be the monopoly of the USAAF.[9] The formulation of a coherent USAAF doctrine of strategic bombardment and preparations for the procurement of a long-range bomber force have been examined elsewhere in this volume.[10] It was a doctrine particularly appropriate to the contemporary security requirements of the USA. After World War II, the United States had no formal commitment to the defense of Western Europe and the only conceivable threat could emanate from the USSR. The Soviet Navy was insignificant and had no blue water capability although it was believed that a submarine building program had started, stimulated by the capture of German U-boats, designs, and tooling. There was no way, however, that the overwhelming numerical superiority of Soviet ground forces could be brought to bear against the North American continent. The only way in which the USSR could threaten the United States was by air. General Arnold testified to the U.S. Senate Military Affairs committee in October 1945:

> New weapons can come through the air unheralded and unannounced. The basic defense against such a plan of attack must lie in the ability to mount rapid, powerful offensive action against the source from which flight begins. . . . The defense has got to be an offensive mission against the source. But, better still, the actual existence of these weapons . . . in sufficient quantities and so located that a potential aggressor knows we can use them effectively against him, will have a very deterring effect.[11]

Three weeks later, General Spaatz testified, "The next war will be preponderantly an air war. . . . Attacks can now come across the Arctic regions, as well as across oceans, and strike deep . . . into the heart of the country."[12]

Within these two comments lay the roots of U.S. air power doctrine which were to permeate the Cold War: the concepts of deterrence and defense by destruction of the enemy's capacity for offensive action, although not necessarily always in that order. National defense based on the long reach of strategic bombers and the power of atomic weapons would capitalize on U.S. technological superiority, exploit the intrinsic difficulty of the USSR to protect its vast territory against air attack, and remove the requirement for U.S. forces to become involved in a ground war of attrition. It remained to be seen whether such a strategic posture would be equally appropriate for an alliance committed to defending territory in Europe.

Strategic bombardment was the most prominent air power issue in 1945, but in the legacy of World War II there were three others which were to impinge heavily on U.S. air operations and the doctrine supporting them. Since Manfred von Richthofen was shot down by Australian soldiers in World War I, command of the air had not been the sole prerogative of the fighter pilot. Anti-aircraft ground fire, for example, had taken a heavy toll of RAF Bomber Command in World War II. Now on the threshold of development, the guided surface-to-air missile (SAM) would complicate the air power offensive-defensive equation still further.

The second innovation, the surface-to-surface missile (SSM), had been extensively used as the V-1 and V-2 by Germany. Among Soviet missile "reparations" were plans for an SSM with intercontinental range. Would such a weapon complement or replace the strategic bomber? Was it a further extension of air power? Or was it simply an extension of the artillery shell? Still further away from significant military deployment was the helicopter. Where did that fit in the air power spectrum? Was it an extension of battlefield mobility and firepower, or a further addition to short-range air power? In a relaxed academic environment, free from budgetary constraints, interservice rivalries and an incipient foreign threat, an objective debate on these legacies from World War II could perhaps have taken place. Like the aircraft carrier/B-36 controversy however, they were inseparable from their associated single service interests.

Between 1945 and 1948, the prospect of war between the United States and the USSR moved from the realm of abstract threat assessment to a dangerous possibility because of events in Europe. As early as October 1945, General Spaatz had expressed prescient unease in a letter to General Arnold: "With the rapid weakening of our forces in Europe and Asia, the USSR is able to project moves on the continent of Europe and Asia which will be just as hard for us to accept and just as much an incentive to war as those occasioned by German policies."[13]

His foreboding was proved to be well founded as the USSR consolidated its military hold on Eastern Europe, encouraged Communist insurrection in Greece, created the Cominform to oppose the implementation of the Marshall Plan and to coordinate the activities of Communist parties internationally in support of Soviet policies, fostered a Communist coup d'etat in Czechoslovakia and finally, in April 1948 began to restrict access to the Western enclave in Berlin.

The Berlin Airlift, 1948–49

Until formerly Soviet documents are released, the exact motives of the USSR in seeking to strangle the Western sectors in Berlin, from June 1948 to May 1949 will remain uncertain. But in the light of subsequent events in the Cold War, the full significance of the Berlin airlift can be evaluated. Four days after the beginning of Soviet interference with rail, barge, and road traffic between the Western zones of Germany and the Western sectors of Berlin, Britain began contingency planning for the air supply of the British garrison only, in case of complete blockade. One day later, on 5 April, a dangerously provocative Soviet Yak fighter collided with a British European Airways passenger plane over Gatow Airport in the British sector, killing all on board both aircraft. The British government immediately allocated RAF fighter escorts to British transport aircraft; an example swiftly followed by the Americans.

Nonetheless, the USSR steadily increased restrictions on traffic throughout April and May, finally halting on 24 June all surface movement from West Germany and rail traffic from the Soviet sector of Berlin. It was estimated that the Western sectors had stocks of food for 36 days, coal for 45 days, diesel for 4–5 months, motor oil for 3–4 months, and gasoline for 7–8 weeks.[14] The Western sectors imported 12,000 tons of supplies daily, of which 75 percent came from the West. They consumed 2,000 tons of food a day. A DC-3 Dakota could lift 2½ tons. There were one hundred U.S. Air Force (USAF) DC-3s and two C-54s in the theater, the RAF had six Dakotas and were about to move another eight over to Germany from Britain. Nonetheless, USAF and RAF relief flights along the air corridors to Berlin began on 25 June. On 26 June, British Foreign Secretary Ernest Bevin issued a statement of British determination to remain in Berlin. Two days later, his position was endorsed unanimously in Parliament, which accepted the risks of war because the only alternative was unacceptable surrender.[15] For good measure, Bevin warned Russia that any interference with the air corridor "would be treated as acts of war."[16] On 30 June, Secretary of State George C. Marshall publicly proclaimed U.S. determination to uphold the right to stay in the city, asserting that air transport would supply Berlin.[17] The U.S. commander in Berlin, General Lucius Clay believed that only a determined show of strength, by convoys with troop protection forcing a passage along the *autobahn* from West Germany, would prevent further Soviet pressure and avert war.[18]

The spirit displayed by the British foreign secretary and U.S. army general was exemplary, but was not consistent with the balance of conventional forces in the region. There were 6,500 allied combat troops in the Western sectors of the city facing 18,000 Soviets, backed by a further 300,000 in the Eastern zone of Germany. The British chiefs of staff advised the cabinet that "Our forces in their present state are not in a position to fight with what we have got."[19] Secretary of State Marshall estimated that the United States would require months to prepare for a war over Berlin.[20]

It seemed therefore that an attempt to force the blockade by land would risk provoking a conflict which even if restricted to Germany would almost certainly result in the military capture of Berlin by the Soviet Union. President Harry S. Truman was not prepared to take that risk, and on 22 July rejected General Clay's proposals and directed that maximum effort be allocated to the airlift. Meanwhile, one other highly visible response by the Allies to the imposition of the blockade had been taken. By the end of 1946, the USAF had one group of 27 aircraft modified to carry atomic bombs. Atomic weapon stocks rose from 2 in 1945, 9 in 1946, 13 in 1947, 50 in 1948, to 250 in 1949.[21] On 25 June, Foreign Secretary Bevin sought the forward deployment of U.S. fighters and heavy bombers to Europe, if necessary based in Britain.²² The deployment of B-29s to Europe was discussed in Washington on 27 June, and on the following day President Truman authorized the deployment of one group to Britain and the completion of a second group's deployment to Germany, where one squadron of 301st Bombardment Group was already on temporary duty. Two squadrons deployed to Germany on 2 July and two more arrived in Britain on 17 July, where their arrival was widely publicized and, with the exception of Communist sympathizers, unanimously welcomed. The B-29s were believed to be nuclear capable, although subsequent research has disclosed that they were not in fact nuclear armed.²³ In any event, the deployment was seen as an unambiguous signal of U.S. commitment to Berlin.

The deployment may well have discouraged any Soviet inclination to interdict the airlift. In the light of the size of subsequent nuclear arsenals on both sides, a handful of atomic bombs under the personal control of a president who was well known to be reluctant to consider their use might not seem a powerful instrument. To the Soviet Union at the time, however, even fifty atomic attacks must have threatened industrial and political obliteration. On the other hand, while the B-29 deployments may have induced extreme caution in Moscow, they did not deter the Soviet Union from maintaining the blockade. After July 1947, only the airlift stood between a continued Western presence in Berlin and humiliating concession to Soviet economic and political control of the city, entailing immeasurable psychological and political damage to the West.

Little by little, the ad hoc flights of June gave way to an increasingly well-coordinated Anglo-American airlift. Daily deliveries gradually increased from 1,404 tons during the last week of June 1948 to a daily average in April 1949 of 7,845 tons. The next generation of transport aircraft, C-97s and C-124s, could have sustained the airlift indefinitely. Every conceivable item was flown in: flour, vegetables, dried milk, coal, fuel oil, machinery, construction equipment, darning and sewing needles—space was even found to fly in a replacement goat mascot for the 1st Battalion Welsh Fusiliers. By the end of April 1949, the Soviet Union realized that it had failed to strangle the Western sectors of the city and on 12 May 1949 formally ended the blockade. For a further four months, the airlift continued to rebuild reserve stocks in the city. The last flight took place on 30 September, by which time 2,325,808 tons of supplies had been airlifted by U.S., British, and latterly, French planes.

At the conclusion of the Berlin Airlift, the city was still a hostage deep inside Soviet occupied territory. The USSR could at any time, in theory at least, have reimposed a blockade or applied other kinds of pressure. Nor had success been easily won. Fifty-four American, British, and German lives had been lost. Heavy maintenance costs were incurred by obsolete aircraft flying continuous short distances for which they had not been designed. Engines were subject to frequent surges rather than long-range cruising, incessant landings with heavy loads wore out airframes, hydraulics, brakes, and tires prematurely. Britain was unable to find enough spares and tools and was constantly short of air crew, drawing upon training crews and thereby compounding the longer term problem. British reservists had to be called up and demobilization stopped to provide sufficient skilled ground crew. The United States, on the other hand, drew upon much greater resources to maintain a flow of air crew and spares, while augmenting ground crew with ex-Luftwaffe personnel. Operationally, the airlift was a reminder that air power depends for its effectiveness on far more than numbers of available aircraft.

With hindsight, the historical significance of the airlift can be fully acknowledged. At the time, air force commanders were naturally concerned about the diversion of transport aircraft to the airlift from their commitment to supporting heavy bomber deployments. The limited number of B-29s equipped to deliver the atomic bomb depended on transports to carry unassembled bombs and supporting equipment to preselected bases in Britain, Germany, the Middle East, and Okinawa. If they had been employed on the airlift when the USSR launched a surprise attack, the USAF's ability to respond would have been severely impaired. In the event, Berlin was sustained and protected by air power to remain an irritant to the USSR for forty more years before its toppling walls presaged the collapse of communism itself.

During the airlift, the Western allies moved from the position of occupying forces in West Germany to become protectors. West Germany itself was enabled to adopt a new democratic constitution on 9 May 1949. The United States demonstrated its recommitment to the freedom of Europe less than three years after the end of World War II. The determination and cooperation shown in the airlift gave additional impetus and confidence to the creation of the North Atlantic Alliance on 4 April 1949. Air power, through its unfashionable, unglamorous transport arm, had won the first decisive battle in the Cold War and checked the spread of Soviet control in Europe.

The Establishment of NATO Strategy

On 17 March 1948, before the Soviet Union imposed the blockade on Berlin, Britain, France, the Netherlands, Belgium, and Luxembourg had signed the Treaty of Brussels, pledging to strengthen economic and cultural ties and to establish a joint defensive structure. On 30 April, the first of many meetings was held in London of the five defense ministers and their chiefs of staff to determine military requirements. From July onward, these meetings were also attended by U.S. and Canadian representatives. In June 1948, the way opened for formal association between the United States and the European allies when the U.S. Senate passed Resolution No. 239, better known as "the Vandenberg Resolution." It included a recommendation for "the association of the United States by constitutional process, with such regional and other collective arrangements as are based on continuous and effective self help and mutual aid, and as affect its national security."[24]

Ten months' negotiations on the content of a treaty of common defense culminated in the signature of the North Atlantic Treaty on 4 April 1949 in Washington by the foreign ministers of Belgium, Canada, Denmark, France, Iceland, Italy, Luxembourg, the Netherlands, Norway, Portugal, the United Kingdom, and the United States.

The treaty included some reassuring sentiments. Article I expressed the signatories' resolve "to unite their efforts for collective defense and for the preservation of peace and security." Article 3 affirmed a collective determination to "maintain and develop their individual and collective capacity to resist armed attack." Article 5 expressed a resolve, in the event of an attack on any signatory, to "assist the Party or Parties so attacked by taking forthwith, individually and in concert with other Parties, such action as it seems necessary, including the use of armed force, to restore and maintain the security of the North Atlantic area."[25]

These agreements were reached during the Berlin Airlift, at a time when both Britain and the United States had privately conceded the difficulties of going to war with the USSR. There was also considerable military disparity between the members of the Alliance. How, then, should NATO prepare to implement the aspirations expressed in Articles 1, 3, and 5?

In December 1948, the U.S. Joint Chiefs of Staff (JCS) had accepted an emergency war plan compiled by Strategic Air Command (SAC). Soviet air defenses were believed to be susceptible to jamming; surface-to-air weapons were ineffective over 30,000 feet; there were no all-weather fighters, no effective ground control system, and no electronic counter-measures. The Soviet bomber force could not yet reach B-29 bases in sufficient strength to check the offensive and even accepting 25 percent attrition all available (50) atomic weapons could be dropped on "the major Soviet urban industrial concentrations."[26] In October 1949, Chairman of the JCS, Gen. Omar Bradley addressed the House of Representatives Armed Services Committee during the hearings on national strategy and unification policies. After referring to the impact of strategic bombing in World War II and to the Soviet Union's creation of a strategic bombing force of its own, he observed:

> Strategic bombing has a high priority in our military planning because we cannot hope to keep forces in being of sufficient size to meet Russia in the early stages of a war. This is particularly true since we are never going to start the war, and the Soviet Union ... can choose the date of starting it. Lacking such forces in being, our greatest strength lies in the threat of quick retaliation in the event we are attacked. Along with many others, I believe that the atomic bomb which has been derided [by US Navy spokesmen], and the Air Strategic Command which has been denounced, have contributed to the avoidance of war during the last couple of years. This combination has been, in my opinion, one of our greatest deterrents to aggression, both here and in Europe.[27]

General Bradley was naturally involved with the contemporary formulation of NATO strategy. In the same testimony to the House of Commons he said, "The basic principle of the North Atlantic Treaty and of the mutual defense assistance program is that each nation shall contribute those things which it can best provide in the collective security plan."[28] Two months later, that "basic principle" was formalized secretly as one of six principles underlying the first North Atlantic Strategic Concept of 1 December 1949.[29] Certain nations, because of their geographical location or because of their capabilities, would be prepared to undertake appropriate specific missions. Specifically, the "Basic Undertakings" included:

Ensure the ability to carry out strategic bombing promptly by all means possible with all types of weapon, without exception. This is primarily a U.S. responsibility assisted as practical by other nations. . . .

Neutralize as soon as practicable enemy air operations against North Atlantic Treaty powers. In this undertaking the European nations should initially provide the bulk of the tactical air support and air defense. . . [and] the hard core of ground forces.

Sea power would be the joint responsibility of the United States and Britain.[30]

Strategic air power, therefore, was to be but one component along with surface forces and tactical air in this first Alliance strategic concept. The implementation of the concept was, however, to prove beyond the inclinations and capabilities of the Alliance membership. Britain had no intention of becoming dependent on U.S. air power for its security. The British Chiefs of Staff had defined their requirement for atomic weapons in 1946, arguing "that Britain, as a great power could not leave her security in the hands of the Americans who, however friendly, could veer so unpredictably from generous international collaboration to self-centered isolationism."[31]

Inevitably, such a commitment impaired Britain's capability to meet its conventional commitments under the NATO concept. In 1949, the Soviet Union exploded its own atomic device. In the same year, Mao Tse-tung completed the Communist seizure of power in China. In June 1950, the Korean War began and was widely seen as further coordinated Communist aggression, probably designed to distract from Soviet destabilization, or worse, of Western Europe. In the United States, the need to increase overall defense provision was argued in the strategic document NSC-68 in April 1950. Paul Nitze, director of the State Department's Policy Planning Staff which actually compiled NSC-68, subsequently explained that their greatest concern had been

the Soviet threat to Europe. Our NATO allies needed reassurance that the balance of power was not tipping in favor of the USSR. Now that the Soviets had acquired an atomic capability, NATO could not for the indefinite future continue to rely primarily on the threat of nuclear retaliation or, if necessary, to repel a Soviet invasion. We had to strengthen our position with other means, and this meant a build up of conventional forces.[32]

For the next eighteen months, the Allied military staffs worked out the conventional force levels necessary to implement the strategic concept. At Lisbon in February 1952, according to the official communiqué,

The decisions taken by the [North Atlantic] Council provided for the earliest build up of balanced collective forces to meet the requirements of external security within the capabilities of member countries. Agreement was reached on the specific defensive strength to be built this year and on a definite program of measures to be taken this year to increase defensive strength in following years.[33]

The unpublicized force requirements for this reassuring build up included ninety-six ground force divisions and 9,000 combat aircraft by 1954. No member of the Alliance would meet its force goals. One commentator observed, "The military authorities called for contributions which the economic conditions of the allies could not sustain. There was a risk that the warrior could be crushed by his own armor. The effort involved in surviving threatened to destroy the survivors."[34] Nor, despite the unanimity of the Lisbon communiqué, were all Alliance partners convinced of the need. Winston Churchill, for one, saw no reason to superimpose a conventional force plan on what seemed a favorable Western balance of power based on atomic weapons and air power.[35] British defense expenditure was cut rather than increased while France, heavily preoccupied in Indochina, was unable to increase its forces in accordance with the Lisbon directive.

Fortuitously, during 1952 the U.S. Atomic Energy Commission proved the feasibility of producing smaller atomic weapons more cheaply. Reduced weight and reduced yield enabled them to be carried by smaller aircraft and employed against armed forces as well as large-scale industrial targets. Their development prompted a new concept of USAF tactical air power in June 1953. Three years previously the mission of tactical air power had been related to the strategy and maneuver of ground forces.[36] The revised regulation defined tactical air operations

> as the application of all airpower, under the command or operational control of a theater or area commander, against an enemy's military potential and capabilities in being. . . . Restricted only by limitations of equipment and capabilities of designated units, tactical air operations may encompass any task necessary in the furtherance of the theater mission.[37]

Twelve months later, the changed responsibilities of tactical air power were publicly and unambiguously specified by the USAF:

> The formidable nature of this new source of firepower in fact reverses the orthodox relationship of air and ground forces. Specifically, it is quite reasonable to say that we should look for a modification in our tactics and in our concepts of war . . . which would point toward the exploitation of tactical air atomic attacks by highly mobile ground forces.[38]

The expansion of the roles of tactical air power reflected a major revision in U.S. defense policy. President Dwight Eisenhower had become concerned about the costs of sustaining large conventional forces and was conscious of domestic opposition to committing American troops to another protracted conventional conflict in Korea. He was a strong advocate of air power and the deterrent value of nuclear weapons. In October 1953, the new basic national security policy was enunciated in NSC 162/2. The Joint Chiefs of Staff were directed to base their planning on the use of nuclear weapons in either general or "local" war, whenever their employment would be "militarily advantageous." Local conventional defense would still be required in Europe, but "the tactical atomic support which can be provided to our allies will become increasingly important in offsetting present deficiencies in conventional requirements."[39]

Secretary of State John Foster Dulles explained the implications of this "new look" to a restricted session of the NATO Council in April 1954. He emphasized the need to deter Soviet exploitation of superior numbers and the option of surprise attack by threatening retaliation by other means at different locations. He argued that NATO would be politically and economically inhibited from matching the conventional force strength of the Soviet bloc and stated that the United States believed that the use of atomic weapons as conventional weapons was essential for the defense of the NATO area. He then recommended that NATO "agree to use the atomic weapons or conventional weapons . . . whenever and wherever it would be of advantage to do so."[40]

NATO ministers approved the concept and nuclear basis for Alliance defense planning in December 1954 as MC 48. Two years later, the North Atlantic Council formally adopted the strategic concept in Military Committee document MC 14/2, known popularly as "Massive Retaliation." The U.S. Army had fired an atomic artillery shell in May 1953, and later that year the first atomic artillery units were deployed to Europe. It was, however, tactical air power which was given the greatest prominence in the revised policy. Again there had been an elision from USAF concepts into NATO strategy. Both would subsequently be complemented by the advent of intercontinental and intermediate-range surface-to-surface missiles.

It may be argued that the circumstances of confrontation have dictated the evolution of air power since 1945, but events in the period 1947–54 suggest that the pattern of confrontation was in fact determined by NATO's adoption of air power and nuclear weapons as the foundation of its strategy of containment, when adequate conventional forces proved impossible to procure. In sum, there would have been no credible NATO strategy without air power.

It was, however, a very constrained form of air power. The nuclear weapons of tactical aviation would complement those of the Strategic Air Command. Previously, the U.S. Army had on several occasions expressed dissatisfaction with USAF procurement for tactical air support. Such arguments were now irrelevant and interservice squabbles overridden. "Lessons" from Korea were not forgotten but simply discarded, because they were operationally and politically so far removed from the "real" environment in Europe. More insidiously, nuclear thinking would come to dominate concepts, tactics, training, aircraft and systems procurement. Air power in a nuclear war did not require precision weapons, or an attritional battle to establish air supremacy, or replacement crews. It made complete strategic sense at the time, but it placed in jeopardy that most prized air power characteristic—flexibility.

The Role of Reconnaissance

When the Strategic Air Command was activated in March 1946, it was allocated three responsibilities. First, it was to be ready to conduct long-range offensive operations independently or in cooperation with land and sea forces. Second, it needed to be able to conduct maximum-range reconnaissance operations. Finally, it was ordered to provide units capable of operations employing the latest and most advanced (that is, atomic) weapons.[41] The command had a front-line strength of 148 B-29 bombers, 85 P-51 escort fighters, and one reconnaissance squadron of modified B-29s.[42] During the next decade, B-29s would be supplemented by B-36s and ultimately both superseded by the B-47 and B-52 as SAC expanded to over 1,300 aircraft.

Under the dynamic leadership of General Curtis LeMay and with priority government funding, early operational deficiencies were corrected and nuclear weapon shortfalls were removed. But however many aircraft, however powerful the weapons, and however accurate the bombing, targets had to be located and reached. In World War II, reconnaissance had been a valuable but secondary complement to strategic bombing. The industrial geography of Germany and Japan was well known to the allies. Not so the USSR after 1945.

The area of the Soviet Union was 8,649,500 square miles. It was a closed society and the only maps available to SAC were based on Luftwaffe photographs largely restricted to the USSR west of the Urals, supplemented by some Tzarist regional maps and even a 1912 Baedeker railway timetable. There was little knowledge of new industrial centers and military installations. To fulfill their mission, SAC crews would be required to bomb from high altitudes at night and in all weather beyond the limits of even unjammed navigation beacons. Bombardiers had to be able to identify their targets from radar and visual photographs. Initially, reconnaissance missions were restricted to the periphery of the Soviet Union, relying on camera and radar slant ranges to look up to 100 miles inside frontiers and coastlines.

Strategic reconnaissance was first directed at the new "no man's land" of the Arctic Circle. In June 1946, RB-29s of the USAAF's 46th (later 72nd) Reconnaissance Squadron began radar and visual photography of the ice cap to identify any undiscovered land areas, of Soviet coastal areas and of airfields, submarine bases, and other military installations from the Kamchatka peninsula in the far east to Novaya Zemlya Island in the northwest. During these flights, techniques of grid navigation were perfected which ultimately were to facilitate transpolar civil aviation as well as routing for SAC bombers. In complete radio silence on missions up to 5,000 miles and over 30 hours, the RB-29s also collected data on the Soviet air orders of battle and air defenses. By the end of World War II, electronic warfare had permeated all aspects of air operations. Now Soviet radar and other electronic emissions were monitored to establish their technical characteristics and source.

Soviet air defenses were still thinly scattered and early USAF Electronic Intelligence (ELINT) flights probed the radars to identify wave-lengths and gaps in their coverage.[43] Thereafter, electronic countermeasure equipment could be designed and fitted to SAC aircraft. A modified B-29 and ELINT B-17s flew in the Berlin Airlift streams gathering data on Soviet radars in Eastern Germany. Such information also disclosed the command structure and deployment of air defense and other Soviet units. The most important single discovery occurred on 3 September 1949, when a specially equipped aircraft detected evidence of a Soviet atomic explosion somewhere in Asia. The impact on U.S. foreign, defense, and procurement policies was dramatic because the USSR had not been expected to develop an atomic bomb for at least four more years. By 1952, the spread of Soviet surveillance radars in Eastern Europe, along the Black and Baltic coasts, and across northern Siberia, plus the mass production of fire control radars was being progressively observed by frequent reconnaissance missions around the Soviet periphery. U.S. naval aircraft, for example, flying from bases in Spanish Morocco monitored emissions as far apart as the

Black Sea and the Gulf of Finland. But as long as flights were restricted to the periphery of the Soviet Union, SAC would be unaware not only of targets but also of the depth and quality of Soviet air defenses. However, President Truman forbade U.S. overflights after the loss of a U.S. Navy Convair PB47-2 near the Latvian coast in April 1950.

British Cooperation

In 1951, Britain's own strategic bombing force was developing. While the RAF's requirement for information on the USSR was not as geographically extensive as that for SAC, General LeMay perceived a common interest. He persuaded the U.S. Chiefs of Staff to invite the RAF to cooperate in a reconnaissance project from which the intelligence product could be shared between the two countries.[44]

By July 1951 an agreement between the two air forces had been reached that RAF crews should fly three RB-45s in RAF markings over the USSR.[45] After training at Lockbourne Air Force Base, Ohio, and at Sculthorpe in Britain, the first sorties were flown over the USSR on 21 March 1952, after one "proving" flight up and down the Berlin corridor. One B-45 made a zigzag flight over the Kola Peninsula, the second covered the Baltic and Moscow and the third flew over Ukraine and the Caucasus.[46] Each flight was supported by air-to-air refueling before leaving and after returning to Western airspace. All were flown in darkness, beyond the range of western navigation beacons. Navigation depended on compass, timing, and H2S radar for ground feature identification plus 1:1,000,000 maps cobbled together from various sources. The lead navigator, Flight Lieutenant (later Wing Commander) R. S. Sanders, with the Special Duties Flight Commander, Squadron Leader J. Crampton, were briefed personally by the Air Officer Commander-in-Chief, Bomber Command, and the Group Captain (Intelligence) shortly before the mission. Each aircraft was equipped with an internal camera to photograph the Plan Position Indicator of the H2S radar. The product of the films was subsequently incorporated in SAC synthetic crew training.

No written instructions were given to the crews but the importance of accurate timing was stressed because the British electronic intelligence gathering "Y Service" and USAF ELINT aircraft on the periphery of the USSR were monitoring the flights. Indeed, it is probable that the responses of the Soviet air defense system to the intruders was of greater significance than the radar derived imagery. It is also probable that the zigzag courses flown by all three aircraft were designed to secure the maximum defensive responses with the minimum risk of interception. The crews were never told why their courses were so designed, nor the product of their missions.

Two years later, however, a second RAF sortie by three B-45s, on similar routes, met very different opposition. In 1952, Flight Lieutenant Crampton and his crew had heard nothing and seen nothing of the air defenses. On 28 April 1954, their route was similar: across the Kiev area, north of the Black Sea to within H2S range of the marshlands north of the Caspian Sea. This time their flight duration was to be 7½ hours at 480 mph flying at 31,000 to 37,000 feet. As before, they had a zigzag track and their flight was coordinated with Western ELINT gathering organizations.

In the space of two years, however, Soviet air defenses had expanded and improved. Near the end of their 20 point radar photography route, they were engaged by dense Soviet flak. Unknown to them, a MiG fighter had been scrambled to intercept and if necessary ram them.[47] The crew were under strict orders to abort their missions and return to base if "compromised." This they did, en route photographing several other ground positions and stimulating continued air defense interest as they did so.

In 1995, some of the agencies involved in exercise "Ju Jitsu"[48] remained tight-lipped about the product of the six RAF flights. The value of the radar photography may have been variable because at least two cameras had been inaccurately set up by the ground crews, but it is highly probable that the ELINT gave invaluable information to the USAF and RAF about the evolution of Soviet air defenses. The RAF crews were personally thanked by General LeMay and each, uniquely, received an Air Force Cross for each mission. No official citation was ever published for the awards, and inquisitive family and friends were informed that, "they were for experiments with inflight-refueling."

Contemporary correspondence between the USAF and RAF reveals the acute sensitivities of Ju Jitsu. The Korean War had not begun when the first sortie was flown. Winston Churchill's personal approval had been given and had any of the six flights ended prematurely over the USSR the political repercussions, especially in Britain, would have been disastrous, particularly had it been learned that President Truman had forbidden USAF aircraft to overfly the Soviet Union. The British Chief of the Air Staff had sought, and was given, reassurance from the USAF that the Soviet air defenses could not reach the B-45s at night. In the event they came dangerously close.

Churchill sought, through the RAF Chief of Air Staff, to persuade the USAF to overfly also, but President Truman remained adamantly opposed, although welcoming the British activity. These missions make it clear that there could have been a "Gary Powers" incident nine years earlier than when it actually occurred. Such an event, with its considerable propaganda value for the USSR, would have come at a time when the Atlantic Alliance was still in hesitant infancy. It is unlikely that a set of innocuous maps and a plea by any captured survivors that they were "off course" would have been very convincing. However, the missions were successful, there were no losses, and the secrets were so well kept that not even the crews' superior officers knew where they had been nor what they had done. As a result, the capabilities of SAC continued to be enhanced.

Meanwhile, an alternative USAF solution was to launch unmanned balloons carrying radar receivers and recording equipment from Western Europe to be carried by prevailing winds across the USSR to the Pacific area, where they would be tracked on U.S. radar and drop their equipment for retrieval on receipt of a radio signal. Known as "Grand Union," the project was completely haphazard and tracks unpredictable. How many were launched and recovered is not known, but some were recovered in the Soviet Union, with embarrassing public displays of the "weather information" equipment laid on in Moscow for the international press.[49] A further attempt was made in 1956 before U-2 operations began. To President Eisenhower's discomfiture, the results predictably repeated those of the previous fiascos.[50]

At least one overflight of Soviet territory was made by SAC in May 1954. An RB-47E from the 91st Strategic Reconnaissance Wing was tasked to photograph nine Soviet airfields near Murmansk and Archangel to discover whether the newest Soviet fighter, the MiG-17 had been deployed to the area, The MiG-17s announced their presence by attempting to intercept the USAF reconnaissance aircraft.[51]

The U-2

On this occasion the crew were able to bring their damaged aircraft back to its base in England. Other crews were not so fortunate. Between 1946 and 1960, at least thirteen U.S. aircraft and ninety air crew were lost on reconnaissance missions around or over Soviet and Soviet occupied territory.[52] While the RB-47 was evading MiG-17s at a height of 40,000 feet, Lockheed had begun developing an aircraft at Burbank, California, in 1954 in response to a joint Central Intelligence Agency/USAF requirement for a high-altitude strategic reconnaissance and research aircraft. The U-2's first test flight took place on 8 August 1955 and made its first Soviet overflight mission, high above Moscow and Leningrad, on 4 July 1956.

The contribution of the U-2 to Western security was immeasurable. For the first time the inner secrets of the Soviet military and military-industrial complex were laid bare. From an unassailable height of 70,000 feet the U-2's Hycon B camera covered a swathe 150 miles wide with high resolution and 750 miles for general reconnaissance.[53] In addition the U-2 carried ELINT collection systems, and brought back invaluable data on Soviet surveillance and fire control radars and communications. The U-2, for example, revealed an air defense radar system which was far denser and much deeper in Soviet territory than had been assumed. Bomber production capacity and deployments were surveyed, of particular importance after Soviet revelations of the M-4 and TU-16 aircraft in 1954 and 1955. It was found that Soviet emphasis in intercontinental bombardment was actually being placed on surface-to-surface missiles, but in 1958 only six ICBM launches had been made—too few to presage the feared deployment by 1960 of one hundred or more.[54]

New Soviet SAM construction and deployments were located. In 1957, the SA-1 became operational in rings around Moscow and Leningrad while the SA-2 was displayed in the November parade in Moscow later in the same year. It was essential that SAC should be given full operational data on which to base ECM equipment and tactical response. As a result of U-2 reconnaissance, the SA-1 was seen to present little threat, but the 60,000 ceiling of the SA-2 prompted the revision of B-47 operations to low level. B-52s were to continue at high altitude, relying on U-2 derived, constantly updated, ECM protection. Prior to the arrival of the U-2, Soviet nuclear weapons development and testing had been monitored on the peripheral flights. Now it regularly monitored the weapons plant at Alma Alta, missile test centers at Tyuratam and Kapustin Yar, and nuclear test sites at Novaya Zemlya, Semipalatinsk, and in China at Lop Nor. Overall, it was estimated that as a result of information brought back by the U-2, SAC's list of potential targets in the USSR grew from three to twenty thousand.[55] U-2 derived ELINT would have been indispensable had SAC aircraft ever had to attack them.

U-2 overflights of the USSR abruptly ceased when Gary Powers was shot down over Sverdlovsk on 1 May 1960. Later that year satellite overflights began. The U-2, however, continued to provide invaluable intelligence in other regions.

The Cuban Crisis

During 1962 the U.S. government became increasingly concerned about Soviet military assistance to Cuba. On 29 August, U-2 photographs revealed the construction of SA-2 sites on the island. By this time U.S. photo interpreters were familiar, from U-2 imagery of the USSR, with the ground configuration of both surface-to-air and surface-to-surface sites. In September a site similar to those identified in the USSR for medium-range surface-to-surface missiles (MRSM) was discovered. On 14 October, a further photograph of the site revealed the presence of MRSM missile transporters, missile tents, missile fuel storage, and other MRSM equipment. On 17 October, several other missile installations were discovered with at least sixteen and possibly thirty-two missiles. On 18 October, Soviet Foreign Minister Andrei Gromyko assured President John F. Kennedy that the USSR would never place offensive weapons or missiles in Cuba.[56] Four days later, President Kennedy broadcast the details of the crisis and the following day released the U-2 photographic evidence which had considerable international impact with its corroboration of the U.S. version of events. Meanwhile, the B-52 force was armed and placed on airborne alert while the B-47 squadrons were dispersed to airfields and airports around the country.

Subsequent U-2 flights and low-level reconnaissance revealed that work on the missile sites was being accelerated and that IL-28 nuclear-capable bombers were being assembled on Cuban airbases. On 27 October, a U-2 was shot down over Cuba and the pilot killed. Robert Kennedy summoned the Soviet ambassador and warned him of the consequences of any further interceptions. He observed succinctly, "If we had not violated Cuban airspace we would still be believing what [Nikita] Khrushchev had said—that there would be no missiles placed in Cuba,"[57] with the near certainty of a Soviet fait accompli and a far more serious risk of unpredictable military conflict. From that day the crisis began to be defused, but it was the occasion when reconnaissance emerged from its supporting air power role to international recognition as a major contributor in its own right to national security. In fact, it had been making such a contribution for several years, but in a highly classified environment.

From the days of the RB-29s of 46th/72nd squadron, through the activities of RAF RB-45s and Canberras, the peripheral operations, and finally to the U-2 overflights, reconnaissance had provided indispensable intelligence to SAC: target locations, air defenses, Soviet ORBATS, and raw material for ECM. In 1975, Central Intelligence Agency Director Richard Helms attributed to the U-2 90 percent of the intelligence required to produce the American estimate of Soviet strength.[58] Aerial

Cold War 253

photography in the 1950s was held to be as important as codebreaking had been in the 1950s,[59] while, according to Helms, "an enemy's electronics order of battle was the most important thing [when SAC was] trying to exploit gaps in the radar cover and missile defenses."[60]

This was an age when it was believed in the West, with a certain amount of justification, that a surprise air attack by nuclear weapons could determine the outcome of a war between the superpowers. U.S. military intelligence assessments and forecasts had to assume the worst case—anything less could have proved disastrous. The military industrial importance of the reconnaissance was that it allowed equipment procurement to be accurately focused on Soviet strengths and weaknesses. The political and psychological benefit resulted from a confident awareness of the Soviet capabilities and level of preparation for long-range air attack, and later, long-range missile attack. The USAF, for example, overestimated Soviet bomber production by 100 percent, while the "missile gap" so feared between 1958 and 1960 and so exploited politically, was known by the Eisenhower administration not to exist, despite the dramatic and debilitating appearance of the first two Sputniks in 1957. Consequently, whatever the public perceptions, the U.S. government, thanks directly to air power in its reconnaissance role, was able to maintain a confident, stable international posture. Unfortunately, the single spectacular and diplomatically disastrous loss of Gary Power's U-2 in 1960, held responsible for the breakdown of détente, was given international publicity. The continuous contribution to international stability of reconnaissance wrapped in secrecy for several years remained unheralded for far too long. Indeed, without the stimulus of nuclear confrontation, it is unlikely that the role, and its benefits, would have advanced so far so quickly. Those benefits were to be extended to subsequent operations and interests in smaller operations and crises down to Desert Storm in 1991.

In-flight Refueling

Nuclear confrontation stimulated a second air power role which, like reconnaissance, grew rapidly from a supporting arm of SAC to a multirole force multiplier. Development of in-flight refueling (IFR) had been suspended during World War II because it was not considered practicable for the numbers of aircraft involved in Allied bombing raids and because the unrefueled range of the B-29 rendered it unnecessary for attacks on Japan. With the end of the war, there was no obvious reason to recommence development and procurement of air-to-air refueling. Its introduction into military operations was by no means inevitable. But even the B-29 and its successors could not reach targets deep inside the Soviet Union. In peacetime, forward air bases were available in Western Europe, North Africa, the Middle East, and along the Pacific Rim. In war, however, many of these could become vulnerable either to Soviet ground forces or to air attack. Range extension by in-flight refueling became a high priority for the USAF.

In 1947, General Spaatz formed a Heavy Bombardment Committee to "study methods of and instrumentalities for air delivery of individual and mass atomic attacks against any potential enemy from bases within the continental United States."[61] After examining several methods of range extension, and assessing the potential range of bombers still under development, the Committee recommended that first priority be given to air-to-air refueling. In 1948, equipment was ordered from Flight Refueling Limited in Britain to convert 100 B-29s to receivers and 60 B-29s to tankers. In 1949, a B-50 flew round the world nonstop, covering 23,108 miles in 94 hours with four refueling from RB-29s. SAC had proved that it now had the reach to discharge its mission of long-range attack against targets worldwide. It was the first step toward the capabilities of the USAF forty years later: "Global Reach—Global Power."

From its initial role as a range extender of SAC, air-to-air refueling reduced dependence on forward bases and enabled deeply penetrating reconnaissance sorties. In the Korean War, another activity was introduced: force multiplication by enabling fighters to sustain extended combat air patrols. That contribution not only ensured continuous air cover but at reduced overall cost. Instead of three combat aircraft—one on station, one returning and one on the ground—being required in a specific period, the overall requirement of airframes, air crew, and ground crew could be cut to one third. Subsequently, in Southeast Asia, the tankers also enabled tactical aircraft to fly longer ranges, carry more munitions and on many occasions to recover to friendly bases after combat damage or fuel shortage. But for political constraints, the tankers could have enabled indirect routing to and from targets, thereby achieving surprise, avoiding the repetitive approaches to and departures from target areas, and reducing the number of casualties caused by predictable tracks.

General LeMay was so convinced of the need for tankers that he sought to procure one RC-135 for every one B-52, but the Air Staff would only authorize two for three.[62] In 1958, SAC placed additional KC-135s ahead of additional B-52Gs, B-58s, B-70s, and all surface-to-surface missiles in its procurement priorities.[63] Five years later, at the height of the Cuban Missile Crisis, tankers sustained the whole of the B-52 fleet on airborne alert,[64] including a detachment, refueled off the Spanish coast, which flew a very high profile holding pattern over the Mediterranean between Crete and Cyprus.[65]

In 1976, tankers deployed to England and enabled air attacks on Libya by F-111s despite being prohibited to over-fly France, Spain, and Portugal. By 1995 air-to-air refueling had become recognized as an indispensable force multiplier for the USAF, U.S. Navy, and many other air forces around the world. Without the stimulus of confrontation and the requirement of a nuclear-equipped SAC, the development and exploitation of air-to-air refueling would have proceeded much more slowly, if at all.

The Soviet Contribution

The Soviet Union contributed to the evolution of air power in the nuclear age in two ways—indirectly by stimulating developments in the USAF, and directly by pursuing policies rooted in the Soviet experience of air warfare and reflecting particular Soviet strategic requirements.

In 1944, three B-29s force-landed in the Soviet Union. Stalin had already asked for B-29s and B-24s under lease agreements without success. Now the B-29s were reverse engineered, reconstructed by the Tupolev Bureau, flown and put into production by March 1945 as the TU-4. For the first time in its history, the United States became vulnerable to hostile air power, albeit on a one-way mission. From 1949, the threat was compounded by Soviet acquisition of atomic weapons. The TU-4 was followed by the twin jet-engined TU-16 and the four-engined M-4, both appearing at the 1954 May Day Red Square fly past. The latter was believed to have a range of 10,000 miles and performance similar to the early B-52. The TU-95 followed one year later. Meanwhile some two dozen bomber bases were built in the Arctic Circle and northern Russia, transport squadrons were allocated to the bomber units, and in-flight refueling was begun.

At the time, there was no reason to doubt U.S. intelligence forecasts about extensive and rapid growth in Soviet strategic air power. For example, between 600 and 700 M-4s were expected to be built, to fly too high and too fast for interception by U.S. fighters.[66] Such assessments strongly and conveniently reinforced the USAF's arguments for reliance on its own bomber counterforce, but had the USSR sustained its priority to bomber production, the forecasts could have proved disturbingly accurate. In fact, such a threat never materialized.

Despite early successes in World War I and for a brief period in the 1930s, the USSR had never given high priority to long-range air bombardment and in World War II only 200,000 sorties were flown by Long Range Aviation. This represented only 6 percent of all sorties flown by the Soviet Air Forces. Of those, 40 percent were flown in support of ground forces and 30 percent against German rail communications.[67]

For the Soviets, strategic air attack was an extension of ground force operations until 1945. Then, it appears that Stalin reciprocated U.S. perceptions of potential threat and accelerated long-range bomber production to be able to reach the United States. But the operational effectiveness of the first generation of Soviet postwar bombers was well below Western evaluation. By the mid-1950s the USSR was beginning to look to surface-to-surface missiles to discharge the strategic bombardment role. In 1955 the commander-in-chief of the Soviet Air Forces, Air Chief Marshal Pavel Zhigarev, forecast the change in emphasis:

> strategic long range bombers are expensive to build, man and maintain and they need to be housed in large airfields where they are vulnerable to air attack, they tie up large numbers of maintenance personnel and need great supplies of fuel. Missiles can be built more easily and cheaply, do not need such a complicated supply and servicing organization and are easily concealable and less vulnerable.[68]

First Secretary Khrushchev ordered a series of highly classified studies on the employment of nuclear weapons and missiles in future wars. A British agent in the Soviet Ministry of Defense reported the progress of the studies and of debates in the Soviet General Committee in the late 1950s which concluded that "the new world must strive to keep war losses to a minimum . . . to keep war short (by swift use of nuclear weapons)" and that it was insufficient simply to modernize old arms, "obviously we must go faster and further both in the theory of using nuclear missile weapons and in their production."[69] In January 1960, Khrushchev summarized the position in an address to the Supreme Soviet:

> Given the present development of military technology, military aviation and the Navy have lost their former importance. This type of armament is not being reduced but replaced. Military aviation is being almost entirely replaced by rockets. We have now sharply reduced and probably will further reduce and even halt production of bombers and other obsolete equipment.[70]

The policy was publicly demonstrated in May 1960 when the Soviet Rocket Forces were formally created as a new and primary armed service. Nevertheless, the potential threat from the Soviet manned-bombers had by 1960 stimulated SAC procedures of dispersal, ground and airborne alerts, as well as billions of dollars of investment in air defense of the North American continent.

Meanwhile, Soviet air defenses, given the highest priority by Stalin, had continued to be extended across the length and breadth of the Soviet Union. They had three components: interceptors, surface-to-air weapons, and the ground environment of early warning surveillance, acquisition, and weapons guidance radars plus associated communications. From the late 1950s onward, increasing priority was given to SAMs and anti-aircraft artillery.

During the 1960s, surface-to-air defenses were spread across Eastern Europe including the SA-3 and SA-5, the radar-guided mobile SA-6, and the heat-seeking portable SA-7. They first complemented the manned interceptors and then began to free them for more offensive operations. By the end of the 1960s, a "third generation" of Soviet tactical aircraft began to emerge. First shown at the 1967 Moscow air display, they included improved models of existing MiG-21, Sukhoi 17/22 fighters with increased range and payload, and the MiG-23/27 family with advanced avionics. In 1974, the SU-24 long-range night all-weather tactical bomber also entered service. It was authoritatively estimated that between 1967 and 1977, "spending for the Air Force increased more rapidly than for any other military service and grew at over 3 times the rate for defense spending as a whole from 1969 to 1973."[71] After periods during which the strategic bomber force and then air defense received resource priority, the emphasis was shifted to tactical aviation. One analyst observed, "Of the total spending for the Air Force between 1967 and 1977, the largest increase by far was for Frontal Aviation, which increased by about 50% its inventory of tactical aircraft over this period."[72]

Hitherto, tactical aviation had remained completely subordinate to Soviet ground forces. "In land theaters, the mission of armed combat will be accomplished primarily by offense. But this will be done by the ground forces, including tactical aviation."[73] By the end of the 1960s, however, Soviet aircraft were increasingly capable of ranging deep into Western Europe well beyond the ground forces in a combined arms offensive. At the same time, new concepts of air operations were formed. By the 1970s, there would be a disturbing degree of convergence between Eastern and Western tactical aviation.

Flexible Response and Convergence

In 1956, the first Soviet medium-range ballistic missiles were deployed in Europe, followed a year later by the shorter range SCUD. Sputniks I and II demonstrated an intercontinental reach. The credibility of NATO Strategy MC 14/2 was undermined. Western tactical nuclear weapons would not offset conventional deficiencies if the opponent was also nuclear equipped. A threat of "massive retaliation" lost its potency when an opponent could retaliate in kind. Moreover, if such mutual evolution were to produce nuclear stalemate, would that not give a potential aggressor freedom to exploit conventional weakness with impunity?

NATO strategy was first questioned, then slowly and laboriously revised to accommodate conflicting interests and priorities among Alliance members.[74] The underlying objective of the revision was to reduce dependence on nuclear weapons. That in turn required increased conventional strength, a reversion to the position sought at Lisbon. The new concept MC 14/3, to become popularly known as "Flexible Response," was defined by the NATO Council in 1967 as

> based upon a flexible and balanced range of appropriate responses, conventional and nuclear, to all levels of aggression or threats of aggression. These responses, subject to appropriate political control are designed, first to deter aggression and thus preserve peace, but, should aggression unhappily occur, to maintain the security and integrity of the North Atlantic Treaty area within the concept of forward defense.[75]

This concept was to remain in being until the end of confrontation and came to depend increasingly on air power for its implementation. NATO announced its new strategy at a difficult time. Increased dependence on conventional force required more personnel, more training, more weapons, more maintenance facilities, more ammunition stocks to sustain conventional warfare, and higher states of conventional force readiness. Yet in 1966, France had withdrawn its forces from integrated NATO command and denied the Alliance French territory and airspace for reinforcement, resupply, and defense in depth. In 1967, the United States was becoming increasingly committed to the war in Southeast Asia and drawing down forces in Europe. Britain was reversing and reducing its defense budget. Overall in the Alliance, there was a reluctance, exactly as at Lisbon fifteen years previously, to meet the increased commitments.

Although tension between East and West had relaxed a little by the end of the 1960s, there was no change in Warsaw Pact deployment in Eastern Europe and, as noted above, the Soviet Air Force was expanding and reequipping. Consequently, in NATO annual exercises held after 1967 to implement the new concept, scenarios assumed a large-scale Warsaw Pact offensive preceded by different durations of warning time. The Pact would always have the advantage of choice of timing, location, and method of attack. Soviet strategy emphasized deception, surprise, and deep combined arms attacks exploited by swiftly moving reinforcements. Usually, in the exercises, the "attacks" were met by NATO conventional forces and after very few days recourse to nuclear weapons was necessary. NATO was no more capable of waging protracted conventional war in Europe after 1967 than it had been fifteen years earlier. Then, dependence was placed on nuclear weapons; now a heavy burden would be placed on NATO air forces, especially in the opening stages of conflict.

In-place NATO ground forces were out-numbered by their opponents and their deployment had been determined by political rather than strategic considerations. For example, the best equipped U.S. forces were located well south of the major east-west threat axis across the north German plain. Elsewhere, the NATO ground troops were spread from north Norway to east Turkey, divided by seas, mountains, and neutral territory, whereas Soviet forces were concentrated in central Europe, across the inner German border. NATO was committed to forward defense, but only a small proportion of national forces were in, or could be easily moved into, appropriate defensive positions. Not surprisingly, by 1989 when the Warsaw Pact collapsed, NATO had come to depend heavily on air power to offset the geopolitical and strategic advantages of its opponents. Less obvious to the public, but well known to Western intelligence, was the equally heavy dependence on air power which had evolved in Soviet doctrine and strategy.

NATO airborne early-warning and control aircraft (AWACs) contributed to surveillance in peacetime, and in crisis or combat would have enabled defensive concentration and counteroffensive control in the manner seen so comprehensively in the Gulf War of 1990–91. The Boeing E-3 AWACs was made possible by the evolution of large airframe, miniaturized, computerized equipment, advanced radar and communications, and engines powerful enough to sustain both the aircraft and its operational systems for extended periods on station. Its contribution to the impact of air power in the later stages of the Cold War is inestimable. It is no coincidence that by 1988, the Soviet Air Force was equipping some interceptor units with high-speed antiradiation missiles expressly to neutralize NATO AWACs aircraft.

Had crisis degenerated into conflict, NATO air forces would have been called upon to protect home bases, reinforcement concentrations, and routes and win local supremacy over battlefields. Warsaw Pact ground force superiority would have been countered by close air support and deeper attack on second echelon and other reinforcements to delay and disrupt the momentum of the Soviet armies and those of their allies. Simultaneously, offensive counter-air sorties could have been flown against the enemy's air bases. Tactical reconnaissance across the theater would have been essential.

First reinforcements within and from outside Europe would have been made by air, committing all NATO air transports plus civilian charter aircraft. Such operations could only take place in secure airspace. In extremis, the Supreme Allied Commander Europe could have been authorized to use some of his strike attack force to deliver nuclear weapons. Without air power the conventional implementation of 14/3 would have been impossible. On this at least, there was no disagreement from the USSR or its allies.

From the mid-1970s the Allies were aware of Soviet air power concepts and strategy.[76] An air operation, mounted by attacks by tactical aviation, medium- and heavy-bomber regiments and escorted by fighters, was the first and essential element in a combined arms offensive. Its objectives included NATO nuclear weapon and storage facilities, headquarters, airfields, reinforcement posts, routes, and assembly areas, and NATO ground forces themselves. By 1985, the Soviet Air Force had the equipment for the task: the MiG-29, SU-27, SU-25, MiG-31, TU-22M, and TU-160 had entered front-line service, complemented by advanced air-to-air and air-to-ground weapons, airborne early warning, and in-flight refueling.

At theater level, the convergence of air power doctrine and capabilities was almost complete. The major difference lay in that whereas NATO saw the threat emanating from the enemy's ground forces, the Warsaw Pact saw, correctly, NATO air power as the major obstacle to success. The latter view was summarized in 1981:

> The experience of the most recent wars has shown that the air forces have always substantively affected the course of the combat action of their own troops. Consequently the problems of combating air forces have been given much attention and deserve still more, because a breaking up or serious weakening of the enemy's air force and nuclear missile groupings leads to a fast decline of his capabilities. By ensuring supremacy in the air, it creates favorable conditions for the action of troops taking part in the operations in the theater ... he who seizes the initiative in the air will dictate his own conditions.[77]

There can be little doubt that had the Cold War ever caught fire, the outcome in Europe would have been determined by whichever side successfully applied its air power concepts. Had NATO's failed, the uncomfortable choice between capitulation and nuclear escalation would have been starkly imminent.

The Final Stimulus

The last stimulus to air power given by the Cold War occurred in its last decade when the West sought to bridge the conventional force gap by heavy investment in advanced technology. Stealth, precision weapons, multirole all-weather aircraft, surveillance, communications, comprehensive electronic warfare capabilities, and satellite interfaces were all procured for use against the USSR and its allies. No amount of Koreas, Vietnams, Libyas, and such could have generated the political will and economic allocation to expand Western, and specifically United States, air power so powerfully in such a comparatively short time. There is strong evidence to suggest that the Soviet military supported President Mikhail Gorbachev's reforms because they had come to realize that the existing Soviet economic system could not compete with Western production of advanced military technology, especially in the air and in space.[78] If so, the final contribution of air power in the Cold War was to accelerate its end and the collapse of the opposition.

Reprise

The contributions of air power to the conflicts of the second half of the twentieth century are to a great extent quantifiable by sortie rate, effectiveness, attrition rates, operational outcomes, and so on. Without exception, they differed considerably from the one feared between East and West. Thus it is productive to trace the evolution of aircraft, weapons, tactics, and operational impact through the campaigns in Korea, Southeast Asia, the Falklands, and ultimately in Desert Storm. Air power had much to contribute in each. In the war which was never fought it was the enemy who capitulated. The outcome of the actual conflicts had regional implications; the outcome of the Cold War changed the direction of the world. In it, air power was multifaceted. Strategic air power dictated the initial Western strategic posture; tactical air power together with ground-based nuclear weapons sustained the Western alliance when its members failed to provide for conventional defense; flexible air power was tasked to underpin the final concept of flexible response. Air power thwarted Stalin and laid the basis for a new Europe in 1948–49; in 1962, it enabled the Cuban crisis to be identified and managed without resort to nuclear war. It elevated reconnaissance and in-flight refueling to primary air power roles. By 1989, the synergy of air power and advanced technology had produced a military instrument beyond the expectations of even the most zealous of air power pioneers.

But there was also a cautionary tale. Role concentration can produce overwhelming success, but it can induce inflexibility. No two conflicts are ever exactly the same. The air power used so powerfully in Desert Storm could not be applied in a similar way in a Bosnian scenario. From 1945 to 1989, it was essential to give priority in preparation to one kind of war. Even that preparation was both complicated and simplified by the possibility of early use of nuclear weapons.

Air power denotes the military exploitation of the air by all its roles, not just by destruction or by the threat of destruction. As a result, "lessons" from the use of air power in the Cold War should not be examined in isolation any more than those from Desert Storm. They should be approached in the political context and priorities of the time, with a flexibility of mind to match that unique quality in air power itself.

Notes

1. General Carl Spaatz, "Strategic Air Power: Fulfillment of a Concept," *Foreign Affairs*, April 1946.
2. General Omar Bradley, "Effects of Air Power on Military Operations: Western Europe," report from 12th U.S. Army Group, 15 July 1945. Excerpts in Eugene Emme, *Impact of Air Power* (New York: Van Nostrand, 1959), 241.
3. Admiral of the Fleet, Lord Hill-Norton, *Seapower* (London: Faber & Faber, 1982), 69.
4. Asher Lee, *The Soviet Air Force* (London: Duckworth, 1961), 72.
5. Ibid., 80.
6. H. D. Hall, *History of the Second World War: North American Supply* (London: HMSO, 1955), 424.
7. Lord Ismay, *NATO: The First Five Years 1949–54* (Brussels: NATO, 1955), 4.
8. See John T. Greenwood, "The Emergence of the Postwar Strategic Air Force, 1945–1953," in this volume.
9. See David A. Rosenberg, "American Postwar Air Doctrine and Organization: The Navy Experience," in this volume.
10. Greenwood, "Emergence of the Postwar Strategic Air Force."

Cold War

11. Cited in David MacIsaac, "The Air Force and Strategic Thought 1945–1951," Woodrow Wilson International Center for Scholars, 21 June 1979.
12. Ibid.
13. Spaatz to Arnold, quoted in R. F. Futrell, *Ideas, Concepts, Doctrine: Basic Thinking in the United States Air Force 1907–1960* (Maxwell AFB, Ala.: Air University Press, 1989), 214.
14. Ann Tusa and John Tusa, *The Berlin Airlift* (London: Hodder & Stoughton, 1988), 141.
15. Report of House of Commons debate, *London Times*, 2 July 1948.
16. Bevin Papers, Foreign Office 800/467, quoted in Tusa and Tusa, 153.
17. U.S. Department of State Bulletin, 30 June 1948, cited in ibid., 157.
18. Clay to Department of Army, 25 June 1995, Clay Papers, cited in ibid., 149.
19. British Ministry of Defence Report, CAB 131/6, 7 July 1948.
20. Robert Murphy, political adviser to Gen. Dwight Eisenhower, in Murphy, *Diplomat among Warriors* (London: Collins, 1964), cited in Tusa and Tusa, 174.
21. Robert F. Futrell, "The Influence of the Air Power Concept on Air Force Manning, 1945–1962," in *Military Planning in the 20th Century* (Washington, D.C.: Office of Air Force History, 1986), 257.
22. Tusa and Tusa, 154.
23. Maj. H. R. Borowski, "A Narrow Victory," *Air Force Magazine*, July/August 1981, 10–27.
24. Cited in Ismay, 9.
25. Text of the North Atlantic Treaty, *Keesing's Treaties and Alliances of the World* (Bristol: n.p., 1968), 69.
26. JCS 1952/1, 21 Dec. 1948, para. 32a, cited in Thomas H. Etzold and John L. Gaddis, eds., *Containment: Documents on American Policy and Strategy 1945–1950* (New York: Columbia University Press, 1978), 358.
27. Excerpt from *Congressional Record*, October 1949.
28. Ibid.
29. Defense Committee of NATO DC 6/1, 1 Dec. 1949, para. 5, cited in Etzold and Gaddis, 337.
30. Ibid., para. 7, 337–38.
31. Sir William Jackson and Lord Bramall, *The Chiefs* (London: Brassey's, 1992), 277.
32. Paul Nitze, Address to the National War College, Washington D.C., 20 Sept. 1993, published in *NSC-68, Forging the Strategy of Containment* (Washington D.C.: National Defense University Press, 1994), 13.
33. Texts of Final Communiqués 1949–1974, NATO Information Service, Brussels, 69.
34. Andre de Staercke in *NATO's Anxious Birth, The Prophetic Vision of the 1940s* (London: Hurst, 1985), 157–59.
35. Jackson and Bramall, 282.
36. Futrell, *Ideas, Concepts, Doctrine*, 347.
37. AFR 23-10 Organization: Air Commands and Air Force Tactical Air Command, 29 June 1953, cited in ibid.
38. Brig. Gen. James E. Ferguson, "The Role of Tactical Air Forces," *Air University Quarterly Review*, Summer 1954. Cited in ibid.
39. Military Strategy to Support the National Security Policy set forth in NSC 162/2, JCS 2101/113, 9 Dec. 1953. Cited in Jane E. Stromseth, *The Origins of Flexible Response* (London: Macmillan, 1988), 13.
40. John Foster Dulles, "Talking Paper," prepared for NATO Council Meeting, Paris, 23 Apr. 1954. Cited in ibid.
41. Norman Polmar, *Strategic Air Command* (Annapolis, Md.: Nautical & Aviation Publishing Co., n.d.), 10.
42. Ibid., 11.
43. See Alfred Price, "The History of US Electronic Warfare," vol. 11, Association of Old Crows, 1989, for a comprehensive operational and technical account of this period.

44. Paul Lashmar, "Skulduggery at Sculthorpe," *Aeroplane Monthly*, October 1994, 10–15. I am indebted to Paul Lashmar for access to the documents from U.S. sources referred to in the following footnotes.

45. Letter from Air Chief Marshal The Hon. Sir Ralph A Cochrane, GBE, KCB, to Air Chief Marshal Sir William Elliot, KCB, KBE, DFC, Head of British Joint Services Mission, Washington, 25 July 1951, in Twining Papers, declassified 15 Dec. 1986.

46. Wing Commander R. S. Sanders, DFC, AFC, and Bar, interview with the author, 2 Aug. 1995. The following account is based on his log book entries and recollections.

47. Lashmar, 15.

48. Marshal of the Royal Air Force Sir John Slessor to General Vandenberg, 12 Sept. 1952, Twining Papers, declassified 27 Feb. 1981.

49. Price, 157.

50. Michael R. Beschloss, *Mayday* (London: Faber & Faber, 1986), 111–12.

51. Col. Harold Austin, USAF (Ret.), "A Cold War Overflight of the USSR," *Daedalus Flyer*, Spring 1995, 15–18.

52. Paul Lashmar, "Shootdowns," *Aeroplane Monthly*, August 1994, 10.

53. Graham Yost, *Spy Technology* (London: Harrap, 1975), 20.

54. Beschloss, 152.

55. Ibid., 156.

56. Robert Kennedy, *13 Days* (London: Macmillan, 1969), 39–45.

57. Ibid., 105.

58. Cited in Yost, 34.

59. CIA Deputy Director Ray Cline, 28 June 1983, cited in Beschloss, 143.

60. Maj. Gen. John B. Marks, Commander Electronic Security Command 1995, cited in Price, 312.

61. Futrell, *Ideas, Concepts, Doctrine*, 232.

62. Ibid., 518.

63. Ibid., 519.

64. Kennedy, 55.

65. Price, 293.

66. Robert Kilmarx, *History of Soviet Air Power*, 253, cited in R. A. Mason and J. W. R. Taylor, *Aircraft, Strategy and Operations of the Soviet Air Force* (London: Jane's, 1986), 132.

67. Maj. Gen. of Aviation B. A. Vasil'ycv, *Long Range Missile-Equipped* (Moscow, 1972), 64, cited in Mason and Taylor, 128.

68. Asher Lee, ed., *The Soviet Air and Rocket Forces* (London: Weidenfeld & Nicolson, 1959), cited in Mason and Taylor, 135.

69. Quoted in Lee, 10–11.

70. As quoted in *Pravda*, 15 Jan. 1960.

71. J. H. Hanson, "Development of Soviet Aviation Support," *International Defence Review*, May 1980, 683.

72. Ibid.

73. V. D. Sokolovsky, "Military Strategy, Soviet Doctrine and Concepts," *Pall Mall*, London, 1962, 281.

74. For a recent European analysis of the genesis of "Flexible Response," see Jane E. Stromseth, *The Origins of Flexible Response* (London: St. Anthony's Macmillan, 1988).

75. North Atlantic Council Communiqué, Brussels, 14 Dec. 1967.

76. *Materials from the Soviet General Staff Academy*, vols. 1 and 2 (Washington, D.C.: National Defense University Press, 1989–90).

77. Col. Alexander Musial, Polish Air Force, "The Character and importance of air operations in modern warfare," *Air Force and Air Defence Review*, Warsaw, December 1981. UKTRANS No. 138, Conflict Studies Research Centre, Sandhurst, England.

78. See, for example, R. A. Mason, *Air Power: A Centennial Appraisal* (London: Brassey's, 1994), chaps. 4 and 7, for an analysis of Soviet perceptions.

Searching for Victory through Air Power: The Conduct of the Air Wars in Vietnam

Mark Clodfelter

In eight years of aerial warfare from 1965 to 1973, American aircraft dropped more than eight million tons of bombs on the landscape of Southeast Asia. Half of that total fell on South Vietnam, America's ally in the perceived struggle against Communist aggression. Roughly three million tons fell on Laos and Cambodia, so-called "neutrals" in the Vietnamese conflict. The remaining million tons landed on North Vietnam, with the bulk of those falling during the "Rolling Thunder" air campaign of 1965–68.[1]

For American air leaders, the million tons dropped on North Vietnamese soil counted far more than those that fell elsewhere. Rolling Thunder, and the 1972 "Linebacker" air campaigns, offered the promise of ending the war independently with air power, while the seven million tons of bombs dropped on Laos, Cambodia, and South Vietnam promised a ground victory that air power had supported. In the end, a victory of any sort that achieved President Lyndon B. Johnson's goal of a stable, independent, non-Communist South Vietnam proved impossible to obtain with American military force. Many air leaders, however, viewed Johnson's political restrictions as the key reason that air power had failed to achieve independent success during Rolling Thunder. They pointed to the 1972 bombing as an example of what unfettered air power could achieve, and many focused specifically on President Richard Nixon's eleven-day December bombing effort known as "Linebacker II." Yet those air commanders who argued that a Linebacker II in 1965 would have "won the war" then failed to see that conditions in 1965 were not the same as those in 1972. Moreover, the objective of Nixon's aerial onslaught—to remove American presence from the war and provide a "decent interval" for South Vietnam—differed significantly from Johnson's goal of an enduring non-Communist South. Regardless of the type of military force used, Nixon's aim was by far easier to achieve.

Vietnam was an especially thorny problem for American air leaders because it did not mesh with their expectations. After finally achieving the holy grail of service independence in 1947, the Air Force had become America's first line of defense in the anticipated "general" war against the Soviet Union. That defense was in fact an offense, in which Strategic Air Command's (SAC) bombers were the centerpiece. "Massive retaliation," the guiding principle of President Dwight Eisenhower's defense policy, promised a tremendous nuclear bomber attack against the Soviet heartland should either the Soviets, or a proxy Soviet state, launch an attack against the United States or its allies. The concept suffered as a credible option, but the Air

Force ingrained the emphasis on an independent aerial victory in its doctrine. Air leaders realized that air power had failed to win a victory in Korea, but they viewed the limited war there as an anomaly, especially given Eisenhower's endorsement of massive retaliation. Although they realized that such a limited conflict might recur, they also believed that readiness for general war—a euphemism for nuclear combat—sufficed for wars of lesser magnitude. "The best preparation for limited war is proper preparation for general war," stated the 1959 edition of the Air Force's Basic Doctrine manual. "The latter is the more important since there can be no guarantee that a limited war would not spread into general conflict."[2]

The Air Force's preparation for general war on the eve of Vietnam hearkened to the prophecy of Billy Mitchell, the teachings of the Air Corps Tactical School, and the perceived effectiveness of strategic bombing in World War II. Mitchell had maintained that air power alone could defeat a nation by paralyzing "vital centers," which included great cities where people lived, factories, raw materials, foodstuffs, supplies, and modes of transportation.[3] All were essential to wage modern war, Mitchell had written in the aftermath of World War I, and all were vulnerable to air attack. Moreover, many of the targets were fragile, and wrecking them promised a victory both quicker and cheaper than one achieved by surface forces. Air power could attack vital centers directly, avoiding the senseless slaughter that had characterized World War I land combat. Bombers would wreck an enemy's will to fight by destroying its capability to do so, and the essence of that capability was not its army or navy, but its industrial and agricultural underpinnings. Eliminating industrial production "would deprive armies, air forces and navies . . . of their means of maintenance."[4] Air power also offered the chance to attack the will to fight directly, but Mitchell thought that bombers did not necessarily have to kill civilians to wreck a nation's will to resist.[5]

Mitchell's conviction that air forces could achieve an independent victory in war by destroying an enemy's capability and will to resist became a hallmark of the coterie of Air Corps officers who subscribed to his beliefs. The names of his disciples reads like a "Who's Who" of American air power: Henry H. Arnold, Carl A. Spaatz, Ira C. Eaker, Frank Andrews, William Sherman, Herbert Dargue, Robert Olds, Kenneth Walker, and Harold Lee George. All not only accepted Mitchell's notion that an autonomous air force was the key to achieving an independent victory through air power, they also spread Mitchell's gospel throughout the close-knit Air Corps community.

Mitchell's prophecy became the fundamental underpinning of the Air Corps Tactical School, the focal point of American air power study during the interwar years. The school provided an intense, nine-month, air power–focused curriculum to the Air Corps' top mid-level officers, and graduated 261 of the 320 generals serving in the Army Air Forces at the end of World War II.[6] Its faculty contained a substantial number of Mitchellites during the late 1920s and early 1930s. Sherman, Dargue, George, Olds, and Walker—the latter two had served as Mitchell's aides—filled key faculty positions, and all promoted Mitchell's vision of independent air power founded on the bomber. Basing many of their assertions on their study of American industry and population centers, they refined Mitchell's notions into an "industrial web theory" that offered the blueprint for how to wreck an enemy state with air power.

The belief, widely shared among Tactical School instructors, that the industrial apparatus essential to a state's war-making capability was also necessary to sustain its populace was the fundamental tenet of the School's industrial web theory. In brief, its main points were: first, in "modern warfare," the military, political, economic, and social facets of a nation's existence were so "closely and absolutely interdependent" that interruption of this delicate balance could suffice to defeat an enemy state; second, bombing, precisely aimed at these "vital centers" of an enemy's industrial complex, could wreck the fragile equilibrium and hence destroy the enemy state's war-making capability; and third, such destruction would also wreck the enemy nation's capacity to sustain normal day-to-day life, which would in turn destroy the will of its populace to fight.[7]

Along with fellow Tactical School instructors Ken Walker, Laurence Kuter, and Haywood Hansell, Harold Lee George relied on the industrial web theory to design an August 1941 plan for a prospective bombing campaign against Nazi Germany and Imperial Japan. Created in the sweltering heat of the Washington Munitions Building, the plan, known as AWPD-1, would guide American bombing in World War II and reflected many of Mitchell's air power notions. It contained the belief that bombing could wreck Germany's war-making capability and will to fight, and that in doing so it could obviate the need for an amphibious assault on Hitler's Europe. Mitchell's followers presented their convictions to Gen. George C. Marshall for review, and the army chief of staff approved AWPD-1 three months before Pearl Harbor.

After World War II, many American airmen viewed the bombing of Germany and Japan as a vindication of Tactical School teachings—as well as a rationale for an independent air force. Although bombing had not single-handedly defeated Germany, the air campaign against the German homeland had significantly damaged its ability to wage war. Moreover, the American bombing effort against Germany had suffered from numerous diversions, such as the need to support ground forces both in North Africa and France, and not until September 1944 did it focus exclusively on bombing Hitler's Reich. Once it did so, the air offensive had a telling impact on German oil and transportation. In the Pacific, the combination of high winds, dispersed "cottage" industries, and a desire to end the war by air power rather than ground invasion led to Maj. Gen. Curtis LeMay's incendiary attacks against Japanese cities. Although those raids certainly hurt Japan's war-making capability, they did not destroy its will to resist, and not until the *Enola Gay* and *Bock's Car* dropped atomic bombs in early August 1945 was that will substantially shaken. Still, American political and military leaders realized that strategic air power had played significant, if not decisive, roles in Allied victory in both Europe and the Pacific, and they noted that the apparent key to future victory was the air-delivered atomic weapon.

The result was an independent air force with a doctrine geared to achieving an independent victory à la Mitchell and the Air Corps Tactical School. Enough similarities remained between the notion of an industrial web and the damage rendered to Germany and Japan for American airmen to make the industrial web theory their doctrinal cornerstone. Published a few months after the Eisenhower administration announced its massive retaliation policy, Air Force Manual 1-8 defined *strategic air operations* as attacks "designed to disrupt an enemy nation to the extent that its will and capability to resist are broken."[8] Such operations would be autonomous, "conducted directly against the nation itself" rather than auxiliary operations supporting friendly land and sea forces against an enemy's deployed armies and navies.[9] The authors of Manual 1-8 proclaimed that destroying the war-making capacity of a nation would "neutralize" its surface forces. They further contended:

> Somewhere within the structure of the hostile nation exist sensitive elements, the destruction or neutralization of which will best create the breakdown and loss of the will of that nation to further resist. . . . The fabric of modern nations is such a complete interweaving of major single elements that the elimination of one element can create widespread influence upon the whole. Some of the elements are of such importance that the complete elimination of one of them would cause collapse of the national structure insofar as integrated effort is concerned. Others exert influence which, while not immediately evident, is cumulative and transferable, and when brought under the effects of air weapons, results in a general widespread weakening and eventual collapse.[10]

The authors concluded that destroying petroleum or transportation systems would cause the most damage to a nation's will to resist. Only "weighty and sustained attacks," however, would succeed in wrecking either system.[11]

Manual 1-8 remained current strategic air doctrine for more than 11 years—Air Force officers did not revise it until 1 December 1965. Its message, along with that articulated in the "Basic Doctrine" manuals of the Eisenhower years, permeated the service. The president's fiscal concerns (a nuclear attack force was cheap, especially compared to funding a large standing army) had meshed with a straightforward policy of massive retaliation to make Gen. Curtis LeMay's Strategic Air Command the dominant military force on the planet. Besides garnering more than half of the nation's meager defense budget during the 1950s, SAC also dominated its service in terms of thought as well as position, and its domination continued when America's active military participation began in Vietnam in early 1965. By then, as LeMay ended his tenure as air force chief of staff, three-fourths of the highest ranking Air Force officers in the Pentagon had SAC backgrounds.[12]

On the eve of sustained combat in Vietnam, the SAC-mindset portended ill for the Air Force. Although their doctrine stated that the industrial web theory applied to "modern" nations, many airmen equated "modern" to "all," and in Vietnam they would futilely try to determine the key industrial component that made agrarian North Vietnam tick. Part of the problem was that the airmen's predecessors had designed the industrial web theory based upon their own vision of the United States, and that vision may not have been accurate. Part of the problem was also that post–World War II airmen had transformed the notion into a guideline for nuclear attack, and the prospects of nuclear bombing did not translate exactly into the conventional arena. Yet Air Force doctrine taught that they did translate, and that preparation for nuclear war sufficed to ready the Air Force for combat at any level. That belief received a significant boost in October 1962 when the threat of SAC B-52s, supplemented by a fledgling intercontinental ballistic missile force, compelled the Soviet Union to back down during the Cuban missile crisis. If the threat of bombing could make the Soviets—America's mightiest potential enemy—retreat, surely that threat would make other, lesser, nations fall into line as well. So believed many American airmen and political leaders as the United States looked for a quick, cheap solution in Vietnam.

Air power appeared to offer the answer in early 1965 when American political and military chiefs agreed to initiate Rolling Thunder. As to *how* bombing North Vietnam would yield an independent, non-Communist South Vietnam, there existed a wide disparity of opinion, ranging from the signal air power would send to the North Vietnamese—that bombs would ultimately destroy their heartland—to the signal that it would send to America's South Vietnamese allies—that it would bolster their fighting spirit, making them fight harder and cause them to prevail against the Viet Cong. Although many airmen believed that North Vietnam lacked the Soviet Union's will to resist, and that the threat of aerial destruction would force Ho Chi Minh to cry uncle, their motivations for bombing the North also subscribed to their doctrine: they believed that the Viet Cong, which formed the vast bulk of the enemy army in South Vietnam, could not fight without the support and direction of the North Vietnamese, and that bombing North Vietnam would deny the Viet Cong the *capability* to keep fighting. This notion was the fundamental premise of Rolling Thunder, and one that American airmen would continue to endorse after Lyndon Johnson's political advisors had given up on it. It was also a premise that was fundamentally flawed.

Despite frequently *stating* that the enemy waged guerrilla warfare, American political and military leaders assumed that the destruction of resources necessary for *conventional* warfare would weaken the enemy's capability and will to fight unconventionally. During the Rolling Thunder era, however, the enemy rarely fought at all. Hanoi had only 55,000 North Vietnamese troops in the South by August 1967; the remaining 245,000 Communist soldiers were Viet Cong.[13] None of these forces engaged in frequent combat, and the Viet Cong intermingled with the Southern populace. Enemy battalions fought an average of one day in thirty and had a total daily supply requirement of 380 tons. Of this amount, they needed only 34 tons a day from sources outside the South.[14] Seven 2½-ton trucks could transport the requirement,[15] which was less than 1 percent of the daily tonnage imported into North Vietnam. Sea, road, and rail imports averaged 5,700 tons a day, yet Hanoi possessed the capacity to import 17,200 tons. Defense Department analysts estimated in February 1967 that an unrestrained air offensive against resupply facilities, accompanied by the mining of Northern harbors, would reduce the import capacity to 7,200 tons.[16] The amount of goods that the Communists shipped south "is primarily a function of their own choosing," the Joint Chiefs remarked in August 1965.[17] Their appraisal remained valid throughout Rolling Thunder.

Initially, though, American military chiefs, and air leaders in particular, believed that the Viet Cong could not function without Hanoi's support, and that conviction received backing from President Johnson and his key political advisors in the spring of 1965. Air Force Chief of Staff LeMay, and Gen. John P. McConnell, who served as LeMay's vice chief and succeeded him as chief of staff on 1 February 1965, called for a concentrated air attack ranging from sixteen to twenty-eight days against transportation centers, bridges, electrical power facilities, and the sparse components of North Vietnamese industry. Admiral U. S. Grant Sharp, the commander in chief of Pacific Command who controlled the forces would conduct such an attack, eagerly endorsed it, as did Lt. Gen. Joseph H. Moore, the commander of the 2nd Air Division in Saigon, who directed many of the Air Force aircraft that would participate. To air commanders, the "sudden, sharp knock" of a three-week air offensive would not only disrupt North Vietnam's war effort, it would also disrupt the fabric of Northern economic and social welfare. Bombing would dissect the web that tied North Vietnam's war-making capability to its way of life, thus compelling Ho Chi Minh to call off the Viet Cong insurgency and eliminating Hanoi's capacity to support it.

In February 1965, President Johnson and his political advisors accepted that air power was an appropriate instrument to bully North Vietnam: it would cost fewer American lives than would sending in American ground troops; they could focus it on key North Vietnamese targets; and, above all, they could control its intensity. Control was essential for Lyndon Johnson. Although he had committed personal as well as national prestige to preserving a non-Communist South Vietnam, equally important to the President was assuring that the war in Southeast Asia did not expand into a larger conflict involving either the Chinese or the Soviets. Remembering Chinese intervention in the Korean War, Johnson was terrified that the Chinese would send troops to support their Communist neighbors in North Vietnam, or, worst yet, that the Soviets would actively join in the conflict—possibly even with nuclear weapons.[18] Both the president's key advisors, Secretary of State Dean Rusk and Secretary of Defense Robert McNamara, warned Johnson not to implement the proposed three-week air campaign against North Vietnam because of the unknown impact that such bombing would have on the Communist superpowers. Rusk, as assistant secretary of state for Far Eastern affairs during the Korean War, had seen first hand the effects of miscalculating Chinese intentions, while McNamara had played a key role in helping resolve the Cuban missile crisis, the world's closest brush with nuclear holocaust. Johnson placed an enormous amount of trust in the opinions of both men. Despite the president's masterful manipulation of domestic politics, "foreign affairs were indeed foreign to him."[19]

Besides his fears that Vietnam might expand into World War III, Johnson's commitment to establishing a "Great Society" caused him to shun a heavy air attack on North Vietnam. His desire to improve the domestic social welfare of poor and underprivileged Americans had been a lifetime goal. "If I left the woman I really loved—the Great Society—in order to get involved with that bitch of a war on the other side of the world, then I would lose everything at home," he recalled. "But if I left that war and let the Communists take over South Vietnam, then I would be seen as a coward and my nation would be seen as an appeaser and we would both find it impossible to accomplish anything for anybody anywhere on the entire globe."[20] He feared that a massive increase in military force in Southeast Asia would advertise the seriousness of the threat to South Vietnam, causing the attention of Congress and the American public to shift away from the social programs that he cherished. A rapid increase in military pressure would have further repercussions. The president hoped to secure a favorable perception of the United States in Third World nations. Too much force in Vietnam might cause those countries to view the American effort as motivated by imperial ambitions or feelings of racial superiority. Johnson also wished to maintain the support of NATO and other Western allies. The greater the effort in Vietnam, the more allies elsewhere would question the ability of the United States to sustain its military commitments.

Johnson's conflicting goals combined to produce the main principle of air strategy against North Vietnam: gradual response. America's political leaders believed that military force was necessary to guarantee the South's existence, yet other goals prevented them from unleashing America's full military power. To assure that the war remained limited, Johnson prohibited military actions that threatened, or that the Chinese or Soviets might perceive as threatening, the survival of North Vietnam. Bombing would begin slowly in the southern part of North Vietnam and incrementally "roll" northward toward the heartland containing Hanoi and Haiphong. Meanwhile, Johnson and his political chiefs would scrutinize bombing's effects, with a wary eye focused on the reactions of Moscow and Peking. Based on those reactions, as well as the response of the American public and the world community at large, they could tighten or loosen the bombing spigot as they saw fit. Rolling Thunder's initial attack on 2 March 1965 struck only one target, an ammunition depot well south of the heartland, and was the only attack of the week. The following week fighters bombed barracks and ammunition depots, again south of the 20th parallel, on a single day.

Johnson and his political advisors hoped that the attacks would signal Ho Chi Minh that ultimately air power would demolish such heartland targets as the North's single steel mill, its one cement factory, or its meager electrical power system. Fighters bombed more targets, on more days, during the third week of March, and the bombs crept northward toward Hanoi. During this span, Admiral Sharp stated that he expected that the limited interdiction would yield success by degrading transportation, diverting manpower to rebuilding roads and bridges, and conveying America's "strength of purpose" that would "make support of the VC as onerous as possible."[21] This faith, also shared by political leaders, that the threat of greater destruction would suffice to make Ho Chi Minh balk probably stemmed from the Soviet retreat during the Cuban missile crisis.

The belief that Ho would cower to air power lasted very briefly, although a meager regard for North Vietnam's tenacity endured throughout the Johnson presidency.[22] By early April 1965, after only six Rolling Thunder missions, National Security Advisor McGeorge Bundy began pressing Johnson to change the focus of American military effort to ground power. Sharp and other air chiefs redoubled their efforts to convince political leaders to bomb key elements of Northern war-making capability directly, but Bundy, Rusk, McNamara, and general-turned-ambassador to South Vietnam Maxwell Taylor persuaded Johnson to emphasize—and enlarge—America's military effort on the ground in the South. To American political leaders after July 1965, bombing the North would serve as a means of supporting American troops in South Vietnam by denying enemy forces unlimited supplies and placing a "ceiling" on the magnitude of the war that they could fight. Many air commanders, however, continued to advocate increased attacks on North Vietnamese heartland targets in the hopes that air power might ultimately wreck the North's capability and will to fight.

Johnson's political restrictions made conducting the systematic air campaign called for by the air chiefs virtually impossible. He initially prohibited B-52s from bombing North Vietnam because he thought that the Chinese or Soviets might deem use of the heavy bomber designed for nuclear missions too provocative.[23] B-52s did not bomb the North until 1966, when Johnson permitted them to attack targets just north of the 17th parallel. Targets in Hanoi remained "off-limits" to all aircraft until the summer of 1966. The president then forbade air commanders from bombing within a 30-mile radius from the center of Hanoi, a 10-mile radius from the center of Haiphong, and 30 miles of the Chinese border without his personal approval. Besides determining *where* his pilots could attack, Johnson decided *how often* they could do so. He stopped Rolling Thunder completely on eight occasions between March 1965 and March 1968, with reasons that varied from giving the North Vietnamese time to negotiate to observing Buddha's birthday. Meeting periodically with key advisors Rusk, McNamara, and Bundy over lunch in the White House on Tuesday afternoons, the president selected specific North Vietnamese targets for attack in weekly or biweekly increments. Not until October 1967 did these luncheons regularly include Army General and Chairman of the Joint Chiefs of Staff Earle Wheeler; prior to that time, no military representative usually attended.

The lack of a military presence during the Tuesday lunches blatantly revealed Johnson's distrust of his generals and caused those wearing the uniform enormous frustration at all levels. Because the luncheon attendees did not publish the results of the sessions, different perceptions of the president's decision-making frequently occurred. Conflicting guidance reached air commanders and produced confusion. Pilots learned that they had authority to attack moving targets such as convoys and troops, but could not attack highways, railroads, or bridges with no moving traffic on them.[24] Moreover, the precise definition of "moving targets" was unclear to those flying Rolling Thunder missions. A wing commander in May, 1965 expressed his uncertainty over permissible interdiction targets to 2nd Air Division Headquarters in Saigon:

> What is a military convoy? How many vehicles constitute a convoy? When a specified number of vehicles covers what length of road is it a convoy? Is a single

vehicle travelling by itself an authorized target? . . . Targets on a "truckable ancillary road" are listed as a target. How far off of a specified route are we authorized to follow a truckable ancillary road? "Troops" are listed as targets. The difficulty of recognizing groups of civilians on the ground from troops is readily apparent. I recognize this as my problem but believe it can be better defined.[25]

In desperation the officer, "pending more definitive guidance from headquarters," defined a convoy as "three or more internal combustion vehicles going the same direction on not more than a one mile segment of a specified route."[26]

Much like the wing commander, air leaders orchestrated the offensive against North Vietnam as best they could given the president's constraints. They refused to surrender the deeply ingrained notion that bombing could break an enemy's capability and will to resist, and their targeting proposals that McNamara carried to the Tuesday lunches adhered to the basics of Air Force doctrine. Johnson, increasingly desperate for a solution to the war, ultimately bowed to many of their targeting suggestions, although he never gave his air chiefs carte blanche to attack North Vietnamese target systems in a coordinated series of "sharp, hard knocks." Raids occurred incrementally over long spans of time, with the effort against Northern transportation counting for 90 percent of all Rolling Thunder missions and running from March 1965 to June 1966 (and during intervals between shifts in bombing emphasis); the attack on oil storage areas occurred from late June to early September 1966; and raids against Northern industry and electric power plants took place in March, April, and May 1967. After those attacks, the objective wavered. At the end of March 1968, in the midst of the domestic furor over the Communist Tet Offensive, Johnson restricted bombing to targets below the 19th parallel in an effort to spur peace negotiations. On 1 November 1968, he halted all attacks on North Vietnam and brought Rolling Thunder to a close.

In the end, bombing could break neither Hanoi's capability nor its will to keep fighting, nor could Rolling Thunder place a "ceiling" on the magnitude of the war that the North Vietnamese and Viet Cong could fight. The 1968 Tet Offensive demonstrated in graphic fashion bombing's failure to limit enemy operations. Rolling Thunder never suited the nature of the Vietnam War, and the inability—or unwillingness—of many air chiefs to recognize that fact produced "military constraints" that further limited bombing effectiveness. Unable to produce telling results against an enemy that rarely fought, air commanders adopted a method of combat score keeping resembling the body count approach used by ground commanders in the South: sortie count. Admiral Sharp's April 1966 division of North Vietnamese airspace into seven permanent bombing zones, or "Route Packages," triggered a competition between Navy and Air Force commanders to produce the most sorties in their respective Route Packages on a given day.[27] The totals flown then became a warped measuring stick of bombing effectiveness. To increase the count, some commanders called for attacks with less than full bomb loads, which in turn endangered additional flyers;[28] one Navy F-4 pilot admitted that he attacked the Thanh Hoa bridge, one of the North's most heavily defended targets, with no bombs at all but was told simply to strafe the structure with 20mm cannon fire.[29]

Besides military limitations, "operational" controls further restricted Rolling Thunder. Those constraints consisted of such vagaries as geography, weather, aircraft types, and enemy defenses. North Vietnam's lush terrain was ideal for camouflage, and the enemy frequently resorted to deception. Hanoi also exploited the proximity of Laos and Cambodia by snaking the sophisticated series of pathways that combined to form the Ho Chi Minh Trail through eastern areas of both countries. Weather was one of the air campaign's most significant operational controls. From September to April, the dense clouds of the winter monsoons made continuous bombing impossible. The monsoons prevented Rolling Thunder from starting in late February 1965 and canceled numerous missions in March, when the president's political advisors had the greatest faith in its success. Most of the raids scheduled during the monsoon season against fixed targets such as bridges became interdiction strikes because clouds obscured the primary objective. In 1966, only 1 percent of the year's 81,000 sorties flew against fixed targets proposed by the Joint Chiefs of Staff,[30] and weather was a key reason for the low total.

Of the aircraft types that performed most Rolling Thunder bombing, none were well suited for North Vietnam's forbidding environment. The Air Force relied primarily on the Republic F-105 Thunderchief, a slow-turning, single-seat fighter designed during the 1950s as a nuclear attack aircraft, and the McDonnell-Douglas F-4 Phantom, developed by the Navy as a high-altitude interceptor and modified for ground attack. It suffered from a vulnerability to ground fire, poor rear cockpit visibility, and engines that emitted thick, black smoke revealing its location. The Navy used the Phantom for bombing as well, but relied mostly on the McDonnell-Douglas A-4 Skyhawk, a diminutive single-seat fighter that could carry only four tons of bombs. Together, these aircraft flew against a defensive array that sported two hundred surface-to-air missile (SAM) sites, seven thousand anti-aircraft guns, a sophisticated ground-controlled intercept (GCI) system, and eighty MiG fighters by August 1967. Hanoi gained the reputation as the world's most heavily defended city, and veteran F-105 pilot Jack Broughton labeled North Vietnam as "the center of hell with Hanoi as its hub."[31] In 1967, the last full year of Rolling Thunder, 326 American aircraft were lost over the North; 921 were lost for the entire three and one-half year campaign.[32]

The air offensive originally envisioned as a quick, cheap alternative to ground war proved to be neither. Moreover, Rolling Thunder spurred both China and the Soviet Union to provide North Vietnam with not only military hardware but also economic backing. As a result of the support received from the Communist superpowers, the North's gross national product actually *rose* during Rolling Thunder.[33] Ho Chi Minh knew that the Chinese and the Soviets vied for influence in his nation, and he adroitly played one against the other to gain maximum support. The bombing also provided Ho with a means to rally the Northern populace behind the war effort. He realized that the constrained bombing would cause little damage, but the American air presence persisted over North Vietnam. Thus, he could consistently point to the air attacks as examples of American barbarism, a claim made time and again by his effective propaganda ministry. "In terms of its morale effects," RAND analyst Oleg Hoeffding argued in 1966, "the U.S. campaign may have presented the [North Vietnamese] regime with a near-ideal mix of intended restraint and accidental gore."[34] Rolling Thunder killed an estimated 52,000 civilians out of a population of 18 million;[35] in contrast, the first B-29 incendiary raid against Tokyo in World War II had killed at least 84,000 Japanese civilians on a single night.[36]

Rolling Thunder destroyed 65 percent of North Vietnam's oil storage capacity, 59 percent of its power plants, 55 percent of its major bridges, 9,821 vehicles and 1,966 railroad cars,[37] but such destruction counted for little in terms of ending the war. The myriad of political, military, and operational controls that plagued the air campaign helped prevent it from achieving Johnson's goal of a non-Communist South Vietnam. Without those constraints, however, its prospects were dim as long as the enemy chose to wage an infrequent guerrilla war. Air Force doctrine had discounted the probability of limited war, especially one in which the enemy rarely fought. That doctrine had also claimed that victory through air power was likely regardless of the nature of the war, and that the keys to victory were identifying and then wrecking the ingredients tying together the enemy's capability and will to resist.

Given the type of war that the Communist army fought, the only two targets that might have hurt the enemy's war effort and the support it received were people and food. None of Johnson's advisors, military or civilian, advocated such attacks. Yet even had raids against population centers or the Red River dikes been conducted against North Vietnam—and succeeded in knocking the North out of the war—in all likelihood they would have had minimal impact on achieving President Johnson's war aim. The key enemy was the Viet Cong in the South, not the North Vietnamese, and the Viet Cong were not dependent on Hanoi's support. Nor were they dependent on Hanoi's direction. Many Viet Cong soldiers and their leaders fought against the American-backed Saigon regime because it was corrupt and mistreated the Southern populace, rather than because of a commitment to North Vietnamese Communist ideology.[38] As long as the Viet Cong kept the war's tempo limited, Rolling Thunder could not prove decisive.

In the final analysis, the war would be won or lost in South Vietnam, where auxiliary air power supported the ground struggle waged by American and South Vietnamese troops. While the U.S. Air Force and the U.S. Navy fought two separate and often unrelated air wars against North Vietnam, in the South no less than six isolated air wars, with disparate arrays of air power, transpired simultaneously. Air Force fighters, based in South Vietnam and Thailand and directed by 7th Air Force Headquarters in Saigon,[39] formed a large part of that combined effort. Yet fighters were not the only U.S. Air Force contribution to the Southern ground battle. Beginning in June 1965, Strategic Air Command's B-52s began bombing suspected enemy positions in South Vietnam in what became a massive operation known as "Arc Light." The Navy bombed targets in South Vietnam as well, flying from carriers positioned at "Dixie" station in the South China Sea in contrast to the more familiar "Yankee" station in the Tonkin Gulf used for attacking North Vietnam.

Naval air often supported U.S. Marine operations in the I Corps section of northern South Vietnam, but the Marines also had their own air component. Flying F-4s, A-4s, A-1 Skyraiders, and numerous helicopters, the Marines supported their ground units from bases throughout the I Corps area, especially Da Nang. U.S. Army helicopters, ranging from the workhorse Chinook to the ubiquitous Bell UH-1 Huey, also dotted the South Vietnamese skies. Although the Army relied on its "choppers" for delivering men, materiel, and firepower throughout the South, the Air Force provided the major airlift support in South Vietnam with its fleet of turbo-prop transport aircraft—the C-123 Provider, C-130 Hercules, and C-7A Caribou, from Pacific Air Forces (PACAF)[40]—and two jet transports—the C-141 Starlifter and giant C-5 Galaxy that were part of Military Airlift Command. Last, and in many respects, least in terms of air forces operating in the South was the South Vietnamese Air Force (VNAF). Created in the image of the U.S. Air Force, the VNAF contained A-1, A-37, and, later in the war, F-5 attack aircraft, as well as a smattering of transports and helicopters. It provided close air support to and transport for the South Vietnamese Army, but suffered from leadership problems, especially after the flamboyant Air Marshal Nguyen Cao Ky left the VNAF to become prime minister in the summer of 1965.

Just as one air chief did not direct the bombing of North Vietnam, no single individual controlled the enormous amount of aircraft flying in the South. The lack of a single air manager made coordination difficult, even among aircraft flying from the same service. B-52 bombers, ultimately refurbished to carry up to thirty tons of conventional ordnance rather than two nuclear bombs, flew to South Vietnam from Andersen Air Base, Guam or from U-Tapao Royal Thai Air Base, Thailand. Initially, President Johnson directed B-52 raids in the South, but in August he transferred that decision-making power to the Joint Chiefs of Staff.[41] After complaints from 7th Air Force about the long-distance control, in April 1966 the Joint Chiefs transferred B-52 targeting approval for raids in South Vietnam to Admiral Sharp in Honolulu, and in November 1966 Sharp gave U.S. Army Gen. William C. Westmoreland, the commander of Military Assistance Command, Vietnam (MACV) in Saigon, the authority to approve all Arc Light targets.[42]

Westmoreland refused to give this control to Air Force Gen. William W. Momyer, the 7th Air Force commander, revealing Westmoreland's view of the B-52 as flying artillery that should be controlled by an Army commander. In contrast, Momyer wanted to coordinate the effort of his fighters with the B-52s and guarantee that the bombers would attack only defined targets of men and supplies.[43] His pleas fell on deaf ears. Westmoreland's MACV staff, predominantly comprised of Army officers focused on fighting the ground war, not only made B-52 targeting decisions but also selected the preponderance of Southern targets for 7th Air Force fighters as well by virtue of directing MACV's intelligence branch.[44] As a result, the air interdiction that occurred in South Vietnam was haphazard and piecemeal.

The Air Force's struggle with the Army for control of the B-52s typified the attitude of many air leaders in regard to how to conduct the air war in the South. Air Force Chief of Staff John McConnell, who pressed vigorously for three weeks of "hard knock" bombing to initiate Rolling Thunder, tried equally hard to obtain an aerial victory in the South. McConnell stated in August 1965 that ground forces alone could not defeat the Viet Cong and that only air power could defeat the enemy.[45] The general meant air power provided by the United States Air Force, however. When the Army prepared to launch its first airmobile offensive into the Ia Drang Valley in November 1965 with the First Air Cavalry Division's 434 helicopters, McConnell ordered Air Force commanders in Hawaii and Saigon to keep detailed statistics on every phase of the operation to show that the Army was incapable of conducting such an offensive without support from Air Force fixed wing aircraft.[46] Many Air Force commanders were also hesitant to condone transformation of the old C-47 transport into the AC-47 gunship because they believed that endorsement of the concept would legitimize the Army's use of armed transport helicopters, which would in turn partly eliminate the Army's need for Air Force air support.[47] McConnell disapproved of using the B-52s in the South because suitable targets were scarce and Westmoreland was reluctant to send ground troops into the bombed areas to determine the exact amount of damage inflicted. Yet, in a twisted bit of parochial logic, in September 1965 he endorsed continued B-52 bombing in South Vietnam "since the Air Force had pushed for the use of air power to prevent Westmoreland from trying to fight the war solely with ground troops and helicopters."[48]

Use of the heavy bombers had mixed results. The initial Arc Light missions produced little damage, as the bombs missed Viet Cong areas, or the enemy, possibly tipped off by agents who had infiltrated the South Vietnamese military, fled before the bombers arrived.[49] On the other hand, if intelligence could pinpoint enemy forces, as occurred during the 1965 Battle of the Ia Drang and the 1968 siege of the Marine outpost at Khe Sanh, then the effects could be devastating. On 14 November 1965, eighteen rapidly dispatched B-52s dropped 344 tons of bombs on two North Vietnamese regiments to wreck their counterattack against the 1st Air Cavalry Division at the Ia Drang.[50] During the seventy-seven-day siege of Khe Sanh, from 15 January to 31 March 1968, B-52s flew 2,548 sorties and dropped 59,542 tons of bombs, and, along with Air Force, Navy, and Marine fighters and Marine artillery, completely destroyed two attacking North Vietnamese divisions.[51] Truong Nhu Tang, Viet Cong minister of justice who survived several B-52 attacks, described a raid as "an experience of undiluted psychological terror." He remembered: "The first few times I experienced a B-52 attack it seemed, as I strained to press myself into the bunker floor, that I had been caught in the Apocalypse. One lost control of bodily functions as the mind screamed incomprehensible orders to get out."[52] Truong later noted that he survived the attacks because of the advance warning provided by Soviet intelligence trawlers that observed B-52 take-offs from Guam and relayed the information to the North Vietnamese and the Viet Cong; flights from Thailand were similarly monitored.[53] Such warnings were of limited value, though, if the enemy chose to wage open warfare, as was the case at the Ia Drang and Khe Sanh. Whenever the North Vietnamese or Viet Cong chose to mass and fight a "conventional" war of movement, they paid the price to American air power. Until the 1968 Tet Offensive, they rarely chose to do so.

The enemy's restrained combat during the Johnson presidency made it difficult for air and ground commanders alike to determine air power's impact on the ground battle. "Unfortunately, for the planners at the time and for subsequent researchers, reliable quantitative indications of results were unobtainable," comments historian John Schlight. "For one thing, the Air Force had no clear-cut objective of its own to measure results in South Vietnam."[54] Most Air Force commanders preferred to use their weaponry in independent operations like interdiction that offered the prospect of large-scale returns, rather than for auxiliary missions like close air support that provided a limited amount of assistance to a single ground unit. To gauge bombing success in the South, Admiral Sharp relied on the arcane sortie count methodology used to evaluate Rolling Thunder. In April 1966, he told assembled commanders in Honolulu that he aimed to complete the planned sortie totals for the year despite a shortage in conventional ordnance that would force aircraft to fly with less than full bomb loads.[55]

Westmoreland further clouded the impact of Air Force attacks by telling air commanders to label all air strikes in the South as close air support missions. His directive displayed his unwavering conviction that the entire country of South Vietnam was simply one large battlefield, and that all air power supported his massive ground campaign on it. Nonetheless, the Air Force continued to track sorties used to assist troops actually engaged in combat with enemy forces, and found that only three percent of all sorties flown in South Vietnam through the end of 1966 went to that end.[56] In addition, the Army requested close air support for only one out of every ten engagements with the enemy. A key reason for the dearth of requests was that half of all ground battles in the South lasted less than twenty minutes, which was too short a span to call upon air power for assistance.[57]

The limited amount of ground combat requiring close air support, along with the enemy's propensity to avoid fighting, led to the development of "free fire zones" in South Vietnam. These areas were "known enemy strongholds . . . virtually uninhabited by noncombatants" where any identified activity was presumed to stem from enemy forces and was thus susceptible to immediate air or artillery strikes.[58] Although the notion seemed to guarantee solid results for air power, in actuality it frequently proved disastrous to oft-repeated goal of the winning the "hearts and minds" of the Vietnamese peasantry. Religious beliefs compelled many villagers to return to ancestral homes they were forced to leave to create the zones, and aerial reconnaissance sometimes mistakenly identified their return as the arrival of Viet Cong soldiers. Such instances had disastrous consequences, and virtually guaranteed that any survivors who might have been apathetic toward the war before the attack would now side with the enemy.

Unlike Rolling Thunder, President Johnson and his political advisors placed few restrictions on the air wars in the South. Johnson deemed that the Chinese or Soviets would raise little outcry over air raids on South Vietnamese territory, and that raids condoned by Southern leaders would not attract the attention of the world press. Until the 1968 siege of Khe Sanh, Air Force aircraft based in Thailand could not attack targets in South Vietnam without first landing in the South and then flying from there.[59] Air commanders also had to receive permission from South Vietnamese province chiefs, who were responsible for the welfare of everyone living in their province, before launching air strikes.[60] Yet, as was the case with free fire zones, obtaining clearance to attack did not guarantee that innocent civilians would not be injured or killed. A favorite technique of Viet Cong units was to fire one or two shots at an American patrol from a South Vietnamese village and then quickly depart the area. The patrol leader might respond by requesting air support, and if the local province chief approved the request, the destruction of an innocent hamlet might result.

In short, applying air power to the struggle in South Vietnam was an enormously difficult proposition. The war was not just a guerrilla conflict, but a civil war as well marked by centuries of animosity. The key to victory was controlling passion rather than position. The location of front lines or the amount of men and equipment lost had meaning only in terms of how those variables affected the remainder of those willing to fight, and the Air Force attempted to eliminate that desire from the enemy through Rolling Thunder. Concurrently, air commanders faced the challenge of how best to provide the Army and Marines with auxiliary doses of air power. To many ground commanders, the answer was simply to provide more firepower sooner. The disparate air forces that flew over South Vietnam could usually reply with large amounts of ordnance, although not always as rapidly as ground commanders would have liked. In a war for the control of hearts and minds, however, more bombs was not necessarily the right answer.

Airlift, a nonlethal form of air power, offered not only the potential to support combat units with men and equipment—as the C-130s, C-123s, and C-7s did in magnificent fashion for the beleaguered Marines at Khe Sanh—but also to carry government officials, food, clothing, and building materials to villages throughout South Vietnam. Such "mercy missions" were in fact conducted, but the "pacification" effort in the South never received the emphasis of the combat airlifts. Moreover, airlift was vulnerable to ground fire, and it was transitory. President Johnson's goal of a stable, independent, non-Communist South Vietnam could be achieved only through a long-term presence on the ground. Accomplishing his war aim would have required a massive outlay of manpower, as well as a fundamental change of mindset about how to use those men. After three years of warfare and the shock of the Tet Offensive, neither the president nor the American public was willing to up the ante, and the war aim changed.

Johnson's successor labeled his goal in Vietnam "peace with honor," but the phrase was a euphemism for a withdrawal that retrieved American prisoners without abandoning the South to an imminent Communist takeover. President Richard Nixon began incrementally removing American ground forces from the South in 1969, and planned to turn the entire combat effort over to the South Vietnamese through a program of "Vietnamization." To buy time for the Southern military to assume control, Nixon turned to air power. On 18 March 1969, in response to a request from Cambodian Prince Norodom Sihanouk, the president began the secret bombing of Cambodia with B-52s. The raids continued through May 1970 and dropped 120,578 tons of bombs on Cambodian soil.[61] Nixon kept the attacks secret partly to maintain Cambodia's official status as a neutral in the Vietnam War, and partly to hide his expansion of the war from an American public that demanded that an end to involvement in Southeast Asia.

Nixon also increased the intensity of the air war in Laos. President Johnson had initiated American bombing there in December 1964 with operation "Barrel Roll," which provided air support to Laotian Army forces in northern Laos battling the Communist Pathet Lao. In 1965 Johnson also approved operation "Steel Tiger," an interdiction effort against those parts of the Ho Chi Minh Trail traversing southeastern Laos. Nixon significantly increased the bombing of the Ho Chi Minh Trail in operation "Commando Hunt." Johnson had halted all bombing of North Vietnam in November 1968 as a result of a perceived agreement with North Vietnamese negotiators in Paris, who seemingly agreed to stop moving men and materiel across Vietnam's demilitarized zone, halt attacks on major South Vietnamese cities, and not shoot at American reconnaissance aircraft in return for an end to bombing the North.[62] The demise of Rolling Thunder caused air commanders to focus on destroying enemy supplies after they departed North Vietnam, and Commando Hunt aimed to accomplish that goal. It began on 15 November 1968, with the number of bombing missions in Laos jumping 300 percent, from 4,700 in October to 12,800.[63] The trend continued with Nixon, who continued Commando Hunt attacks until April 1972.

Although the Air Force claimed impressive totals for raids, the actual impact of Commando Hunt was far less certain. In December 1971, Secretary of the Air Force Robert C. Seamans reported that bombing the Ho Chi Minh Trail had prevented all but 21,000 tons of materiel from arriving in South Vietnam out of the 68,000 tons shipped during the 1969–70 season, while in the 1970–71 season, only 9,500 tons arrived at their final destination out of 68,500 tons shipped.[64] For the enemy, however, such totals more than sufficed to carry on the war. Even though the 1968 Tet Offensive had been a tremendous political victory for the North Vietnamese and Viet Cong, in strictly military terms the offensive had been a disaster. American and South Vietnamese firepower decimated the Viet Cong units that came out of hiding to launch the attack, and the Viet Cong never fully recovered as an effective fighting force. North Vietnamese units, like those that massed to attack Khe Sanh, suffered fearfully as well. After the summer of 1968, the enemy again reverted to infrequently waged guerrilla warfare, in large measure to recover from the wounds of Tet, and the North Vietnamese became the dominant enemy in the South. Because they fought only "at the gnat-swarm stage," North Vietnamese supply requirements remained minimal until they launched the Easter Offensive in 1972.[65] For air commanders hoping to achieve an independent success through Commando Hunt, once again the nature of the war severely limited their prospects.

Those prospects improved at the end of March 1972 when the North Vietnamese began a massive, three-pronged offensive against South Vietnam. Except for a lack of air cover, the assault resembled the German blitzkrieg of World War II—more than 100,000 men, backed by T-54 tanks and 130mm heavy artillery, stormed across the South Vietnamese border. The Easter Offensive transformed the war into a fast-paced conventional conflict. The North Vietnamese drive needed vast logistical support to sustain it, and their supply lines proved highly vulnerable to air attack and airborne mining. The transportation links that American pilots had bombed with sparse results during Rolling Thunder had suddenly become worthwhile targets.

Nixon probably never appreciated the air power implications of the change in North Vietnamese strategy,[66] but he had known that an offensive was imminent, and he did not intend for Hanoi to claim victory while he was president. The continued withdrawal of American troops prevented him from responding to the threat with ground power; by May 1972, only 69,000 American troops would remain in the South. If he intended to use military force, air power was his sole option. In late 1971, after learning that the North Vietnamese had begun stockpiling goods for a potential attack on the South, he launched operation "Proud Deep Alpha," in which Air Force fighters flew more than one thousand sorties from 26 to 30 December against North Vietnamese supply areas south of the 20th parallel.[67] The Air Force further attacked North Vietnam continually for forty-eight hours in mid-February 1972.[68] More significant than the bombing was the start of a tremendous transfer of aircraft to Southeast Asian bases. Nixon sent 207 Air Force F-4s to South Vietnam and Thailand between 29 December 1971 and 13 May 1972, giving him a total of 374. He also ordered 161 B-52s to Andersen Air Base, Guam, and U-Tapao, Thailand, between 5 February and 23 May, creating a total of 210 B-52s in the Far East—more than half the bomber fleet of Strategic Air Command.[69] In addition, he dispatched the carriers *Constellation* and *Kitty Hawk* to join the *Coral Sea* and *Hancock* in the Tonkin Gulf, which gave the Navy a force of 300 attack aircraft by mid-April.[70]

When the North Vietnamese attacked at the end of March, Nixon was primed for action, and he took those around him by storm. After military chiefs reported that monsoon weather would temporarily prevent flying over the North, the president told aide H. R. Haldeman and Attorney General John Mitchell on 4 April to "try and get the weather, damn it, and if you know any prayers, say them. . . . The bastards have never been bombed like they're going to be bombed this time."[71] Two days later, Nixon met with Air Force Gen. John W. Vogt, Jr., en route to Saigon to take command of 7th Air Force. Vogt described the President as wild-eyed as he berated air commanders for lacking aggressiveness. "He wanted somebody to use imagination—like Patton," Vogt remembered.[72] Nixon concluded that he would have to be that individual. He had long admired the World War II hero, a California native born in Nixon's Twelfth District, and was especially fond of watching the George C. Scott movie version of the general's World War II career. The president relished Patton's tough demeanor, and he emulated it. "The US will not negotiate at the point of a gun," he scrawled on 10 April in notes for a projected speech on the Vietnam situation.[73]

After the weather cleared on 5 April, Nixon ordered fighter attacks against supply concentrations south of the 18th parallel. When those raids failed to slow the North Vietnamese offensive, the president ordered B-52 raids against targets in the Northern heartland, including Haiphong. Nixon also wanted to send the giant bombers against targets in Hanoi, but National Security Advisor Henry Kissinger warned against it, claiming that such attacks would produce great domestic criticism. Although the president relented, he wanted to make certain that the North Vietnamese feared his willingness to use military force. As Kissinger left Washington in early May for a negotiating session in Paris with chief North Vietnamese negotiator Le Duc Tho, Nixon barked: "Henry, you tell those sons of bitches that the President is a madman and you don't know how to deal with him. Once reelected I'll be a mad bomber."[74]

Unlike Lyndon Johnson, who agonized over the prospects of bombing North Vietnam and was ill at ease in the foreign policy arena, Nixon could rely on his substantial expertise in foreign affairs, plus he had master diplomat Henry Kissinger by his side. Moreover, the world situation had changed significantly from the way that it had appeared to Johnson and his political advisors, and the Nixon-Kissinger duo aimed to take advantage of it. An obvious Sino-Soviet split enabled the president to travel to Peking in February 1972 with the promise of diplomatic recognition, and then fly to Moscow three months later to sign the Strategic Arms Limitation Treaty and provide the Soviets with desperately needed grain. Nixon went to Moscow after he had begun heavy air attacks on North Vietnam and mined Northern harbors, gambling that the Soviets' desire for arms reductions and wheat would override their loyalty to Hanoi's war effort. His intuition proved correct. Although neither the Soviets nor the Chinese, who only responded with verbal protests to Nixon's bombing, were eager to forsake North Vietnam, their desire for détente eclipsed their commitment to Northern victory.

Besides gaining a free hand from the Communist superpowers, Nixon had no conflicting domestic agendas like the Great Society. He also had the advantage of the Easter Offensive's overt aggression that appeared as a legitimate threat to South Vietnam and demanded a forceful response, rather than the pinpricks of guerrilla warfare that had faced Johnson. Thus, Nixon had a tremendous freedom of action that his predecessor had lacked, and the situation was especially ripe for the type of action taken.

On 10 May, two days after aircraft mined Northern ports, Nixon initiated operation "Linebacker" against North Vietnam. The campaign, consisting primarily of Air Force and Navy fighter attacks against many of the same targets bombed during Rolling Thunder, continued until 23 October 1972. Nixon placed few political controls on Linebacker, and soon after it began air commanders could attack targets in Hanoi or Haiphong on their own initiative. The Air Force also made widespread use of "smart" munitions—laser and electro-optically guided bombs that had come into service just before Rolling Thunder ended and that had devastating effects against fixed targets like bridges. For the most part, the weather cooperated, as the North Vietnamese had launched their offensive near the start of the North's dry season. The five months of Linebacker, plus the month of attacks preceding it, deposited 155,548 tons of ordnance on North Vietnam, reducing overland imports via rail and truck from 160,000 tons to 30,000 tons a month, while mining reduced seaborne imports from more than 250,000 tons a month to zero.[75] These reductions had a dramatic impact on the North Vietnamese Army in the South, which persisted in waging a conventional war of movement. By July, the Communist offensive had sputtered to a halt, unable to receive supplies from the North and ravaged by American and VNAF air strikes in the South as well as vigorous South Vietnamese Army counterattacks.

Kissinger believed that the time had come to restart the peace negotiations terminated after he had met with an intransigent North Vietnamese delegation in Paris on 2 May. The president concurred. Although the public talks begun during the Johnson presidency had continued nonstop since 1968, Nixon put his faith in Kissinger's private sessions with Le Duc Tho, which, until the May meeting, had been conducted in secret. Hanoi agreed to renew the private talks and they resumed on 19 July. At them, Kissinger stressed a peace proposal that he had carried from Nixon in late April when he visited Moscow to make final preparations for the May summit—the North Vietnamese could keep troops in the South after a negotiated settlement that included a prisoner exchange, provided that they agreed to withdraw the soldiers who had entered since the start of the invasion.[76] Le Duc Tho made several concessions during July and August, including dropping the demand for South Vietnamese President Nguyen Van Thieu's immediate removal. Finally, on 8 October, Tho dropped his demand for a coalition government in the South and agreed to an in-place ceasefire followed by the withdrawal of remaining American troops.

Prospects for peace appeared high, but those prospects had excluded the South Vietnamese. Not until 19 October did President Thieu first see the draft agreement's text, and when he did, he was outraged.[77] Several provisions in the draft caused him great concern, particularly those allowing Northern troops to remain in the South and creating a "National Council of Reconciliation and Concord," containing Communist representation, that he feared would be a stepping stone to a future coalition government. Thieu's fears prevented him from acquiescing to the accord hammered out between Tho and Kissinger and forced Kissinger back to the bargaining table. "As a token of good will," Nixon ended Linebacker and suspended air attacks in the North Vietnamese heartland above the 20th parallel. "But," he remembered, "there was to be no bombing halt until the agreement was signed. I was not going to be taken in by the mere prospect of an agreement as Johnson had been in 1968."[78]

Nixon now found both Hanoi and Saigon blocking his goal of an "honorable" disengagement from the war. Thieu presented Kissinger with a list of sixty-nine changes to incorporate into the agreement. When Kissinger presented the list to the North Vietnamese, they balked and began withdrawing concessions that they had made earlier. Inconclusive negotiations continued in November and December, with both Tho and Thieu refusing to back down. Now, however, Nixon had a new concern—the Democratic Congress elected during his landslide reelection in November threatened to cut funding for South Vietnam if Nixon did not end American involvement in the war by the time that Congress convened in early January. Nixon believed that he could not afford to delay, and after a disappointing session on 13 December, he ordered Kissinger home from Paris. The president would rely once more on air power.

This time, he focused that weapon on the enemy's will to continue the war—as well as the will of his ally to oppose an agreement. In late November, Nixon had told the Joint Chiefs of Staff to prepare to bomb Hanoi with B-52s if the enemy got "hardnosed" and refused to negotiate,[79] and on 14 December he ordered a massive air offensive to begin against targets in the North Vietnamese heartland on the 18th. Dubbed "Linebacker II," the eleven-day air campaign from 18 to 29 December (no attacks occurred on Christmas Day) dropped 20,000 tons of bombs on the North, with B-52s delivering three-fourths of that total on rail centers and storage areas.[80] Air commanders had originally designed the air offensive to include large numbers of fighter attacks, but monsoon weather prevented the fighters from flying smart bomb missions during all but twelve hours of the campaign.[81] Nixon, however, was pleased by the large-scale participation of the B-52s. He aimed to send a signal to both Hanoi and Saigon that he was willing to risk a key part of America's nuclear triad to secure peace in Vietnam. Although the Air Force lost fifteen of the giant bombers to surface-to-air missiles, and leading congressional Democrats and the world press condemned the bombing, on 28 December the North Vietnamese agreed to restart negotiations in early January. They also accepted Nixon's caveat that they would not deliberate on matters already covered by the basic settlement accepted in October.

Once Kissinger and Le Duc Tho met, talks progressed rapidly, and the two drafted an agreement on 13 January 1973. That accord contained only cosmetic differences from the one negotiated in October. Thieu agreed to endorse it only after Nixon sent him a series of ultimatums stating that Congress would definitely cut funds to South Vietnam if he refused to accept the settlement. The South Vietnamese president probably appreciated Linebacker II, but whether the bombing induced him to back the agreement remains a matter of conjecture.

Nixon believed that his December air offensive had succeeded in extracting America from its most difficult war, and many air chiefs agreed. After receiving word on 28 December that Hanoi had accepted his conditions for a return to Paris, the president told advisor Charles Colson: "The North Vietnamese have agreed to go back to the negotiating table on our terms. They can't take bombing any longer. Our Air Force really did the job."[82] Admiral Thomas Moorer, chairman of the Joint Chiefs of Staff, noted that "air power, given its day in court after almost a decade of frustration, confirmed its effectiveness as an instrument of national power—in just nine and a half flying days."[83] Air Force Gen. J. C. Meyer, commander in chief of Strategic Air Command, and Lt. Gen. Gerald W. Johnson, the 8th Air Force commander whose B-52s had conducted Linebacker II, both shared Moorer's opinion.[84] Many air chiefs likened Linebacker II to the plan for an intensive, three-week campaign in the spring of 1965, and concluded that such an effort then would have won the war.[85]

Yet Nixon's Christmas air offensive had not "won" the war, nor had it achieved success alone. The peace treaty signed in Paris by Secretary of State William P. Rogers on 27 January 1973 enabled the United States to recover its prisoners and permitted a resupply of South Vietnam after the American departure, but it did not guarantee the survival of a non-Communist South. By allowing North Vietnamese troops to remain inside South Vietnam, Nixon made that survival problematic. Air power had certainly hurt the North Vietnamese Army, and its desperate straits were likely one reason that Hanoi agreed to sign an accord in January.[86] Yet North Vietnamese leaders had no intention of surrendering their goal of a unified Vietnam, and America's departure from the conflict, along with the sanctioned placement of communist troops in the South, boded well for their prospects of overthrowing the Saigon regime.

Air power's ability to help achieve Nixon's war aim stemmed from several factors: Hanoi's decision to wage a fast-paced, conventional war of movement, which made its army dependent on vulnerable supply lines in the North and susceptible to a tremendous aerial pounding in the South; the widespread development of precision guided munitions, which had maximum effect against an enemy waging conventional war; the skill of the president and Kissinger in foreign affairs, which enhanced air power's effectiveness as a political tool; Nixon's freedom to apply air power without having either the domestic or foreign policy concerns that plagued his predecessor, which in turn resulted in an air campaign having minimal political controls; the happenstance of suitable flying weather throughout most of Linebacker I; and, most importantly, the fact that Nixon's objective in Vietnam was vastly more limited than Johnson's, and hence easier to obtain.

Many air leaders failed to see these distinctions, however, and pointed to Linebacker II as the proper way to apply air power in any future conflict. General LeMay, who had pushed for a Linebacker-type attack during his last year as chief of staff, remarked in July 1986 that America could have won the Vietnam War "in any two-week period you want to mention" had it relied on an intensive air campaign against the North.[87] The Air Force's "Basic Doctrine" Manual produced in 1984 reflected a similar conviction, noting that destroying "a selected series of vital targets" that comprised a nation's industrial or military establishment was likely to wreck an enemy's war-making capacity and will to fight.[88] For many air commanders, Linebacker II became a vindication of the doctrine spawned by Billy Mitchell and Air Corps Tactical School instructors, as well as proof that the doctrinal tenets of massive retaliation did indeed suffice for wars of lesser magnitude. Historian Donald J. Mrozek observes that for many strategic bombing enthusiasts, "Linebacker II was the contemporary equivalent of the Battle of New Orleans—an event of ambiguous pertinence to the war, but an indispensable mythic force for reasserting long-held beliefs."[89]

The Air Force has not necessarily abandoned those beliefs since the Vietnam War, although the publication of its latest "Basic Doctrine" Manual in the aftermath of Desert Storm reveals much less dogmatic thinking. Dated March 1992, the current manual notes that "there is no universal formula for the proper employment of aerospace power in a campaign" and that "optimum use of aerospace forces depends upon a host of dynamic circumstances peculiar to the conflict at hand."[90] Still, the manual hearkens to earlier convictions about the primacy of independently applied air power despite couching the notion that air power can win a solo victory. "While powerful synergies can be created when aerospace, land and naval forces are employed in a single, integrated campaign," the manual declares, "it is possible that aerospace forces can make the most effective contribution when they are employed in parallel or relatively independent aerospace campaigns."[91]

With one eye on Vietnam, air planners for Desert Storm designed an "Instant Thunder" air campaign for the Gulf War that sought to avoid Rolling Thunder's gradualism. That plan originally called for heavy attacks on industrial, transportation, and communication targets in Baghdad, and was later expanded to include simultaneous raids against deployed Iraqi troops in Kuwait. The desire for an independent air victory remains strong among many in the Air Force, and against Saddam Hussein's Iraq the focus on attacking the nation's perceived military-industrial web was perhaps appropriate. The Iraqi leader tightly controlled all aspects of Iraq's military; Iraqi forces waged conventional war in a desert environment where they were especially susceptible to America's sophisticated air weaponry; Saddam Hussein gave the Allied forces five and a half months to prepare their assault; and President George Bush succeeded in completely isolating Iraq on the world stage, allowing airmen to conduct an air offensive with minimal political controls. In addition, American military chiefs remembered the problems caused by the lack of a single air commander in Vietnam, and they empowered Air Force Lt. Gen. Charles Horner with the authority to direct multiple air forces as the Joint Forces Air Component Commander.

The forty-two-day air war against Iraq was exceedingly effective, but air leaders should be wary of making the campaign a prototype for future conflicts. No blueprint exists for the proper application of air power. Many factors during Desert Storm combined to produce success—much as they had during the Linebacker campaigns—and the Iraqis fought the ideal type of war for American air power. The next enemy may not prove so accommodating. A determined opponent waging guerrilla warfare, especially one fighting infrequently like the Viet Cong, is extremely difficult to defeat with air power applied either as an independent force or in support of ground operations. Should air commanders again face a guerrilla enemy, they may find, like those conducting Rolling Thunder, that the application of air power actually helps the enemy's war effort more than it hurts it. Air chiefs wedded to the notion of an independent air victory may also find that the concept is illusory.

Notes

1. Ralph Littauer and Norman Uphoff, eds., *The Air War in Indochina* (Boston: Beacon Press, 1972), 11, 168–72; Earl H. Tilford, Jr., *Crosswinds: The Air Force's Setup in Vietnam* (College Station: Texas A&M University Press, 1993), 109.

2. Air Force Manual 1-2, 1 December 1959, 4.

3. William Mitchell, draft of *War Memoirs*, 2, File: Diaries, May 1917–February 1919, box 1, William Mitchell Papers, Library of Congress, Washington, D.C.; William Mitchell, *Skyways* (Philadelphia, Pa.: Lippincott, 1930), 253.

4. William Mitchell, "Aeronautical Era," *Saturday Evening Post*, 20 Dec. 1924, 3.

5. "It may be necessary to intimidate the civilian population in a certain area to force them to discontinue something which is having a direct bearing on the outcome of the conflict," Mitchell observed in his bombing manual. Achieving that goal might cause some civilian deaths, but the number would pale compared to the deaths produced by a ground war between industrialized powers. Moreover, once bombed, civilians were unlikely to continue supporting the war effort. "In the future, the mere threat of bombing a town by an air force will cause it to be

evacuated and all work in munitions and supply factories to be stopped," he asserted. See ibid.

6. Robert T. Finney, *History of the Air Corps Tactical School, 1920–1940* (Maxwell AFB, Ala.: Air University, 1955; reprint ed., Washington, D.C.: Office of Air Force History, 1992), 43. Of the three-star generals in the Army Air Forces at the end of the war, eleven of thirteen were Tactical School graduates, and three four-star generals—Joseph McNarney, George Kenney, and Carl Spaatz—graduated from the school.

7. Wesley Frank Craven and James Lea Cate, *The Army Air Forces in World War II,* 7 vols. (Chicago: University of Chicago Press, 1948–1958; reprint ed., Washington, D.C.: Office of Air Force History, 1983), 1:50–52. See also Donald Wilson, "Origin of a Theory for Air Strategy," *Aerospace Historian* 18 (March 1971); and Michael S. Sherry, *The Rise of American Air Power: The Creation of Armageddon* (New Haven: Yale University Press, 1987). Wilson, who served as chief of the "Air Force" section from 1931 to 1934, and director of the Department of Air Tactics and Strategy from 1936 to 1940 at the Air Corps Tactical School, was highly influential in the development of the industrial web theory.

8. Air Force Manual 1-8, 1 May 1954, 6.

9. Ibid., 2.

10. Ibid., 4.

11. Ibid., 5, 8.

12. USAF Oral History Interview of Brig. Gen. Noel F. Parrish by Dr. James C. Hasdorff, 10–14 June 1974, San Antonio, Tex., Air Force Historical Research Agency (hereafter cited as AFHRA), Maxwell AFB, Ala., file K239.0512-744, 204.

13. "Meeting with Foreign Policy Advisors on Vietnam," 18 Aug. 1967, Meeting Notes File, box 1, Lyndon Baines Johnson Presidential Library (hereafter Johnson Library), Austin, Tex. In July 1965, Defense Department analysts had estimated that 192,000 Viet Cong and three North Vietnamese Army regiments (7,500 men) fought in the South. See McNamara to the president, 3 Nov. 1965, National Security Files, Country File: Vietnam, folder 2EE, box 75, Johnson Library.

14. Headquarters USAF, *Analysis of Effectiveness of Interdiction in Southeast Asia, Second Progress Report,* May 1966, AFHRA, file K168.187-21, 7. The study further noted: "The present low requirement of 34 tons/day, though made up largely of ammunition, provides much less than is usually calculated for North Vietnamese forces. Thirty-six percent of the supply support for a soldier in a North Vietnamese light division consists of ammunition. When he is deployed to the south this drops to 18%. Only 6% of the supplies furnished Viet Cong Main Force soldiers is ammunition. Only a 13% firepower utilization rate is presently being experienced by the VC/NVA troops in South Vietnam." McNamara acknowledged in 1967 that Communist forces fought an average of one day in thirty and remarked that they needed fifteen tons of supplies daily from external sources. The Joint Chiefs had estimated in August 1965 that the enemy needed thirteen tons per day of "external logistical support." See U.S. Congress, Senate, Committee on Armed Services, Preparedness Investigating Subcommittee, *Air War against North Vietnam,* 90th Cong., 1st sess., 25 Aug. 1967, pt. 4, 299, and Annex A to JCSM 613-65, 27 Aug. 1965, National Security Files, Country File: Vietnam, folder 2EE, box 75, Johnson Library. Truong Nhu Tang, *A Viet Cong Memoir* (New York: Vintage, 1985), 156–64, details the specific supply requirements of the Viet Cong.

15. The standard military 2½-ton truck could transport 5 tons of goods over roads and 2.5 tons overland.

16. Rostow to the president, 6 May 1967, National Security Files, Country File: Vietnam, folder 2EE, box 75, Johnson Library, and *The Pentagon Papers: The Defense Department History of United States Decisionmaking in Vietnam,* Senator Gravel edition, 5 vols. (Boston: Beacon, 1971), 4:146.

17. Appendix A to JCSM 613-65, 27 Aug. 1965.

18. Lyndon Baines Johnson, *The Vantage Point* (New York: Holt, Rinehart & Winston, 1971), 66–67, 153.

19. Bernard Brodie, *War and Politics* (New York: Macmillan, 1973), 138.

20. Doris Kearns, *Lyndon Johnson and the American Dream* (New York: Signet, 1976), 263.

21. Message, 04030Z April 1965, CINCPAC to JCS, in *Commander-in-Chief, PACOM, Outgoing Messages, 22 January–28 June 1965,* AFHRA, file K712.1623-2.

22. In a 15 July 1985 interview with the author at Athens, Ga., Rusk stated that underestimating North Vietnamese determination was one of his greatest mistakes regarding Vietnam: "I thought the North Vietnamese would reach a point, like the Chinese and North Koreans in Korea, and Stalin during the Berlin airlift, when they would finally give in."

23. Goodpaster to the president, "Meeting with General Eisenhower," 12 Oct. 1965, National Security Files, Name File: President Eisenhower, box 3, Johnson Library.

24. Senate Preparedness Subcommittee, *Air War against North Vietnam,* pt. 5, 27–29 Aug. 1967, 478.

25. Message, 190935Z May 1965, 41st ADIV ADVON to 2 AD CP, in *PACAF Outgoing Messages, 3 April–24 December 1965,* AFHRA, file K717.1623.

26. Ibid.

27. USAF Oral History Interview of Lt. Gen. Joseph H. Moore by Maj. Samuel E. Riddlebarger and Lt. Col. Valentino Castellina, 22 Nov. 1969, AFHRA, file K239.0512-241, 17–18.

28. John Morrocco, *Thunder from Above* (Boston: Boston Publishing, 1984), 125; Lt. Col. William H. Greenhalgh, interview with the author, Maxwell AFB, Ala., 17 May 1985.

29. Statement to the author in July 1989 by a retired Navy pilot who preferred to remain anonymous.

30. *Pentagon Papers,* Gravel ed., 4:138.

31. Jack Broughton, *Thud Ridge* (New York: Bantam, 1969), 24.

32. Littauer and Uphoff, 283.

33. By January 1968, Hanoi had received almost $600 million in economic aid and $1 billion in military assistance. See Jason Summer Study, "Summary and Conclusions," 30 August 1966, *Pentagon Papers,* Gravel ed., 4:116, and Department of Defense Systems Analysis Report, January 1968, ibid., 225–26. The Systems Analysis Report stated: "If economic criteria were the only consideration, North Vietnam would show a substantial net gain from the bombing."

34. Oleg Hoeffding, *Bombing North Vietnam: An Appraisal of Economic and Political Effects* (December 1966), RAND Corporation Memorandum RM-5213, 17.

35. NSSM 1 (February 1969), *Congressional Record 118,* pt. 13 (10 May 1972), 16833.

36. Sherry, 277.

37. U. S. Grant Sharp and William C. Westmoreland, *Report on the War in Vietnam (as of 30 June 1968)* (Washington, D.C.: Government Printing Office, 1969), 53.

38. See Truong, esp. chaps. 13 and 16.

39. In early 1966, 2d Air Division became 7th Air Force.

40. In October 1966, 7th Air Force gained the new 834th Air Division, which contained the 315th Air Wing's C-123s that had previously operated in South Vietnam under PACAF's jurisdiction. The new 483rd Air Wing at Cam Ranh Bay, also a part of the 834th Air Division, gained the C-7s that had formerly belonged to the Army as a result of a March 1966 agreement between Air Force Chief of Staff McConnell and Army Chief of Staff Gen. Harold K. Johnson. Seventh Air Force never gained full control of PACAF's C-130s, and had operational control of them only when they deployed to Vietnam from other Pacific bases. See Ray L. Bowers, *The United States Air Force in Southeast Asia: Tactical Airlift* (Washington, D.C.: Office of Air Force History, 1983), 174–82, 241–47, 353–76.

41. John Schlight, *The United States Air Force in Southeast Asia: The War in South Vietnam: The Years of the Offensive 1965–1968* (Washington, D.C.: Office of Air Force History, 1988), 83.

42. Ibid., 148–51.

43. For Momyer's personal thoughts on his efforts to gain control of the B-52s, see his *Air Power in Three Wars* (Washington, D.C.: Government Printing Office, 1978), 99–104.

44. The chief of MACV plans was the only Air Force general on the MACV staff, and he was excluded from daily planning and focused on long-range concerns, like Southeast Asia Treaty Organization planning. The J-3 Army general handled operational matters. See Schlight, 10–11, 126–27.
45. Schlight, 76.
46. Ibid., 103, 106.
47. Ibid., 90.
48. Ibid., 82.
49. Ibid., 52.
50. Carl Berger, ed., *The United States Air Force in Southeast Asia, 1961–1973: An Illustrated Account* (Washington, D.C.: Office of Air Force History, 1984), 150. For the whole of the Ia Drang Battle, B-52s flew 96 sorties and dropped 1,795 tons of bombs.
51. Bernard C. Nalty, *Air Power and the Fight for Khe Sanh* (Washington, D.C.: Office of Air Force History, 1988), 88.
52. Truong, 167–68.
53. Ibid., 168.
54. Schlight, 136.
55. Ibid., 119.
56. Ibid., 216.
57. Ibid.
58. Quoted in Sean A. Kelleher, "Free Fire Zones," in James S. Olsen, ed., *Dictionary of the Vietnam War* (Westport, Conn.: Greenwood, 1988), 163.
59. Schlight, 39, 277.
60. Ibid., 38.
61. Berger, 141.
62. The willingness of the North Vietnamese to agree to such an arrangement remains a point of controversy. President Johnson wrote in his memoirs: "Before I made my decision [to halt the bombing], I wanted to be absolutely certain that Hanoi understood our position. . . . Our negotiators reported that the North Vietnamese would give no flat guarantees; that was in keeping with their stand that the bombing had to be ended without conditions. But they had told us that if we stopped the bombing, they would 'know what to do.' [American negotiators] were confident Hanoi knew precisely what we meant and would avoid the actions that we had warned them would imperil a bombing halt." See Johnson, *Vantage Point,* 518.
63. Tilford, 109.
64. Littauer and Uphoff, 73.
65. Douglas Pike, *PAVN: People's Army of North Vietnam* (Novato, Calif.: Presidio, 1986), 223.
66. Nixon stated in his memoirs that an intensive air campaign against North Vietnam in 1970, in concert with the Cambodian invasion, would have brought the war to a close. See Richard Nixon, *RN: The Memoirs of Richard Nixon,* 2 vols. (New York: Warner, 1978), 2:79. In April 1988, he told NBC's "Meet the Press" that his failure to bomb and mine North Vietnam immediately after taking office was the greatest mistake of his presidency. "If we had done that then," he remarked, "I think we would've ended the war in 1969 rather than in 1973." See Transcript of NBC's "Meet the Press," 10 Apr. 1988.
67. Headquarters, 7th Air Force, *7 AF History of Linebacker Operations, 10 May 1972–23 October 1972,* n.d., AFHRA, file K740.04-24, 1.
68. Lavelle to Ryan and Clay, "Daily Wrap-Up," 10 Feb. 1972, in *Pave Aegis and Other Miscellaneous Messages, June 1971–June 1972,* AFHRA, file K717.03-219, vol. 5.
69. Ibid., 3–6; *Air War—Vietnam* (New York: Arno, 1978), 115–25; James R. McCarthy and George B. Allison, *Linebacker II: A View from the Rock* (Maxwell AFB, Ala.: Air War College, 1979), 11.

70. "The New Air War in Vietnam," *U.S. News and World Report,* 24 Apr. 1972, 15.
71. "Nixon Bombing Recorded in Tape," *New York Times,* 30 June 1974.
72. Quoted in Seymour M. Hersh, *The Price of Power: Kissinger in the Nixon White House* (New York: Summit, 1983), 506.
73. Nixon's handwritten notes, "Vietnam Points to Emphasize," 10 Apr. 1972, Folder: "Monday, 10 April 1972," box 74, President's Personal File—President's Speech File 1969, Nixon Presidential Materials Project, Alexandria, Va.
74. Quoted in Hersh, 568.
75. Guenter Lewy, *America in Vietnam* (New York: Oxford University Press, 1970), 411; U.S. House, Committee on Appropriations, Subcommittee on DOD, *DOD Appropriations: Bombings of North Vietnam,* Hearings, 93rd Cong., 1st sess., 9–18 January 1973, 43; Robert N. Ginsburgh, "North Vietnam—Air Power," *Vital Speeches of the Day* 38 (15 Sept. 1972): 734.
76. Tad Szulc, "Behind the Vietnam Cease-Fire Agreement," *Foreign Policy* 15 (Summer 1974): 36; Allan E. Goodman, *The Lost Peace: America's Search for a Negotiated Settlement of the Vietnam War* (Stanford, Calif.: Hoover Institution, 1978), p. 120; and Hersh, 512–13.
77. Tad Szulc, *The Illusion of Peace: Foreign Policy in the Nixon Years* (New York: Viking, 1978), 629.
78. Nixon, *RN,* 2:193.
79. Elmo R. Zumwalt, Jr., *On Watch: A Memoir* (New York: Quadrangle, 1976), 412–15.
80. PACAF Study Group, "Linebacker II Air Operation" (briefing given 18 January 1973) in *Department of Air Force Letters Concerning USAF Air Operations in Southeast Asia, 10 October 1972 to 31 January 1973,* AFHRA, file K168.06-232; Lt. Gen. Gerald W. Johnson, *End of Tour Report* (15 Sept. 1973), AFHRA, file K416.131, 80.
81. House Appropriations DOD Subcommittee, *DOD Appropriations: Bombings of North Vietnam,* 4.
82. Charles W. Colson, *Born Again* (Old Tappan, N.J.: Chosen Books, 1976), 78.
83. "What Admiral Moorer Really Said About Airpower's Effectiveness in SEA," *Air Force,* November 1973, 25.
84. Howard Silber, "SAC Chief: B-52s Devastated Viet Air Defenses," *Omaha World Herald,* 25 Feb. 1973; USAF Oral History Interview of Lt. Gen. Gerald W. Johnson by Mr. Charles K. Hopkins, 3 Apr. 1973, Andersen AFB, Guam, AFHRA, file K239.0512-831, 11–13.
85. See, for example, USAF Oral History Interview of Gen. John W. Vogt by Lt. Col. Arthur W. McCants, Jr. and Dr. James C. Hasdorff, 8–9 August 1978, AFHRA, file K239.0512-1093, 69; U. S. Grant Sharp, *Strategy for Defeat: Vietnam in Retrospect* (San Rafael, Calif.: Presidio, 1978), 252, 255, 272; and Momyer, 339.
86. Gen. Tran Van Tra, commander of Communist forces in the southern half of South Vietnam during the Easter Offensive, later wrote after having undergone nine months of continual bombing: "Our cadres and men were fatigued, we had not had time to make up for our losses, all units were in disarray, there was a lack of manpower, and there were shortages of food and ammunition. . . . The troops were no longer capable of fighting." See Tran Van Tra, *Concluding the 30-Years War* (Ho Chi Minh City, 1982 [in Vietnamese]; reprint ed. [in English], Arlington, Va.: Joint Publications Research Service, 1983), 33, quoted in Gabriel Kolko, *Anatomy of a War: Vietnam, the United States, and the Modern Historical Experience* (New York: Pantheon, 1985), 444–45.
87. Interview of Curtis LeMay by Mary-Ann Bendel, printed in *USA Today,* 23 July 1986, 9A.
88. Air Force Manual 1-1, 16 Mar. 1984, 3-2.
89. Donald J. Mrozek, *Air Power and the Ground War in Vietnam: Ideas and Actions* (Maxwell AFB, Ala.: Air University Press, 1988), 165.
90. Air Force Manual 1-1, March 1992, 1:9.
91. Ibid.

Air Power in the Gulf War: Plans, Execution, and Results

Thomas A. Keaney

Aerial bombing has had a controversial history. From the first time powered flight revealed the potential effects of air power on the surface battle, aerial bombardment has been the most controversial of its applications. Criticism from various camps has questioned bombing's accuracy, efficiency, effectiveness, and morality. In 1991, however, the Gulf War offered the opportunity to make good on past claims and set the record straight. The introduction of smart bombs and aircraft with precise bombing apparatus, reconnaissance and communications support from space, superb crew training, and an array of command and control systems brought significant new capabilities to air attacks. Technology had provided the potential to remedy previous shortcomings in accuracy and to make good on past claims. Iraq's seizure and occupation of Kuwait in August 1990 provided the opportunity.[1]

Circumstance afforded air power in general and aerial bombardment in particular a special importance. The military planning by the United Nations coalition formed to combat Iraq counted on extensive use of air power. Though Saddam Hussein would have wished otherwise, the disposition of his forces played directly into that strength. Bombing operations that began in January 1991 thus faced a particularly favorable setting. The Iraqi army sat in a well-delineated area, for the most part away from noncombatants, in a climate with generally clear skies, with no jungle or terrain-masking of targets, no neighboring sanctuary from the bombing, and with long, exposed lines of communication and supply. Iraq's air force and air defense system, while a good regional force, could not hope to contest command of the air against the overwhelming coalition air forces. And, finally, whereas in an earlier time the fear of escalation might have constrained operations, the loss of Soviet sponsorship had removed such restraints, and all of Iraq's governmental and economic infrastructure lay open to all-out air attack by the coalition.

Because of frequent misunderstandings concerning the purpose and doctrine behind the air attacks, a few definitions are in order. Air attacks on surface targets fall into several categories, or missions. Under current United States Air Force terminology, these missions are strategic attack, offensive counterair, interdiction, and close air support.[2] Considerable philosophical disagreement concerning priority, control, and effectiveness of these missions exists within the Air Force and between the services.

Strategic air attack encompasses those missions carried out against enemy command elements, war production, and supporting infrastructure. Offensive counterair operations are those that seek to destroy, disrupt, or limit enemy air power as close to its source as possible; this essay will consider that part of offensive counterair involved in attacking ground-based air defense installations, airfields, and surface-to-air missile launch facilities. Interdiction, actions to disrupt, delay, or destroy enemy potential before it can be used against friendly forces, in the Gulf War involved two distinct operations: air attacks against the lines of communications and attacks on the Iraqi fielded forces themselves. Both operations are discussed separately, though both meet the definition of air interdiction. Close air support, air attacks made nearby to friendly forces and which require integration into the fire and movement of these forces, is used here to consider only those attacks made during the four-day ground offensive in February 1991 and during the Iraqi incursion that resulted in the capture of the Saudi town of Al Khafji in late January.

Planning

Before the crisis of August 1990, the Soviet threat shaped the expected roles of air power and the strategy for its employment in the Southwest Asia region, though changes in the planning were already underway. Until 1989, Southwest Asia was a secondary theater for United States; planning called for a limited defensive campaign, conserving resources for the more important struggle with the Warsaw Pact on Europe's central front. In October 1989, the emphasis shifted: the planning scenario posited Iraq as the primary threat, with no Soviet involvement anticipated. Furthermore, Washington planners allocated to Central Command (CENTCOM) double and sometimes triple the forces previously available. Within the first two weeks of a deployment, Air Force tactical fighter squadrons, two Navy carriers, and brigade-sized Army and Marine aviation units were among the first scheduled to arrive, followed shortly thereafter by a B-52 squadron and a third carrier. Central Command prepared a draft of this new plan, Operations Plan 1002-90, "Defense of the Arabian Peninsula," in the spring of 1990.[3] Notably missing from the forces available in the draft operations plan were the Air Force's stealth aircraft, the F-117s. Their existence had only become public in November 1988.

While still a draft plan, Operations Plan 1002-90 became the starting point for the Desert Shield deployment in August 1990, though the size and mission of the deploying forces soon made the plan obsolete. Fortunately, Central Command had an opportunity to examine the flow of the draft operations plan in July 1990 in a command post exercise, Internal Look 90. The exercise focused the command's attention on a potential conflict with Iraq just weeks before the real thing. The exercise uncovered two problems that were to figure prominently in the war: a shortage of precision-guided munitions and difficulty obtaining timely bomb damage assessment for use in planning further air attacks.[4] An important addition to the exercise force list was the F-117s, which had by then been made available to the theater commanders for regional operations plans.

Iraq's invasion of Kuwait on 2 August 1990 created a situation even more stressful than the scenario envisioned in Operations Plan 1002-90, and if anything made the use of air power more crucial. On 8 August 1990, two days after King Fahd of Saudi Arabia approved the deployment of American forces to defend his country, President George Bush outlined U.S. objectives in the region. They were to (1) secure the immediate, unconditional and complete withdrawal of Iraqi forces from Kuwait, (2) restore the legitimate government of Kuwait, (3) assure the security and stability of the Persian Gulf region, and (4) protect American lives.[5] The first step, however, was defensive, establishing a secure position in Saudi Arabia—the purpose of Desert Shield. Only then could further operations begin to secure the President's objectives—the purpose of Desert Storm.

Internal Look 90 and the 1002-90 draft plan it exercised anticipated the defensive planning of the early days of Desert Shield. The offensive actions of the Desert Storm air plan, however, evolved not from prewar planning but from special and ad hoc organizations whose existence no one had anticipated before the war.

Shortly after the Iraqi invasion of Kuwait, a group of officers on the Air Staff in the Pentagon began planning an air campaign designed to eject Iraqi forces from Kuwait. Colonel John A. Warden III, the deputy director of Air Force Plans for Warfighting, organized and supervised the effort. The group, which grew to more than one hundred officers from the Air Force and other Services, operated out of offices that previously housed an Air Staff division called Checkmate, and this ad hoc group itself became known as Checkmate. Warden and his group developed military objectives, a concept of operations, and a targeting scheme designed to accomplish the goals of the campaign using air power alone.

The plan they developed, named Instant Thunder, called for an intense six-day air campaign designed to incapacitate Iraqi leadership and destroy the country's key military capabilities. Warden organized around "centers of gravity"—key elements of the enemy state and armed forces, the destruction or disabling of which would compel the enemy to yield. The most important center of gravity was Saddam Hussein's ability to lead and control his country, so attacks on telecommunications sites and command centers would isolate him from the Iraqi people and his armed forces. Together with a psychological warfare effort directed against the Ba'athist regime, these attacks would disable or even fatally weaken the regime.

Instant Thunder also targeted Iraq's nuclear, biological, and chemical facilities, and its national air defense system and airfields. Other targets included electric power, oil production, railroads, and military production. Always, however, the United States would strenuously avoid civilian casualties and, indeed, any long-term damage to the Iraqi economy. The plan envisioned six days of operations, striking eighty-four targets (nineteen in the telecommunications set), all in Iraq, not Kuwait. Initially, Warden planned to use only Air Force assets, including F-117 aircraft. Further development of the plan included the Navy and Marine Corps aircraft deploying to the region and the Navy's Tomahawk land attack missiles (TLAMs) launched from ships and submarines.[6]

Only under unusual circumstances would the Central Command Commander-in-Chief, Gen. H. Norman Schwarzkopf, have approved of such a plan, but these were not normal circumstances. As a theater commander, he was not predisposed to ask for assistance from a Service staff, and he was not looking for an option to eject Iraq from Kuwait using air power alone, certainly not by the means of strategic bombing. He was looking for an air option to retaliate on short notice against Iraq, however, and his own staff was fully occupied managing the deployment. He considered the air options already prepared by his own command as only "symbolic" in nature. Schwarzkopf's request for help from the Air Staff brought Warden's plan before him, and Schwarzkopf was favorably impressed by the scope and offensive nature of Warden's plan. He directed Warden to brief the plan to Lt. Gen. Charles A. Horner, Commander, Air Force Component, Central Command (CENTAF), then in Saudi Arabia.[7] Horner, who was serving as Commander, Central Command, Forward, in General Schwarzkopf's place, gave Warden's plan an entirely different reception.

Lieutenant General Horner thought the plan seriously flawed in its operational aspects and disagreed with its focus on Baghdad at the expense of dealing with the Iraqi forces in Kuwait. Horner sent Colonel Warden back to Washington, though Horner retained in the theater several of Warden's planners who had accompanied him. To produce a more acceptable plan, Horner selected Brig. Gen. Buster Glosson, an Air Force officer then on another assignment in the region, to direct a planning effort for an offensive air campaign.[8] While Horner may have dismissed Warden, however, he did not dismiss his concepts, for the plan that emerged from Glosson's efforts retained the same target sets, the same focus on Iraqi leadership, and the same intent—to isolate Saddam Hussein from the Iraqi people and his forces. Instead of being an entire campaign in itself, however, the revised plan became the first phase of a more general plan to eject Iraqi forces from Kuwait.

On 25 August 1990, General Schwarzkopf briefed Gen. Colin Powell, chairman of the Joint Chiefs of Staff, on a four-phase plan, code-named Desert Storm, to eject Iraqi forces from Kuwait. The first phase, called the strategic air campaign, was essentially the Instant Thunder plan with an added aim of preventing reinforcement of Iraqi forces in Kuwait; the second phase would gain air superiority over Kuwait; the third phase consisted of air operations to reduce Iraqi ground forces' capability before the ground attack; and the fourth phase was a ground attack into Kuwait. This planning concept was identical to the one executed the following January and February, though much work remained to be done on the development of the ground attack. The timing for the attacks was a vital consideration, however, because while the briefing estimated that the first three phases would be ready to execute by early October, the ground phase would not be possible until December.[9]

Planning for the strategic air attacks, now captured in Phase I of a larger plan, went on as a matter of high priority. A Special Planning Group was set up with General Glosson as director, and the group's activities were kept secret, even from many on the CENTAF Staff. The high level of secrecy responded to concerns of the friendly Arab governments that threats of offensive military action might impede a negotiated settlement. The makeup of the Special Planning Group, nicknamed the Black Hole, included Army, Navy, and Marine Corps representatives, with most of the Air Force officers drawn from outside the CENTAF Staff. By mid-September, representatives from Great Britain's Royal Air Force and later from the Royal Saudi Air Force had joined the effort. By 2 September, the Black Hole planners had prepared and General Horner had approved a CENTAF operations order for Phase I. Briefings of the plan took place through the chain of command, reaching President George Bush and members of the National Security Council on 10–11 October. Instant Thunder had envisioned approximately 150 strike aircraft; the plan briefed to the president in October called for over 400, with another 300 aircraft—half of them helicopters— withheld in order to defend against an Iraqi invasion and to initiate the third phase.[10] Even with increases in numbers of aircraft, the concepts of the first phase remained remarkably constant.

According to the operations order for Desert Storm, there were six military objectives, and Phase I was to focus on three "centers of gravity":

Theater Military Objectives
1. Attack Iraqi political/military leadership and command and control.
2. Gain and maintain air superiority.
3. Sever Iraqi supply lines.
4. Destroy chemical, biological, and nuclear capability.
5. Destroy Republican Guard forces.
6. Liberate Kuwait City.

Centers of Gravity
1. Iraqi National Command Authority;
2. Iraq's chemical, biological, and nuclear capability;
3. The Republican Guard Forces Command.[11]

Phase I, the strategic air campaign, directed attacks against twelve sets of targets[12] in order to "result in disruption of Iraqi command and control, loss of confidence in the government, and significant degradation of Iraqi military capabilities."[13] First, command of the air was to be gained by attacks on the Iraqi *strategic air defense system* and *airfields*. The most important centers of gravity were *leadership* and *command, control, and communications* facilities. To eliminate long term Iraqi offensive capabilities, the *nuclear, biological, and chemical weapons research, production, and storage* facilities, along with the *Scud missiles, launchers, and production and storage* facilities were targeted. The key elements of the Iraqi armed forces and their supporting industries made up the remainder of the target sets: the *Republican Guard forces; military storage and production sites; naval forces and ports; railroads and bridges; electrical production;* and *oil refining and distribution* facilities.

The original Instant Thunder plan of August 1990 contained a total of eighty-four targets to be struck over a six-day period, and although the target sets remained essentially the same, the total number of targets grew significantly during the fall of 1990. By December, targets numbered 237, with the largest increases being in leadership and military production facilities. The growth in numbers reflected the much greater knowledge of the Iraqi military forces and leadership structure gained after the United States focused its reconnaissance capabilities on Iraq in the fall of 1990. Growth also came about as an indirect consequence of the increased number of bombing aircraft available and the consequent ability to target a larger portion of the Iraqi air defense and military support structure. The intensity of the operations also grew, because although the duration remained constant at six days, the number of aircraft available continuously increased.[14]

First priority of the air planners was gaining command of the air. This goal was a basic tenet of air doctrine, and achieving it would generate at least three specific advantages in the war. First, incapacitating airfields and the air defense system would allow sustained prosecution of attacks against the other target sets. Second, command of the air would prevent Iraqi offensive strikes against coalition forces, in particular strikes delivering chemical weapons. Third, the coalition would prevent Iraqi reconnaissance flights that might uncover the planned shift of coalition ground forces to the west, the surprise to be sprung at the start of the ground offensive.[15] Planners therefore directed their most intense and immediate attention to destroying the Iraqi air defense system through the use of F-117s, other aircraft employing antiradiation missiles to attack radar systems, and an array of electronic countermeasures.

The attacks against the nuclear, chemical, and biological weapons, and Scud facilities served short and long term objectives. In the short term, their destruction would prevent their employment in the Gulf War. Planners believed Iraq fully capable of using both chemical weapons and Scuds.[16] In the long term, the objective of security and stability in the Persian Gulf required eliminating Iraqi weapons of mass destruction along with the Scuds that could serve as their delivery means.

Planners knew that the Iraqi ballistic missile force had mobile launchers, some numbers of which would escape destruction and fire their missiles. Although the Black Hole had planned since August 1990 to attack the fixed Scud sites, neither that group nor anyone else had devised, before the war, a search-and-destroy scheme for dealing with the mobile launchers. Planning focused on reducing the offensive threat the launchers represented by attacking potential concealment sites and support facilities, but not the mobile launchers themselves.[17] The planners in the Black Hole, like CENTCOM's leaders, regarded Iraqi ballistic missiles chiefly as nuisance weapons when used tactically against military forces, but realized that the launching of missiles against Israel would cause political difficulties for the alliance.

Military and naval support facilities, such as ports, were obvious target sets. Several other of the target sets—railroads and bridges, electric power production, and oil facilities—while undeniably forming a part of any country's military power, also served its nonmilitary economic power and its civilian populace.[18] Here, planners attempted to diminish the military support provided by this infrastructure while limiting the damage in other respects. The Black Hole planners did not want to spare the Iraqi population completely. Rather, they wished to inflict disruption and a feeling of helplessness on the Iraqi public without causing severe suffering, all in the hope

Air Power in the Gulf War

of weakening Saddam Hussein's grip on power. As a result, planning for attacks on the industrial power of the country had a dual nature. On the one hand, the objectives were to "cripple production" and "complicate movement of goods and services."[19] On the other hand, planners harbored an "intent to convince the Iraqi populace that a bright economic and political future will result from the replacement of the Saddam Hussein regime..." and specified that "execution planning will emphasize limiting collateral damage and civilian casualties and preserving the Iraqi and Kuwaiti capability quickly to reconstitute their economies."[20]

To comply with this guidance, targeting attempted to distinguish between short and long term damage to electric power generation and oil facilities. For oil targets, this meant that coalition aircraft would hit oil refining and storage facilities, but not oil production facilities. Within the refinery target subset, aircraft would hit distribution points, not cracking towers. For electric power targets, they would strike transformers, which were thought to take months to repair, instead of the generator halls, which were thought to require years.[21]

Attacks against leadership and command and control had both political and military dimensions. Separating the national leadership in Baghdad from the military forces in the field would delay the coordination of military operations and show the Iraqi forces the powerlessness of their leaders. Planners also hoped for a more direct political effect: if Saddam Hussein could not communicate with the Iraqi people he could not propagandize against the United States and its allies, or mobilize the country for war. As a result, the air campaign targeted radio and television transmitters, relay stations, telephone and telegraph facilities, and military command posts. Also attacked, besides facilities that might house Saddam Hussein, were the buildings of the Ministry of Defense, Ba'ath Party headquarters, and similar sites.

Planners considered these strikes as a way to end the war by air power alone. The strikes, in coordination with others, would not just neutralize the government, but change it by inducing a coup or revolt that would result in a government more amenable to the coalition demands. The final Central Command operations plan of December did not stress these intentions, but the CENTAF operations order of September 1990 stated them forcefully: "When taken in total, the result of Phase I will be the progressive and systematic collapse of Saddam Hussein's entire war machine and regime."[22]

The Republican Guard received particular attention in Central Command planning, enough to have it specified as one of the target sets in the Phase I plan. Planners identified the Republican Guard as a center of gravity of the campaign and a priority target of the air campaign. Not only did this elite force serve as the strategic reserve of the Iraqi forces in Kuwait, it also provided essential support to Saddam Hussein's regime. Schwarzkopf's planners intended to rout them so that they could not help Saddam Hussein retain order in the country. The operations order directed that the roads and rail lines south of Basra should be blocked to prevent the withdrawal of the Republican Guard forces.[23] While these forces were seen as a target that had to be dealt with in Phase I of the air campaign as well as in Phase III, that issue of priority became moot when, because of the number of coalition aircraft available, the first three phases of the air campaign began at essentially the same time.

The planning for the second, third, and fourth phases of the air campaign dealt with the Iraqi forces in the Kuwaiti theater. Planners made Phase II, Air Superiority in the Kuwaiti Theater, a separate phase only at General Schwarzkopf's suggestion—most of the air planners tended to see the Iraqi air defense system as an integral whole that included the Kuwaiti theater—and this phase received little special elaboration. Phase III, Battlefield Preparation, however, called on air power to accomplish significant attrition of ground forces to a degree not heretofore planned for by any air force. In August, an analysis group working for General Schwarzkopf had concluded that for a coalition offensive to be successful with a single corps, the air campaign would have to achieve 50 percent attrition of enemy ground forces.[24] Initial planning had therefore sought attrition of troops and all major pieces of equipment, but CENTCOM planners later narrowed the requirement to attrition of tanks, armored personnel carriers, and artillery. Moreover, the attrition was left as a theaterwide goal, not attached to specific divisions or areas within the theater.

A controversial aspect of the planning concerned control of the coalition air attacks in both Iraq proper and in the Kuwaiti theater, which had been delegated to a Joint Force Air Component Commander (JFACC) in the person of General Horner. A convoluted command structure during the Vietnam War had divided the geographic area into multiple regions and command of the aircraft among several services; recent doctrinal changes had attempted to avoid repeating such a structure again. The planning documents had given General Horner the functions of "planning, coordination, allocation, and tasking based upon USCINCCENT (Schwarzkopf) apportionment decisions."[25] An initial issue concerned how many aircraft were subject to JFACC control. General Horner settled it by negotiating agreements with each Service: the Navy would continue to control sorties for fleet defense; the Marines would give up aircraft beyond those needed for Marine requirements; the Army would be able to fly helicopters with little outside control.

A second set of issues concerned who would make the decisions on target priority; these issues were not addressed directly and continued to fester throughout the war. General Schwarzkopf gave Horner and his staff a relatively free hand in selecting targets for strategic bombing in Iraq, and the ground commanders had little interest in these decisions. For targets in the Kuwaiti theater, however, the ground commanders sought to direct the targeting of the Iraqi army and questioned whether the Air Force was giving proper attention to the ground officers' priorities. Army and Marine Corps commanders and staffs sought to focus the attacks on Iraqi army targets that would help "shape" the battlefield for the ground offensive, but those priorities often conflicted with the targeting of mobile Scuds, lines of communications, or other targets within the theater.[26]

The JFACC's tool for exercising control of air operations was the air tasking order, or ATO, and this document itself came under considerable criticism. Hundreds of pages in length, it was cumbersome to prepare (daily), to transmit and receive. Critics also assailed the three-day process for the ATO's construction as too long to allow needed battlefield flexibility. Despite the criticism, the ATO was an essential mechanism for controlling up to 3,000 sorties a day, and claims of inflexibility were often not warranted or overstated. While a cumbersome document, many units had already received the information from the relevant portion of the ATO via secure telephone long before the actual document arrived. Moreover, as the campaign unfolded, many ATO sorties received targeting from an airborne controller on the scene, not from a three-day-old document; the document served more to control the flow of aircraft to the scene.

CENTAF planners, supported by Air Staff calculations, divided air attacks on the Iraqi ground forces into two parts: those directed against the Republican Guard and those aimed at the remainder of the army in the Kuwaiti theater. The planners earmarked higher performance aircraft (F-16s, F/A-18s, F-15Es, F-111Fs) for attacks against the Republican Guard, reserving the A-10s, AV-8Bs, and attack helicopters for dealing with the Iraqi divisions in Kuwait. Attrition calculations assumed 600 sorties a day against each of these two parts of the Iraqi army, relying on precision munitions such as Maverick missiles (but not laser guided bombs—that was an innovation made during the air campaign) and various cluster munitions to create the attrition.

Based on the above criteria, General Glosson briefed General Schwarzkopf in December 1990 that the desired 50 percent attrition could be attained in five days against the Republican Guard and in approximately twelve days against the remainder of the forces in Kuwait.[27] General Schwarzkopf not only accepted these attrition estimates, he reduced the time available, issuing an operations plan two weeks later that specified eight days for Phase III. Some flexibility returned in the operations order for the campaign issued on 16 January: here the length of Phase III was left "to be determined."[28] The requirement for 50 percent attrition made the ground campaign depend on unprecedented air power success in destroying an army. Air power had an enormous task in Phase III alone—to destroy approximately 5,000 pieces of dug-in and defended Iraqi equipment.

If all went as envisioned, planners estimated that the final phase, Phase IV, Ground Offensive Operations, would commence several weeks after the launching of Desert Storm. The objectives for this phase were to liberate Kuwait, cut critical lines of communication into southeast Iraq, and destroy the Republican Guard in the Kuwaiti theater. The ground attack would be "combined with continuous B-52 strikes, TACAIR (tactical air) attacks, and attack helicopter operations." In anticipation of the main attack, "The bridges, roads and rail line . . . will be cut to block withdrawal of RGFC and to form a kill zone north of Kuwait."[29]

Planners intended to provide close air support (CAS) for ground forces by using a "Push CAS" system. Flights of aircraft would arrive at locations within the anticipated target areas continuously, sometimes as frequently as every seven minutes. In other words, without waiting for a ground commander to request support, the aircraft sorties were "pushed" to his location. If the commander did not need the aircraft immediately, they would orbit for a short time, then go on to attack a planned back-up interdiction target, and another flight of aircraft would then arrive to fly in orbit at the commander's location. Aircraft were to strike ground targets under the control of the tactical air control party, naval gunfire liaison team, or an airborne forward air controller.[30]

By 15 January, the coalition air forces comprised more than 1,000 fixed-wing attack aircraft and another 800 air defense fighters and electronic combat aircraft to prosecute the air campaign, more than double the numbers available in October.[31] The air forces came from ten different countries and flew from bases in Saudi Arabia and the other Gulf States, from six U.S. Navy aircraft carriers (three in the Red Sea and three in the Persian Gulf), from Incirlik Air Base in Turkey, and, for the B-52s, from bases far distant from the region. When the air campaign began on 17 January 1991, air planners and commanders were confident. At a relatively low cost—one hundred or so aircraft losses maximum—they thought that the ambitious objectives set in the operations plan could be met. Planners had little doubt that within a month, the Iraqi army would flee Kuwait or, more likely, lie shattered in place, that Iraqi military industry and the Iraqi air force would be destroyed, and that Saddam Hussein's grip on Iraq would be, if not removed, weakened beyond repair.

Execution

Early on 17 January, the coalition launched a concentrated air campaign against strategic military, leadership, and infrastructure targets in Iraq. Even before the first shots were fired at 0238 local time by helicopters attacking an early-warning radar site in southern Iraq, B-52s from Barksdale Air Force Base in Louisiana were en route with conventional cruise missiles, Navy ships had fired salvos of Tomahawk land-attack missiles (TLAMs), and F-117 stealth aircraft were approaching Baghdad. Waves of aircraft followed, rapidly sweeping into Iraq, attacking airfields, nodes of the integrated air defense system, and leadership command and control systems, known Scud sites, nuclear/chemical/biological production sites, and electrical power facilities. By dawn, the attacks spread to include ground forces in the Kuwaiti theater. The second night and day saw more of the same attacks, with oil production and storage facilities and naval sites also coming under assault. The first two days of air operations were the most thoroughly planned and the most complex of the war, as coalition commanders attempted to dismember Iraqi air defenses while attacking targets across the entire spectrum of strategic target sets. The coalition hit virtually every target set on the initial strikes, although the greatest weight of effort was directed against air defenses, airfields, and command elements of the Iraqi regime.

As was inevitable with any planning, unanticipated Iraqi reactions and operational complications of coalition forces forced some adjustments. Almost from the beginning, circumstances dictated diversions from the planned strategic air campaign. The first change occurred on the second day when the first Scud missiles launched from western Iraq landed in Israel. As noted earlier, Iraq's attempts to split the coalition by firing Scud missiles at Israel had been anticipated before the war, and for that reason the fixed Scud sites in western Iraq were targeted on the first night's raid. These strikes failed to neutralize what became the true Scud threat—mobile Scud launchers capable of moving from hidden sites, firing, then hiding again before aircraft could attack them. Intensive combat operations began in an attempt to find, destroy, or simply suppress the mobile missiles; these activities continued throughout the war. The Scud hunt included continuous airborne surveillance of western and southern regions of Iraq, positioning strike aircraft within the Scud launch areas for more immediate targeting, attacks on communication links thought to be transmitting Scud launch authorization, attacks on suspected sites, and strikes against Scud production and storage facilities. By war's end, nearly every type of strike and reconnaissance aircraft employed in the war participated in the attempt to bring this threat under control, but with scant evidence of success.

A second redirection of targeting involved digging the Iraqi air force out of its shelters. Subject to almost immediate engagement by coalition aircraft ranging over the Iraqi bases, the Iraqi aircraft elected not to contest control of their airspace and sought protection in hardened aircraft shelters that were thought immune from coalition attack. On 23 January, airfield attack operations shifted from attacking runways to destroying aircraft shelters. Attacking the nearly 600 Iraqi shelters required a substantial shift of coalition air power resources, mainly F-117s and F-111Fs dropping laser-guided bombs (LGBs). For two weeks, F-111Fs devoted approximately 60 percent of their strikes to these shelters. They were then drawn off for use against tanks and other ground force equipment. Twenty-eight percent of the total British precision-bombing effort was against hardened shelters. Meanwhile, F-117s continued to prosecute shelter attacks until the last week of the war.

Attacks against leadership and command and control facilities in Baghdad peaked on the first night, then diminished substantially until the fourth week of the war. Just as the intensity resumed, however, the F-117 strikes carried out against the Al Firdos district bunker in downtown Baghdad brought intense media attention. Unknown to coalition air planners, the upper level of the bunker was, according to the Iraqis, being used at night by families. Iraqi sources claimed that 200–300 civilians, including more than 100 children, died in the bunker.[32] In the wake of television coverage, a sharp reduction of attacks on leadership targets ensued, with General Schwarzkopf personally reviewing any targets selected for attack in downtown Baghdad.[33] From then on, no significant number of attacks took place on Baghdad until the last week of the war.

Several days after beginning the air attacks, coalition air commanders changed bombing tactics to help reduce losses of aircraft and crews. Although some crews initially had tried NATO-style, low-level ingress tactics during the first few nights of Desert Storm, the sheer number of widely dispersed antiaircraft artillery, combined with the ability of Stinger-class infrared surface-to-air missiles (SAMs) to be effective up to 12,000–15,000 feet, quickly persuaded most everyone on the coalition side to abandon low altitude flying, especially for the weapon release maneuver. Commanders restricted bombing missions to medium altitude (approximately 10,000 to 15,000 feet).[34]

This decision had a price. For aircraft such as the F-16 and F/A-18, which principally employed unguided or "dumb" munitions during Desert Storm, the decision meant a definite sacrifice in bombing accuracy. A consequence of the altitude restriction was the need for higher weather ceilings to deliver ordnance visually; thus a higher-than-anticipated incidence of mission changes occurred because of weather. Air Force planners estimated that by 6 February, three weeks into the war, approximately half of the attack sorties into Iraq had been diverted to other targets or canceled because of weather-related problems.[35] The weather not only impeded accuracy but also hampered accurate bomb damage assessment.

With the degradation of overall bombing accuracy, LGBs, laser-designating aircraft, and other precision weapon systems became ever more important. First, the nature of the strategic targets in Iraq required an extensive use of LGBs to achieve the desired accuracy and effects, to avoid excessive collateral damage, and to reduce the risk of delivery aircraft having to conduct repeat attacks. Since the risk of flying aircraft over Baghdad during daylight was high, TLAMs were used to keep the pressure on during daylight. Some target systems succumbed quickly—electrical power, for example. Other target sets consisting of many individual targets (for example, almost 600 hardened aircraft shelters) required repeated visits throughout the war. Most targets in these targeting categories required precision-guided weapons with a capability to penetrate hardened buildings. Precision weapons were at a premium, so much so that aircraft capable of using them could not be spared to attack bridges until the second week in the war. Moreover, a further large, and unplanned, number of LGBs were used against Iraqi ground forces, mainly during direct attacks against Iraqi armor.

In early February, as CENTCOM's focus shifted to the Kuwaiti theater in preparation for the ground offensive, questions arose about whether the Iraqi army's effectiveness was being reduced to the extent claimed. Planning was based on destroying 50 percent of Iraqi tanks, artillery, and armored personnel carriers. While Washington and the theater leaders disagreed over what level of success was being achieved, a greater effort was clearly called for. To increase the lethality of the attacks, A-10s, thought to be the most effective aircraft against armor, decreased their attack altitude to between 4,000 and 7,000 feet.[36]

A second adjustment, one that led to far more immediate results, was employing LGBs against Iraqi armor. F-111Fs conducted night tests during the first week of February using infrared sensors to detect the hot skin of the tanks (or any other metal equipment) contrasted against the cooler sand that surrounded them. Following these tests, F-111F, F-15E, and A-6 aircraft flew laser-guided bomb attacks against Iraqi armor in a procedure known as "tank plinking." From that point, the number of recorded armor and artillery kills climbed rapidly.[37] Later, in preparation for the ground attack, the weight of effort shifted from the Republican Guard and other heavy divisions in the rear to direct attacks on the Iraqi front line divisions.

When the ground offensive began at 0400 local time on 24 February, a surge in the number of aircraft sorties, including those aircraft allocated under the "push-CAS" system, put more than 3,000 sorties in the air over the battlefield. The generally light opposition to the rapid ground advance, however, generated few targets for close air support aircraft. The surplus of aircraft available often could be put to use attacking other targets well ahead of the rapidly moving ground advance, but not always. In poor weather or facing restricted visibility caused by burning oil wells, aircraft equipped with radar-aimed release systems could be useful, while many aircraft not so equipped had to return to bases with their bombs.

A limited amount of close air support did occur, but most of the air effort, and the subsequent destruction, fell on the heavy reserve divisions and the retreating columns of the Iraqi army as it fled Kuwait. The highway proceeding northwest out of Kuwait City and Al Jahra and over Mutla Ridge, the "highway of death," was one such bottleneck of traffic that came under attack during this retreat. Only the ceasefire declared on 28 February prevented more such scenes of destruction.

Results

Coalition air attacks had dominated Iraqi actions in the war and had put much of that country in disarray, but at war's end there remained questions concerning air power's effectiveness against certain targets. Later analysis could resolve some of the issues, but unknowns still remain. There were several reasons for each condition. First, since only a small part of Iraq was occupied by coalition forces, no direct assessment of target damage in Iraq proper was possible, with one significant exception: inspection teams under United Nations auspices gained extensive access to the country in order to monitor and destroy Scud missiles and launchers as well as nuclear, chemical, and biological weapons programs. These teams brought back information that greatly revised downward the damage levels estimated earlier. Second, even in the Kuwaiti theater itself no adequate, theaterwide, assessment of damage occurred, leaving much to conjecture or anecdotal evidence. Finally, interrogation of Iraqi prisoners and refugees and more systematic analysis of wartime imagery permitted a more complete understanding of what had occurred. Within these limitations, evidence gathered both during and after the war permits some judgments of the effectiveness of the attacks against the target sets outlined earlier: offensive counterair, strategic attack, interdiction, and close air support.

Counterair attacks against Iraqi air defenses and airfields attained overwhelming success, enabling nearly complete freedom of action over Iraq and Kuwait. To explain the meaning of this success, one needs to look at qualitative, not quantitative, measures. During the first six days of Desert Storm, radar surface-to-air missiles downed or damaged eight coalition fixed-wing aircraft; for the rest of the campaign, these SAMs accounted for only another five aircraft. Neither during nor after the war was it usually possible to prove whether a SAM site was destroyed or just silent (or abandoned) because of fear of attack if its radars were employed. Whether destroyed or not, the radar SAMs were not used effectively, and that was the effect sought. Similarly, while many Iraqi runways remained intact, 375 of the nearly 600 hardened shelters were destroyed by coalition aircraft during the war. Because of the shelter attacks, some aircraft were destroyed in the shelters, while many more either attempted to flee to Iran or were dispersed in the open in Iraq, both on and off airfields. In either place they were of no use as a fighting force. As a result, the Iraqis were able to retain by war's end nearly half their aircraft (an estimated 300 to 375 combat aircraft) but at the expense of that air force's entire combat capability during the war.

Actually, the most telling measures were those things that did *not* happen: the number of coalition aircraft *not* shot down or damaged during more than 118,000 combat and combat-support missions; the role *not* played by the low-altitude antiaircraft and SAMs because coalition aircraft could safely fly at higher altitudes; and the reconnaissance and strike missions *not* flown by Iraqi aircraft, allowing coalition surface forces to operate with little fear of attack and preserving the element of surprise for the coalition ground forces shift west. The freedom from air attack, however, must be qualified because of the threat posed by Scud missiles, a subject addressed later.

Air planners hoped and even anticipated that the attacks against the core strategic targets[38] in Iraq would achieve the coalition's military objectives without the need for a ground invasion. The actual campaign, however, focused on attaining more pragmatic objectives against these target sets, objectives that differed from those sought in previous wars. Where attacks in previous wars targeted industrial production to diminish the support of the fielded forces, the Desert Storm strategic attacks instead sought to disorganize, intimidate, and disorient Iraq's political and economic structure. Only the attacks on nuclear, biological, and chemical (NBC) capabilities and Scuds aimed at outright destruction—and these strikes turned out to be the least effective. Strategic target attacks accounted for less than 15 percent of the coalition strikes, though they absorbed nearly 30 percent of the precision guided munitions. Even considering these more restrictive conditions, the attacks achieved rather ambiguous results.

Air attacks on the Iraqi nuclear weapons research program and Scud launchers and support systems produced an apparently good score card in terms of aim points hit but ultimately poor results in terms of objectives sought. Reports during and immediately after the war indicated a high level of destruction against the entire nuclear program, based on analysis of damage to the *known* facilities. Later information showed there was only incomplete destruction of the known facilities; and, more importantly, those facilities were only a small portion of the entire Iraqi nuclear program. Coalition intelligence had underestimated both the size of the program and also the Iraqis' determination to protect it. Whereas the coalition planners began the war certain of only two sites, postwar analysis by United Nations inspectors revealed 16 "main" nuclear facilities and another five nuclear-related sites.[39] Furthermore, the Iraqis went so far as to remove both nuclear fuel and machinery from buildings engaged in nuclear research and bury these items in fields, essentially making them relatively invulnerable to precision air attacks.[40] Attacks on known facilities, in other words, at times were hitting nearly empty structures. One could look on the poor results simply as an intelligence failure, but the more explicit lesson is that targeting complex systems such as a research program (as opposed to simply aiming at facilities) requires a comprehensive understanding of the program.

Targeting and damage assessments of Iraqi Scud launchers and support facilities had much in common with the experience against the nuclear program. As with the nuclear program, coalition intelligence lacked a full understanding of the target base and then during the war misinterpreted the actual damage being inflicted. Iraq was known to have both fixed-launch sites and a mobile launch capability for their Scuds. Coalition planners thought the Iraqis would use the fixed sites initially, and that the mobile launchers, though much more difficult to target, would be vulnerable because of their estimated prelaunch setup times (among other assumptions). Not anticipated was the dispersal of the mobile Scud force to unknown locations even before the beginning of the air campaign and the presence of decoy mobile launchers that could not be distinguished from the real thing even from a distance of twenty-five yards.[41] Moreover, during the early days of the air campaign, pilot reports and pictures of what were described as destroyed mobile Scud launchers tended to mask the actual lack of success in destroying them.[42] Although the Iraqi mobile Scud force was no doubt disrupted, harassed, and to a degree suppressed, coalition aircraft succeeded in destroying few, if any, mobile Scud launchers during the war.

What, then, are the best measurements of the anti-Scud attacks, and what do the results show? The number of fixed sites destroyed appears to have little relevance in this case, and if only the number of mobile Scud launchers is considered, the attacks were a failure. There were other indicators, however, which suggest part success, or at least make the operations worthwhile. One indicator is the launch rate for Scud missiles during the war, and a second is the degradation of Iraq's longer term offensive capabilities based on the extensive attacks on production and storage facilities. The mobile Scud launchers if not destroyed were at least suppressed after the first ten days; a final surge of launches toward the end of the war, however, also showed that the threat had not been completely dealt with. In addition, the suppression both cut down the number of missiles launched and diminished the accuracy of those actually launched because of a shortened setup time and rushed procedures.

The attacks against the Scud missiles and nuclear program call for many subjective judgments concerning cause and effects to determine success. At least in these two cases, there were postwar United Nations inspections to throw further light on the levels of actual damage done. In examining the evidence of attacks on Iraqi leadership and communications, there is far less such postwar information, and the available measures are just as indistinct. Complete success would have entailed the removal of the Iraqi leadership, particularly Saddam Hussein, in the one instance, and the inability of Iraq to control its forces in the Kuwaiti theater from Baghdad, to communicate with the outside world, and to maintain internal control of the population in the other. While complete success would have been readily apparent, there are few objective measures for determining how far short of, or near, that goal the coalition's air campaign may have been by 28 February 1991.

One can, of course, make a compelling case for the tactical effects of the air strikes against leadership targets. Detonation of 2,000-pound bombs within feet of aimpoints on governmental ministries, national command and control facilities, and telecommunications centers forced the surviving elements of Saddam Hussein's government to relocate and shift to backup communications and undoubtedly caused governmental officials to fear for their lives. More far reaching effects of the bombing are less distinct. On the one hand, Saddam Hussein and his Ba'athist regime remained in power, able to communicate with the field commanders in the Kuwaiti theater, and to continue to launch Scuds until the final days of the war. On the other hand, the Iraqi leadership was forced to relocate often, had its communications severely disrupted, and its control of the Iraqi people severely shaken. There were rebellions by the Kurds in the north and the Shiite Moslems in the south, and Saddam Hussein was being criticized openly in Baghdad.[43] Little more can be said. Estimating more precisely the degree of dislocation that occurred requires access to Iraqi officials or records. Estimating these effects during the war itself was and probably will remain more difficult for these target sets than for any others. Even more than nine years after the war, information from Iraq has not given much greater insight into the level of instability produced.

For electric and oil facilities, fairly complete information is available on the measures and degree of success of these attacks in attaining immediate, tactical results. For electrical power, there was a rapid shutdown of commercially generated power throughout the country, the bulk of the loss occurring in the initial days of the air campaign. The shutdown took place even faster than anticipated, in part because the Iraqi engineers at times chose to shut down the plants themselves to protect the systems from overload. Some residual power, perhaps 12 percent of capacity remained available in isolated areas from smaller power plants that were not attacked.[44] The immediate objective of shutting down the national grid was, therefore, quickly attained. The degree to which this measurable result led to the desired effects on the Iraqi military and the country overall is, however, less clear. Some friction was undoubtedly imposed on the Iraqis by forcing national-leadership and military systems countrywide to switch to backup power. But how much friction was induced remains difficult to quantify based on available evidence, but the national electric power grid remained out of action.

Coalition air strikes rendered 90 percent of Iraqi petroleum refining capability inoperative, caused mainly by the employment of a relatively few precision strikes against distillation towers. Air strikes destroyed a much smaller percentage of oil storage capacity because those targets were spread over extensive areas and were less vulnerable to rapid destruction. The lack of distilled petroleum caused few problems for the Iraqi military, however, through no fault of the attack plan. A fuel shortage may have affected the Iraqi air force if it had not chosen to remain on the ground.

Similarly, Iraqi ground forces in the Kuwaiti theater had access to Kuwaiti fuel and in any case used only minimal petroleum while dug into static positions. They had more than enough diesel fuel for a 100-hour ground war, but would have soon run into difficulty finding even transportation to supply the fuel within the theater. The impact of strikes against oil facilities must, in other words, be linked with the interdiction attacks to and within the Kuwaiti theater.

United States forces had experience prosecuting air interdiction operations in World War II, Korea, and Vietnam, but this war provided a few new twists. The objectives were to cut the flow of supplies to the theater, stop the movement of forces within the theater, and, a special concern, stop the Iraqi forces from leaving the theater intact. Since most Iraqi ground forces were already in place when the air war began, the need to block reinforcements was limited. Geography provided the attackers some advantages: terrain consisted of broad plains and farmland, providing little cover for vehicle traffic; the principal lines of communication between Baghdad and the theater generally followed and frequently crossed rivers, and the system of roads narrowed on the approach to Basra. Bridges, therefore, became the key targets.

Bridge destruction began in earnest at the beginning of February 1991 and progressed quickly after that. Overall, 75 percent of the bridges along the route to and from the Kuwaiti theater were damaged or destroyed.[45] Despite the prominent role of bridges in the transportation system, however, their destruction alone was not enough. The Iraqis did not attempt to repair any of their permanent bridges, but did mount a massive effort to build earthen causeways and use ferries and pontoon bridges to bypass downed bridges. The Iraqi skill in coping with the loss of bridges led General Horner after the war to caution:

> Anybody that does a campaign against transportation systems [had] better beware. It looks deceivingly easy. It is a tough nut to crack. [The Iraqis] were very ingenious and industrious in repairing them or bypassing them. . . . I have never seen so many pontoon bridges. [When] the canals near Basra [were bombed], they just filled them in with dirt and drove across the dirt.[46]

Within the theater itself, there were few bridges or chokepoints to target, but attacks on Iraqi supply trucks had devastating effects on the transportation system. Information from enemy prisoners of war indicated that more than half the trucks were destroyed or out of service for lack of parts, and that the drivers were no longer willing to travel the roads.

With all of the evidence on the success of air operations against the bridges, trucks, and entire route system, what conclusions can be drawn about the effectiveness of air interdiction? The coalition aircraft performed well, greatly reducing the flow of supplies, if not completely severing the supply lines. Nevertheless, the inaction of the Iraqi army itself served to mask the actual impact of the damage to its logistical support system. As air interdiction efforts in past wars show, those operations work best when the enemy is engaged in high tempo operations on the ground and is thus consuming supplies at a high rate. The Iraqi army was essentially inert during the air campaign, so that the limited supplies that got through, combined with the large stocks positioned in the theater from August 1990 to January 1991, were enough for the army to remain in place. Whether several more days or weeks of air interdiction alone would have eliminated all resupply and shattered what was left of the distribution system is a matter of speculation. What is certain is that the outbreak of large-scale ground combat increased the demand for supplies (especially ammunition and petroleum) to the point where the residual supply flow was insufficient for any prolonged effort.

While not a centerpiece of the war, operations against the Iraqi navy consumed a significant amount of the air effort, particularly for carrier-based aircraft in the Gulf, and showed the difficulties of operating in confined waters in the presence of even a small enemy force. Coalition aircraft attacked Iraqi naval targets to secure freedom of action in the northern Persian Gulf, both to bring the carrier and battleship firepower closer to the action and to allow the amphibious force to be in position for the deception plan, and landings (if necessary). Just as with targets on land, a lack of bomb damage assessment made the threat picture unclear; the U.S. Navy antisurface warfare commander could not declare the threat defeated until 17 February, two weeks after later analysis would show the last of the Iraqi missile boats had been destroyed.[47] Even with the Iraqi surface navy all but entirely sunk, however, a serious threat remained for coalition naval forces: mines and Silkworm antiship missiles. An unknown number of mines remained, and missile boat destruction removed only one launch method for missiles; the threat of ground launched Silkworms continued to affect coalition naval operations until the end of the war. Repeated strikes against seven suspected Silkworm sites did not remove this threat. Only two Silkworm launches took place during the war, both from a site south of Kuwait City on 25 February, fired, obviously, just before the site was overrun.[48] Just as with the anti-Scud operations, it is difficult to judge whether further launches were in fact suppressed or whether the Iraqis simply chose to retain the missiles until an amphibious attack occurred.

Air attacks against the Iraqi army in the Kuwaiti theater comprised much more than half the coalition air effort. All forty-three Iraqi divisions in the theater received some attention, but the three Republican Guard heavy (armored or mechanized) divisions received the most, followed by the other eight heavy Iraqi divisions that made up the tactical and operational reserves. Attacks against the Iraqi front line divisions, all infantry divisions, peaked just before the ground offensive.

In the opening two weeks of the war, attrition of the Iraqi army by air attacks fell far behind the projected rate, in part because of a combination of poor weather and lower-than-planned sortie rates, but principally because of poorer bombing accuracy and weapon performance from the high release altitudes employed. After the adjustments described earlier, attrition rates increased, often rather dramatically, but the 50-percent goal for the theater was not attained by the beginning of the ground offensive. At that time, Central Command estimated that equipment attrition as 39, 32, and 47 percent for tanks, armored personnel carriers, and artillery, respectively.[49]

The amount of equipment attrition suffered by the Iraqi army became a contentious issue at the time, and postwar reconstruction can only partly resolve the controversy. First, postwar analysis has revealed that the number of Iraqi tanks in the theater just before the war was 800 fewer than the more than 4,200 originally estimated. This error in the estimate, however, is offset by the subsequent overcounting of the tank attrition during the war, making the Central Command wartime estimates of attrition *percentages* on the eve of the ground offensive (23 February) approximately correct. Second, imagery showed that more than 800 Iraqi tanks escaped from the theater at the war's end, making the number of tanks destroyed by both air and ground action around 1,000 less than the more than 3,800 Central Command claimed at the end of the war. And finally, tank attrition on the eve of the ground offensive was approximately 40 percent for the Iraqi army overall and 20 percent for the Republican Guard units; the total wartime Iraqi tank attrition was approximately 75 percent and 50 percent, respectively.[50] In other words, the most important units got off with the least damage. Although far from the least attacked, Republican Guard tanks were better dug-in and defended. Another contributing factor was that these tanks were far enough to the rear to escape the theater; tanks farther forward, even if functioning, had to be abandoned.

Even if the number of tanks had been correctly estimated, too much attention was attached to that weapon. When the attrition goals were set in September 1990, tanks represented what would have been the vanguard of an Iraqi attack into Saudi Arabia. By January 1991, no ground commander set a particular premium on destroying tanks from the air. By then, the key target was artillery, since Iraqi artillery represented the chief danger for blunting the ground attack through use of chemical weapon shells during the breaching effort. The U.S. Army corps commanders, Lt. Gens. Gary Luck and Frederick Franks, and Lt. Gen. Walter E. Boomer, Commander, Marine Forces, Central Command, pointed to artillery as the chief obstacle; General Boomer ordered Marine pilots to direct their energies to attacking artillery instead of tanks.[51] Tanks were not in the front lines in any numbers, and the corps commanders were confident of being able to handle Iraqi tanks in a war of movement, both by air, since tanks on the move were more vulnerable, and by using the superior range of the M1A1 tank's gun.

Ironically, the loss of equipment, the key index of damage assessment during the war was not decisive in any direct way. The key to the defeat of the Iraqi army was not the specific targets destroyed, but the combination of targets attacked and the intensity with which the attacks took place. The soldiers were affected by the bombs that hit their targets as well as by those that missed. The air interdiction effort, the damage to communications and supply systems, along with the equipment attrition, affected the Iraqi soldier beyond the direct inflicting of casualties. The Iraqis did not defect or surrender in droves during the air and ground war because their armor and artillery were being destroyed (in fact, statements by prisoners of war indicate they appreciated the discrimination of the air forces in aiming at the equipment and not them), but because they were short of food, water, and confidence that the equipment was going to do them any good. The Iraqi army did not run out of tanks, or armored personnel carriers, or artillery; in fact, much of the equipment remaining intact at the start of the ground offensive was abandoned, or at least unoccupied, when the coalition ground forces arrived. The total number and operability of the tanks had less meaning under these conditions.

Close air support received a great deal of attention both before and after the Gulf War, but any lessons learned have limited value since the lack of determined resistance during the ground offensive made close air support by fixed-wing aircraft a peripheral aspect of the war. All the front line Iraqi divisions crumbled quickly, often with no resistance at all, and as the coalition corps advanced, they reported only light resistance throughout the theater. Except for isolated instances of determined resistance, possibly two in the Marine area of operations and several more in clashes by Army forces with units of the Republican Guard, rarely was the opposition not handled easily by Army or Marine ground weapons alone. There were, in sum, few situations of "troops in contact" to test how well close air support by fixed-wing aircraft could be synchronized with ground fire support systems. Air power's greater effectiveness was in attacking the forces deeper in the Iraqi defense areas, in the regions where these attacks blended in with interdiction strikes; sorties reported as either interdiction or close air support were quite similar, often occurring just a few miles apart.

No single target or tactic brought on the crumbling of the Iraqi army, rather it was the cumulative pressure of air attacks applied throughout the forty-three-day campaign. Lack of communications, destruction of the theater distribution system, and equipment attrition combined to make the Iraqi forces in some cases unwilling and in other cases unable to maneuver or mount an effective defense. Air power had destroyed not only large amounts of equipment, it had destroyed the confidence of Iraqi soldiers that the equipment would do them any good—on the contrary, the equipment was seen as a magnet for air strikes. In short, even before the ground attack took place and coalition forces swept through the theater, air attacks had already made Iraqi resistance disorganized and ineffective.

In Retrospect

Air power's success in the Gulf War brought forth a variety of interpretations. Perspectives varied based on whether one thought the war represented the end of an era or the beginning of one. Evidence for each perspective existed because the political, technological, and doctrinal contexts of air power were all in a state of flux. First, coming as it did at the end of the Cold War, air power doctrine for bombing found itself just breaking from the confined mindset of employing "tactical" aircraft (fighters and fighter bombers) to halt the advance of ground forces and "strategic" aircraft (long-range bombers) to deliver nuclear weapons. Second, the joint planning system, still in the sorting-out period of the Goldwater-Nichols legislation of 1986, had only begun to resolve some of its most basic issues, particularly the position of the joint force air component commander. Finally, some of the most critical technologies of the war were just making their appearance in U.S. operational forces: stealth systems (F-117 aircraft and Tomahawk cruise missile) and Joint Surveillance and Targeting Attack Radar System (JSTARS) aircraft for tracking moving ground targets were two of the most noteworthy additions.

Only the circumstance of favorable timing can explain the rise to prominence of Col. John Warden's air planning concept for attacking Iraqi "centers of gravity." In ordinary times, General Schwarzkopf would have likely tossed aside the suggested plan of an Air Force staff officer in the Pentagon as unwarranted interference, let alone sought such assistance from the Air Staff himself. Both Generals Schwarzkopf and Horner were alike in their disdain for Washington meddling in operations in their theater. These were not ordinary times, however, as the shift in emphasis from the Soviet Union to a regional aggressor such as Iraq changed many planning assumptions. Furthermore, a strong joint planning system was not yet in place to pursue such planning options during the crisis.

The Desert Storm air campaign itself shows the relative state of development for the various concepts for employing air power. The planned attacks on the Iraqi airfields and air defense system sit at one extreme: precise coordination of actions, sophisticated analysis of the target system, and a search for functional effects. The air attacks on the Iraqi army appear in stark contrast: a planned goal of fifty percent attrition of Iraqi army equipment, theaterwide, a figure arrived at in August or September of 1990 and never refined for either the subsequent changed conditions in the theater or the later-determined scheme of the coalition ground attack. The inescapable conclusion: air forces know well how to attack other air forces, and armies know how to attack armies; neither air forces nor armies, however, possess as sophisticated an understanding of how to attack an army from the air (other than as part of a ground-air assault).

During the war, coalition air forces had to make a series of adjustments to the air plan that demonstrated both the flexibility of the air power instrument and the capacity of the airmen to make it work. The changes required flexibility by not just the commanders, but by all levels of the organization. Invariably, too, both commanders and crews showed their ability to improvise to be an absolute strength. For the commanders, one change involved forfeiting the skill of their crews in low-level bombing, built up over years of training, in order to shift night bombing operations to the relatively safer medium altitudes, where the crews had much less experience. A second and even more dramatic adjustment involved simultaneous changes in tactics, aircraft, and weapons over a matter of days: the shift to employing LGBs in night attacks against Iraqi armor. For these and other of the multiple adjustments, crews had to handle inflight target changes as a matter of course, fashion procedures for coordinating with the JSTARS aircraft, and experiment with various methods in the ultimately fruitless search for mobile Scuds. Even though results of every operation were not successful, the potential value that accrued from possessing a well-trained, adaptive crew force was inescapable.

Based on the results described in this essay, did bombing operations succeed in escaping the controversies of the past? No. While the combination of air attacks dominated the conduct of the war to a degree not seen before, both during and after the war familiar arguments reappeared concerning bombing accuracy, target selection, and relative value of close air support. Moreover, some questioned whether Iraq posed a rigorous test of bombing's capabilities.

Added to the more familiar controversies, successes in the Gulf War raise two further propositions. The first concerns whether the devastating effects that aircraft achieved against ground forces might presage a new relationship between air forces and ground forces. The second concerns whether control of the air (and space) over the battle area might now have become so critical a requirement in the conduct of military operations that it requires the integrated effort and cooperation of all the military services. These added issues could, in fact, become central to the future debate. A sterner test against a more capable adversary may be required, but if air attacks can again exert a similar dominance, the conclusion would follow that a threshold in the relationship of air to surface forces was first crossed in Desert Storm.

Reviewing the results of attacks against all the targets sets, one is left with suspicion of the value of relying on any single discrete indicator to measure results. Even in a campaign as brief, recent, and militarily lopsided as Desert Storm, the broader problems of assessing (not measuring) effectiveness in attaining higher level military objectives, as opposed to achieving more immediate tactical effects, remain as difficult, controversial, and infected with subjectivity as ever. While tactical effects seem more amenable to quantitative measures, results can be deceiving. The need to consider enemy countermeasures and deception as well as the actual, real-world objectives of operational commanders and planners argues that effectiveness is, in the end, essentially a qualitative issue. For example, the supposed destruction of the *known* Iraqi nuclear weapons sites told only a small part of a much grimmer story. Similarly, though the coalition air forces failed to destroy the promised 50 percent of Iraqi armor and artillery in the Kuwaiti theater before the beginning of the ground offensive, coalition air power still created the circumstances under which the 100-hour blowout on the ground was possible.

Planning air strikes that employ very accurate munitions requires similar understanding of the system of which the target is a part. The improved capability to hit within 8–12 feet of a target with precision-guided weapons represented an unprecedented advance in bombing accuracy by comparison with past conflicts. The problem of understanding the vulnerability and functioning of entire target systems in relation to more far-reaching objectives, however, may have been as riddled with uncertainty in the cases of the Iraqi nuclear program and mobile Scud missile capability as were target systems like ball-bearings in World War II. Hitting targets is getting easier, but knowing what targets (or precise aim points) to strike across an array of targets remains, in general, open to uncertainty when facing a dedicated, reactive adversary.

Finally, many commentators on the Gulf War point to the results and lessons as heralding a revolution in warfare. What sort of revolution? Much evidence suggests that the air strikes only in a crude way began to exercise their potential of combining precision intelligence, precision strike, and instantaneous command and control to create extremely lethal strike systems. Others argue that the Gulf War was as good as it gets for air power: that no future adversary will allow such a one-sided advantage in space systems or be as unprepared to combat or neutralize the effects of air attacks. Both conclusions are likely true. The Gulf War experience indicates, however, that uncertainty continues to frustrate the calculations of those who would rely on quantitative measures alone to define success in the application of military power. Without doubt, those calculations will continue to require skilled, adaptive leadership, on the ground and in the air, able to change tactics, adjust plans, and interpret the results.

Notes

1. Bombing as used here is the common parlance for all attacks against surface targets from the air. The term as used in this essay includes not just dropping gravity bombs, but also air-to-ground missiles, cruise missiles, and other ordnance dropped or fired from aircraft.

2. United States Air Force, Air Force Manual 1-1, Volume I, *Basic Aerospace Doctrine of the United States Air Force,* March 1992, 6–13. The U.S. Navy and Marine Corps do not use the term strategic attack. U.S. Marine Corps terminology divides air attacks into two parts: deep air support (interdiction) and close air support.

3. Notes, Col Bryan A. Sutherland, USA, CENTCOM, J-5, 3 Oct. 1990, IRIS-881768, Air Force Historical Research Agency, Maxwell AFB, Ala. (AFHRA); fact sheet, "USCINCCENT OPLAN 1002-90-Arabian Peninsula," USCENTCOM J-5-P, 1 June 1990, IRIS-881768, AFHRA; and USCINCCENT OPLAN 1002-90 Concept of Operations, 16 Apr. 1990, 19–20, Checkmate Historian's Files, Chain of Command (CHC), AFHRA.

4. USCINCCENT, Desert Shield/Desert Storm and Internal Look 90 After Action Reports, 15 July 1991, New Acquisitions (NA) 9, AFHRA.

5. Address to the Nation Announcing the Deployment of U.S. Armed Forces to Saudi Arabia, in *Public Papers of the Presidents of the United States: George Bush, 1990* (Book II) (Washington, D.C.: Office of the Federal Register, National Archives and Record Administration, 1991), 1108.

6. Briefing, "Instant Thunder," "Iraqi Air Campaign," and like names, August 1990, contained in Checkmate Historian's Files, Desert Shield (CHSH) 5 and 7, AFHRA.

7. General H. Norman Schwarzkopf, *It Doesn't Take a Hero* (New York: Bantam, 1992), 313–20.

8. Transcript, Lt. Gen. Horner's taped responses to written questions by CMSgt John Burton, CENTAF Historian, March 1991, Checkmate Historian's Files, Planner (CHP) 13A, AFHRA.

9. Briefing viewgraphs, Headquarters Central Command, "Offensive Campaign: DESERT STORM," 24 Aug. 1990, NA 208, AFHRA.

10. Operations Order, COMUSCENTAF Operations Order, "Offensive Campaign-Phase I," 2 Sept. 1990, Black Hole Files (BH) 8-133, AFHRA; Briefing slides in "General Glosson Briefings," Gulf War Air Power Survey (GWAPS), folder 60, box 3, AFHRA.

11. USCINCCENT OPORD 91-001 for Operation Desert Storm, 16 Jan. 1991, paras. 1D, 3B, and 3C, NA 357, AFHRA. Note the linkage of objectives, centers of gravity, and targets in the planning.

12. The term "strategic air campaign" as used in the operations plan, means simply those

activities planned as part of Phase I. The term was a controversial one at the time. Joint doctrine states that there is only one campaign, the theater campaign, not separate ones for air, naval, or ground.

13. OPORD 91-001, para. 3C.

14. The operations order stated six to nine days. OPORD 91-001, para. 3A.

15. Memorandum for Record, Lt. Col. David A. Deptula, USAF, subj.: Observations on the Air Campaign Against Iraq, Aug. 1990–Mar. 1991, 29 Mar. 1991, 3, GWAPS, Safe #12, D-01, AFHRA.

16. Briefing, Maj. Gen. Robert Johnston, CENTCOM Chief of Staff, to Joint Staff and National Command Authority, "CENTCOM Offensive Campaign," 10 and 11 Oct. 1990, in Report, CENTCOM J-5 Plans, Argumentation Cell, *After Action Report* [Vol. IX SAMS], Tab C, 28 Feb. 1991, NA 259, AFHRA.

17. Target List (with objectives for each target category, including Scuds), BH, Other Documents, folder 8, AFHRA.

18. Many of the telecommunications targets would also fall into this category.

19. Briefing, Joint Chiefs of Staff, "Iraqi Air Campaign Instant Thunder," 17 Aug. 1990, CHSH 5-3, AFHRA; Operations Order, COMUSCENTAF Operations Order "Offensive Campaign-Phase I," 2 Sept. 1990, 4.

20. Ibid., 3–4.

21. Weapons and aircraft accuracy limitations sometimes did not allow such discrimination, and when Iraq began to dump oil into the Gulf and employ oil-fired trenches as part of its defenses, some of the pumping stations in southern Iraq and Kuwait were attacked. (Glosson interview, 12 Dec. 1991, GWAPS, Cochran files, AFHRA.)

22. COMUSCENTAF Operations Order, 2 Sept. 1990, 4.

23. Operations Order 91-001, para. 3.E.2.b.2.

24. Report, Combat Analysis Group, 21 Mar. 1991, in Vol. VI of CENTCOM J-5 Plans, *After Action Report*. The president's November decision to double the forces for a two-corps ground offensive did not change the calculations because intelligence reported that Iraq had also deployed more forces to the Kuwaiti theater.

25. JFACC authorization appeared in draft CENTCOM OPLAN 1002-90, 18 July 1990, 28. The term JFACC had already been used in Joint publications, including Joint Pub 1-02, *Department of Defense Dictionary of Military and Associated Terms*, 1 Dec. 1989.

26. Richard M. Swain, *"Lucky War" Third Army in Desert Storm* (Fort Leavenworth, Kans.: U.S. Army Command and General Staff College Press, 1994), 184–87.

27. "Briefing to the CINC, Phases II and III," 1 Dec. 1990, in "General Glosson Briefings." Subsequent to this briefing, the Iraqi army in the Kuwaiti theater continued to increase in amount of equipment, but so too did the coalition air forces. Coalition air did not attain a level of 600 strikes a day in the Kuwaiti theater on a sustained basis until three weeks into the air campaign. For the first two weeks, air strikes against Iraqi ground forces averaged just more than 300 a day.

28. OPORD 91-001, 6.

29. Headquarters, U.S. Central Command, Combined OPLAN for Offensive operations to Eject Iraqi Forces from Kuwait, 17 Jan. 1991, 6–7, CHC 18-1, AFHRA; USCENTCOM OPLAN Desert Storm, 16 Dec. 1990, 13–14, IRIS-269602, AFHRA.

30. Coalition Combined OPLAN, 17 Jan. 1991, C-6-2.

31. Viewgraph, General Horner's briefing to the secretary of defense, 20 Dec. 1990, in "General Glosson Briefings."

32. Middle East Watch, *Needless Deaths in the Gulf War* (Washington, D.C.: Middle East Watch, 1991), 128–29.

33. Lt. Col. David A. Deptula, USAF, interview with the author, 20 and 21 Dec. 1991.

34. Maj. Gen. Buster C. Glosson, USAF, interview with the author, 14 Apr. 1992. As would be expected, the exact flight and weapon release "floors" for many aircraft fluctuated during the

course of the war in response to tactical conditions and mission requirements. Before the war, planners anticipated using the medium altitude bombing tactic only during daylight.

35. Briefing viewgraph, prepared by AF/XOXWF for the secretary of defense's visit to Checkmate, 6 Feb. 1991, CHC 1-6, AFHRA.

36. During this same period, the employment of F-16 killer scouts (Pointers) began, and F-16s were directed to release bombs below 8,000 feet (Lt. Col. Lewis point paper, "Corps Air Support at Desert Storm," 3 July 1991, Checkmate Historian's Files, Desert Storm (CHST) 22-15, AFHRA); 23/354 TFW(P), "Battle Staff Directive No. 26," 31 Jan. 1991, GWAPS Microfilm Roll 26554, AFHRA.

37. U.S. Department of Defense, *Conduct of the Persian Gulf War, Final Report to Congress,* April 1992, 138.

38. The eight core strategic targets referred to are leadership; command, control, and communications; electric power; oil refining; key bridges and railroads; ballistic missile launchers and support; NBC capabilities; and military support facilities. In the discussion of the strategic targets, I am particularly indebted to the analysis done by Barry Watts. Barry D. Watts and Thomas A. Keaney, *Gulf War Air Power Survey, Vol. II, Part II: Effects and Effectiveness* (Washington, D.C.: Government Printing Office, 1993).

39. Master Target Lists, Master Target Folder, BH 53, AFHRA; United Nations, Security Council, Report S/23215, 7th IAEA on-site inspection (11–12 Oct. 1991), 14 Nov. 1991, 8, 63.

40. David Kay, "Arms Inspections in Iraq: Lessons for Arms Control," 12 Aug. 1992, 3, NA 375, AFHRA.

41. GWAPS discussions with Defense Intelligence Agency (DIA) analysts, 30 Sept. 1992; DIA, "Mobile Short-Range Ballistic Missile Targeting in Operation Desert Storm," OGA 1040-23-91, Dec. 1991, 1.

42. "A-10 Mission Results: Targets Destroyed—Confirmed," in NA 292, AFHRA; Department of Defense, "Special Central Command Briefing," Riyadh, Saudi Arabia, 30 Jan. 1991, transcript #672561.

43. Chris Hedges, "After the War: Iraq in Growing Disarray, Iraqis Fight Iraqis," *New York Times,* 10 Mar. 1991, 1, 14.

44. Walid Doleh, Warren Piper, Abdel Qamhieh, and Kamel al Tallaq, "Electrical Facilities Survey," October 1991, 7–9, in International Study Team, *Health and Welfare in Iraq After the Gulf Crisis: An In-Depth Assessment,* October 1991.

45. DIA, *Final BDA Status Report,* 90–91, NA 519, AFHRA.

46. Interview, Perry Jamison, Richard Davis, and Barry Barlow, Center for Air Force History, with Lt. Gen. Charles A. Horner, Shaw AFB, S.C., 4 Mar. 1992, 49–50, NA 303, AFHRA.

47. Peter P. Perla, *Desert Storm Reconstruction Report, Vol I: Summary 24* (Alexandria, Va.: Center for Naval Analyses, 1991), 78.

48. Robert W. Ward et al., *Desert Storm Reconstruction Report, Vol VIII: C3/Space and Electronic Warfare* (Alexandria, Va.: Center for Naval Analyses, 1992), 4-6 to 4-10.

49. USCINCCENT Situation Report, 23 Feb. 1991, Hq USAF Operations Center Files, Contingency Support Staff (CSS) 29, AFHRA.

50. Directorate of Intelligence, Central Intelligence Agency, "Operation Desert Storm: A Snapshot of the Battlefield," Report IA 93-10022, September 1993.

51. U.S. Marine Corps Research Center, Quantico VA, "Fire Support Coordination During Operation Desert Storm," Research Paper #92-0007 (Part 1), 23.

An Afterword

David MacIsaac

For those stalwart readers who have made it all the way to here, let me begin these brief remarks by noting that this volume has a long history. The majority of its present content derives from a symposium at the Air Force Academy that took place 18–20 October 1978, an event that was initially recorded in Colonel Alfred F. Hurley and Major Robert C. Ehrhart, editors, *Air Power and Warfare* (Washington, D.C.: Office of Air Force History, 1979).[1] Among such volumes, few have been cited more often, leading to incessant requests that it be reprinted—a situation the Government Printing Office seemed to find more entertaining than imperative. Then someone came up with a better idea, and thus, the book you hold in your hands.

The story of the 1978 symposium, the eighth in a series that began in 1967, is one thing; fond memories attendant thereto are another. Among those distinguished scholars and airmen no longer here to share the new edition are: Noel Parrish, Charles H. Gibbs-Smith, Eugene Emme, Al Coox, Ken Whiting, Hal Deutsch, Generals Curt LeMay, O. P. Weyland, Ed Lansdale, John B. McPherson and the then "Grand Old Man of the Air Force," Ira Eaker.

No less importantly among those now lost to us are two brilliantly enthusiastic young assistant professors from those days: Maj. John F. Shiner and Capt. Robert E. Wolff. Colonel Shiner, who later went on to tell the Bennie Foulois story also later served as Acting Head, Department of History, before moving to the Air Staff in Plans and Operations before becoming Deputy Chief, Office of Air Force History, prior to being medically retired owing to the onset of multiple sclerosis.[2] As Executive Director for the symposium here immortalized, Fred was a whirlwind of energy, enthusiasm, and competence. Fred's string ran out early on the morning of 19 March 1995, whereupon he joined the crowd at Arlington with full military honors, on 23 March. His eulogy, delivered by his great and good friend Elliott V. Converse, III, Colonel, USAF, retired, will never be forgotten by those present.

Among Fred's "best lieutenants" in October 1978 was Captain Wolff, sometimes "Bobbie" or "Woofie," but always the brightest, happiest face, all day long.[3] Colonel Wolff had flown the toughest days of LINEBACKER II, seemed to think that Texas Tech was something we had all heard of, and then went on to great success on the Air Staff and in the Strategic Air Command. He was serving as Vice Commander of the 92nd Bomb Wing at Fairchild AFB when he was lost in a truly tragic B-52 aircraft accident in June of 1994.

Planning for the 1978 Military History Symposium began early in the fall of 1976, when Fred Shiner was designated by Colonel Hurley as "Executive Director Presumptive." To the occasional amazement of many beyond the Academy's gates, only two of the previous seven Military History Symposia had placed any particular emphasis on air power or on air warfare.[4] But 1978 would mark the seventy-fifth anniversary of Kitty Hawk, so there was never any real question regarding primary focus. By February 1977, Fred had drawn up a basic outline and by April the Superintendent had given his blessing. The overriding basic premise was that the symposium speakers represent a mix of academic specialists and veteran practitioners. As matters turned out in the end, the mix was almost exactly equal at about fifteen each, although several—like I. B. Holley, Jr., and Noel Parrish—had a foot in each camp.

General Parrish got the affair off to a dynamic start on the evening of 18 October with his Harmon Memorial Lecture on "The Influence of Air Power upon Historians."[5] Such influence, he suggested, was as yet still difficult to discern. With engaging and occasionally acerbic wit, he decried what he saw as the ahistorical bent of U.S. defense policy, the dearth of biographies of air leaders,[6] and a tendency among younger academic historians to look upon airmen as little more than "heralds of the Apocalypse." Two short excerpts capture his concerns and tone:

> Our purpose in meeting here, as I understand it, is to enjoy the living elements of air power history, to mourn for the missing, the departed, and the ill-conceived, and to speculate hopefully on those elements yet unborn. Since the influence of air power upon most historians is largely negative, I shall also discuss the influence of historians on air power, which, by contrast, is practically non-existent. Despite the commendable efforts of many, our traditions and the memories that made them have been neglected, our costly lessons from the recent past are in danger of being forgotten before they are really learned.

The symposium reconvened on Thursday morning, 19 October, with three papers on the period 1903 to 1941 (the latter date involving a peculiarly American concept of when the war started). All are included in the present volume and are now joined by Brig. Gen. Lucien Robineau's essay on the French experiences before and during World War I. This welcomed addition adds a note of balance often lacking in British—especially British—and American accounts, many of which tend to ignore the contributions of men like Clément Ader, especially in his book *L'aviation militaire*, published in 1908, and Col. Maurice Duval.

On Thursday afternoon, separate sessions treated the Japanese, Russian, and German experiences during World War II, and U.S. air leadership during the war. In the new edition, the papers by Professors Coox, Whiting, and Boog are retained, now joined in Part II by what was originally General Eaker's banquet address of Thursday evening. Regrettably, much of the panel discussion on U.S. air leadership—of which we now know much more than we did in 1978—had to be omitted in the new edition to make room for the new essays that make up Part IV.[7] As an example, consider one of General LeMay's comments from the original volume which vividly recalls the early days of the 305th Bomb Group in England:

> When I was given command of a bomb group [the 305th], it consisted almost 100 percent of inexperienced people. I had one major, who had been commissioned from the rank of master sergeant, an administrative clerk, and he was my group adjutant. I had two pilots, besides myself, who had flown B-17s before, and we

three had to check out the other pilots, who came directly from single-engine school. The armament officer was an ex-Marine corporal who had been a captain in the Nicaragua National Guard for a while. He knew something about machine guns, so he was my ordnance officer. My prize was a first lieutenant who had been a line chief in B-17s as a technical sergeant.

The navigators I got two weeks before we went overseas. They had had one ride in a B-17 before they navigated across the Atlantic; the first time half of them had ever *seen* the Atlantic was when they navigated across it. The bombardiers had never dropped a live bomb. They had dropped some practice bombs over a desert on a nice white circle you could see for fifty miles, something entirely different from trying to hit a factory in the midst of a built-up area in the industrial heart of Europe. The gunners had been to gunnery school, supposedly, but had never fired a gun from an airplane.

We never got to fly formation until we got to England. The first day we could fly I got up in formation and it was a complete debacle. The next flight, I got up in the top turret on the radio and positioned each aircraft until the gaggle I had around me approached the formation I wanted to fly. The third time we flew, we went across the Channel. That was our start in combat.

On the morning of Friday, 20 October, the symposium reconvened in separate sessions of two major papers each, three of which are included in Part III of the present volume. In each case the essays represented, in 1978, major steps forward in our understanding of the evolution of air policy and doctrine in the postwar decade. For just one example among the three, David Alan Rosenberg's contribution reflected his yeoman work in breaking down the secrecy barriers surrounding nuclear weapons, and presaged his prize-winning articles in the *Journal of American History* and *International Security*, among others.[8]

The final session on Friday afternoon was chaired by Professor Theodore Ropp and included presentations by Gen. T. R. Milton and Col. Ray L. Bowers, with commentaries by Lt. Gen. John B. McPherson and Maj. Gen. Edward G. Lansdale—all four USAF, retired. The symposium concluded with a discussion period capped by a brief "instant evaluation" of the symposium as a whole by Professor Ropp. This section is not included in this edition, but can be found on pages 299–343 of the first edition.

Part IV is composed of four essays commissioned especially for this edition. Gen. Bryce Poe, former president of the Air Force Historical Foundation, who had participated in the symposium (see original edition, pp. 410–17) focuses here on the war in Korea, but, as always, brings an informed and wide-ranging perspective to the task. Air Vice-Marshal Tony Mason's association with the Academy goes back to 1969–71 and has remained particularly helpful ever since. His truly exceptional speaking and publishing record over the past two decades has led to his reputation as the leading air power analyst of our age. His essay here reflects both the long study revealed in his two best known books and adds new materials from unusual sources.[9]

Lieutenant Colonel Mark Clodfelter's essay on air power in Vietnam is of course informed by his ground-breaking book *The Limits of Air Power: The American Bombing of North Vietnam* (New York: The Free Press, 1989), now recognized as the best treatment of its topic yet published.[10] Colonel Thomas A. Keaney's essay on the Gulf

War of 1991 is informed by his service on the Secretary of the Air Force's Gulf War Air Power Survey (GWAPS), for which he co-authored, with Eliot A. Cohen, the Survey's Director, the GWAPS' *Summary Report* (Washington, D.C.: Government Printing Office, 1993). His conclusions should be read carefully, since not all of them mirror the enthusiasm that attended the initial coverage of the air campaign.

That three of the four contributors to Part IV are former members of the Department of History, yet were chosen for their tasks for reason of their later achievements, must be a point of special pride for Cols. Carl Reddel and Mark Wells—indeed the whole Department of History roster, past and present. The list of department veterans who share that pride is long indeed, but pride of place among them must go to Brigadier General Alfred Hurley, USAF, retired, who not only originally hired all three, but is also responsible for General Poe's long association with the Academy.

All that said about the 1978 symposium, I cannot here refrain from a few short comments about the series as a whole—actually less about the series than the people involved who have made it the premier gathering of military historians. Unknown to most is that the twenty-five to thirty people assigned to the Department of History during the symposium make it happen as "an additional duty." By now, I suspect the numbers of these symposium veterans are in the hundreds. They include not only those who meet flights at Colorado Springs airport, arrange perfect banquets, cajole local innkeepers into favorable rates, serve as personal aides to the distinguished speakers, keep the doors of H-1 from squeaking—the list could go on forever—but also those whose efforts during one or the other of the symposia presaged their many later contributions to the Air Force. Herewith only three examples: (1) when Ken Alnwick, way back in 1969, convinced the Air Force Academy Association of Graduates to help underwrite the series; (2) when Alan L. Gropman, as a senior colonel in charge of a key division of the Air Staff, corralled at least eight former members of the Department of History, to keep him on track; and (3) when Ronald R. Fogleman, Executive Director of the 1972 Symposium, was appointed Chief of Staff, United States Air Force, in October 1994.

Notes

1. From 1966 to 1982 Alfred F. Hurley was Professor and Head, Department of History, and guiding light for the Academy's series of Military History Symposia. He is now Chancellor of the University of North Texas and Brigadier General, USAF, retired. Robert C. Ehrhart is a Lieutenant Colonel, USAF, retired, a Ph.D., and currently a cowboy running loose in the wilds of Montana—a.k.a. Wrangler Bob.

2. John F. Shiner, *Foulois and the U. S. Army Air Corps, 1931–1935* (Washington, D.C.: Office of Air Force History, 1983) treats more than its title implies. Fred also later co-authored, with Bernard C. Nalty and George M. Watson, *With Courage: The U.S. Army Air Forces in World War II* (Washington, D.C.: Air Force History & Museums Program, 1994).

3. And, like Fred, one of our most talented instructors. He was surely the only B-52 driver of his era who could hold cadets spellbound throughout a lecture on "The Relative Decline of the Papacy between the 11th and 13th Centuries" (History 438, Western Ideas & Institutions, Spring 1978).

4. As the editors noted in their preface, the "very limited amount of scholarly work in this field [had previously] stymied proposals to present a symposium on the topic." For the two partial exceptions, see William Geffen, ed., *Command and Commanders in Modern Warfare*, 2nd ed., enlarged (Washington, D.C.: Office of Air Force History, 1971), 253–324. This edition includes several commentators whose remarks were solicited on the contents of the original edition (USAF Academy, 1969). The original participants were Noble Frankland, Robin Higham, Robert F. Futrell, and Gen. Ira C. Eaker (with Gen. Carl A. Spaatz at his side). Later comments include those of Marshal of the RAF Sir John Slessor and Air Vice-Marshal E. J. Kingston-McCloughry, and from the Luftwaffe, Field Marshal Erhard Milch, and Gens. Adolph Galland and Johannes Steinhoff. See also Monte D. Wright, ed., *Science, Technology and Warfare* (Washington, D.C.: Office of Air Force History, 1971), 85–185. The session reproduced in these pages was chaired by the late Bernard Brodie, featured papers by I. B. Holley, Jr. and Melvin Kransberg, and comments by Robert L. Perry, Clarence G. Lasby, John C. Fisher, Francis X. Kane, and Eugene M. Emme.

5. Not included in this edition, but separately published (USAF Academy, 1969) and included in Harry R. Borowski, ed., *The Harmon Memorial Lectures in Military History, 1959–1987* (Washington, D.C.: Office of Air Force History, 1988), 25–42.

6. A situation much improved in the intervening years; see Col. Phillip S. Meilinger, *American Air Power Biography: A Survey of the Field* (Maxwell AFB, Ala.: Air University Press, 1995). As Colonel Meilinger notes in his introduction, "there is still much work to be done."

7. Indeed, the stream of books on air power during World War II seems unending. Among the best of the more recent offerings, our own editor's *Courage and Air Warfare: The Allied Aircrew Experience in the Second World War* (London: Frank Cass, 1995) stands out. But cf. Geoffrey Perret, *Winged Victory: The Army Air Forces in World War II* (New York: Random House, 1993); Walter J. Boyne, *Clash of Wings: Air Power in World War II* (New York: Simon & Schuster 1994); Brereton Greenhouse et al., *The Crucible of War, 1939–1945* (Vol. III, Official History of the RCAF, Toronto, 1994); Stephen L. McFarland, *America's Pursuit of Precision Bombing, 1910–1945* (Washington, D.C.: Smithsonian Institution Press, 1995); and, on a too long-neglected topic, Thomas A. Hughes, *Over Lord: General Pete Quesada and the Triumph of Tactical Air Power in World War II* (New York: Free Press, 1995). Latest into the list is Kenneth P. Werrell, *Blankets of Fire: U.S. Bombers over Japan during World War II* (Washington, D.C.: Smithsonian Institution Press, 1996) which treats far more than its modest title implies.

8. See David A. Rosenberg, "American Atomic Strategy and the H-Bomb Decision," *Journal of American History* (June 1979); "'Smoking, Radiating Ruin at the End of Two Hours': Documents on American Plans for Nuclear War with the Soviet Union, 1954–1955," *International Security* 6 (Winter 1981/82); and "The Origins of Overkill," ibid., 7 (Spring 1983).

9. Tony Mason, *Air Power: A Centennial Appraisal* (London: Brassey's 1994) and (with Air Marshal Sir Michael Armitage) *Air Power in the Nuclear Age*, 2d ed. (Urbana: University of Illinois Press, 1985).

10. And not only by this proud father-in-law!

Contributors

Horst Boog. B.A. (1950), Middlebury College; Ph.D. (1965), University of Heidelberg. Has been a German Air Force reserve officer, military analyst for NATO, and for over twenty-five years a historian at the Militärgeschichtliches Forschungsamt (German Armed Forces Office of Military History) in Freiburg. Retired in 1993 having been the office's Senior Director of Research and Head of the Research Department for the Second World War.

Mark Clodfelter. Lieutenant Colonel, USAF (Ret.). B.S. (1977), U.S. Air Force Academy; M.A. (1983), University of Nebraska; Ph.D. (1987), University of North Carolina. Commissioned in the USAF (1977); served as an air weapons controller at Myrtle Beach Air Force Base, South Carolina, and Osan Air Base, Korea. Instructor and Associate Professor of History, USAF Academy (1983–84; 1987–91). Professor of Airpower History, School of Advanced Air Power Studies, Maxwell Air Force Base, Alabama (1991–94). Professor of Aerospace Studies, University of North Carolina (1994–97). Professor of Military Strategy, National War College.

Alvin D. Coox. B.A. (1945), New York University; M.A. (1946), Ph.D. (1951), Harvard University. Had been a member of faculty or visiting professor at Harvard University, Johns Hopkins University, Shiga National University, Japan, University of Maryland (Far East Division); Historian, Japanese Research Division, U.S. Army, Japan; Analyst, U.S. Air Force, Japan; Associate/Managing Editor, *Orient/West* magazine; since 1964 had been at San Diego State University. Professor and Director, Center for Asian Studies. Died 1999.

Ira C. Eaker. Lieutenant General, USAF (Ret). B.S. (1917), Southeastern Normal College, Oklahoma; A.B. (1933), University of Southern California; law student at Columbia University; University of Philippines, George Washington University; Air Corps Tactical School (1936). Commissioned U.S. Army (1917) and completed flying training (1919); command and staff assignments in the Philippines and the U.S. (1920–42); Commander, 8th Bomber Command, 8th Air Force, Mediterranean Allied Air Forces (1942–45); Deputy Commanding General and Chief of the Air Staff, USAAF (1945–47); retired 1947. Wrote regularly on national security affairs for military and civilian journals. Died 1987.

Eugene M. Emme. A.B. (1941), Morningside College; M.A. (1946), Ph.D. (1949), University of Iowa. Served in U.S. Navy (1943–45) and U.S. Air Force Reserve (1948–73). Taught at University of Iowa (1946–48); civilian professor of history at Air University, Documentary Research Division (1949–51); Air War College, Director of Graduate Study Group and Research Advisor (1952–58); Operations Research, Office of Civil and Defense Mobilization (1958–59); NASA Historian (1959–78). Died 1985.

Ronald R. Fogleman. General, USAF (Ret). B.S. (1963), U.S. Air Force Academy; M.A. (1971), Duke University. Commissioned in the USAF (1963); served in variety of flying, staff, and command assignments in the United States and overseas. Instructor of military history, USAF Academy (1970–73). A command pilot and parachutist, he amassed more than 6,500 flying hours in fighter, transport, tanker, and rotary wing aircraft during his thirty-four year career. He flew 315 combat missions and logged 806 combat hours in Southeast Asia. Commander, 7th Air Force (1990–92); Commander, Air Mobility Command (1992–94); Air Force Chief of Staff (1994–97).

John T. Greenwood. B.A. (1964), University of Colorado; M.A. (1966), University of Wisconsin; Ph.D. (1971), Kansas State University. Taught at Wisconsin State University, Whitewater (1966–67); Historian, Strategic Air Command (1970–73); Chief Historian, Space and Missile Systems Organization, Air Force Systems Command (1973–75); Historian, Office of Air Force History (1975–78). Chief, Historical Division, U.S. Army Corps of Engineers.

Edward Homze. B.A. (1952), M.A. (1953), Bowling Green State University; student, Free University of Berlin (1957–59); Ph.D. (1963), Pennsylvania State University. Active duty with the USAF (1954–56). Visiting Professor of Strategy, Naval War College (1975–76). Professor, University of Nebraska.

Thomas A. Keaney. Colonel, USAF (Ret). B.S. (1962), USAF Academy; M.A. (1971), Ph.D. (1975), University of Michigan. Served in a variety of Air Force flying and staff assignments including B-52 Squadron Commander, Forward Air Controller in Vietnam, planner on the Air Staff, and Associate Professor of History at the USAF Academy. Base Commander, Wurtsmith Air Force Base, Michigan (1983–85); researcher and author, Gulf War Air Power Survey, Office of the Secretary of the Air Force (1991–92); faculty member, Department of Military Strategy and Operations, National War College, Ft. McNair, Washington, D.C. (1992–98). Executive Director, Foreign Policy Institute, School of Advanced International Studies, Johns Hopkins University, Washington, D.C.

David MacIsaac. Lieutenant Colonel, USAF (ret). A.B. (1957), Trinity College, Connecticut; A.M. (1958), Yale University; Ph.D. (1970), Duke University. Commissioned USAF (1958); numerous assignments including Spain (1961–64), Vietnam (1970–71), USAF Academy (1964–66, 1968–70, 1972–75, 1976–78). U.S. Naval War College (1975–76). Fellow, Woodrow Wilson International Center for Scholars (1978–79). Senior Research Associate, Air Power Research Institute, Center for Aerospace Doctrine, Research, and Education (1982–84). Associate Director of the Air Power Research Institute and Professor of Military History, Air University, Maxwell Air Force Base, Alabama (1985–91).

R. A. Mason. CB CBE MA DSc. Air Vice Marshal, RAF (ret). M.A. (1956), St. Andrews University; M.A. (1967), University of London; U.S. Air War College, Associate Program (1971); Royal Air Force Staff College (1972). Commissioned RAF (1956); served in several RAF staff appointments and on exchange assignments to History Department, USAF Academy; Director of Defence Studies, RAF; visiting lecturer on strategy and air power at ten universities in Great Britain; Air Secretary, RAF. Retired 1989. Defense analyst, lecturer and author.

Contributors 321

Richard J. Overy. B.A., M.A., Ph.D. Caius College, Cambridge University. Research Fellow, Churchill College (1972–73); Fellow and Assistant Lecturer in History, Queen's College (1973–79); Professor of History, King's College, University of London.

Robert Perry. A.B. (1947), Marshall University; M.A. (1949), Ohio State University. Historian, Air Force Historical Program, Wright-Patterson Air Force Base, Ohio (1951–61); Historian, Air Force Systems Command (1961–74); RAND Corporation, Program Chairman, Research and Development and Systems Acquisition programs; consultant to numerous government agencies; lecturer, Ohio State University, Wittenberg University, and elsewhere.

Bryce Poe II. General, USAF (Ret). B.S. (1946), U.S. Military Academy; M.A. (1964), University of Omaha; M.A. (1965), George Washington University; National War College (1965). Completed flying training while a cadet and commissioned U.S. Army Air Corps (1946); numerous flying and command assignments (1946–69); Headquarters USAFE (1970–73); Commander, Odgen Air Materiel Area (1973–74); Vice Commander, USAFE (1974–76); Commander, Air Force Acquisition Logistics Division, AFLC (1976–78); Commander, Air Force Logistics Command (1978–80); President, Air Force Historical Foundation (1987–96); Publisher, *Air Power History*.

Lucien L. Robineau. Brigadier General, French Air Force (Ret). Graduated Ecole de l'Air (1953); Commissioned and completed flying training (1953). Served in various flying a command assignments including Squadron Commander (1960–63); Wing Commander (1966–68); Base Commander (1976–78); Professor, French Air Force War College (1973–76); Chief of Staff of the Territorial Command of Bordeaux (1979–81). Director, French Air Force Office of History (1985–94). Head, History, Arts and Letters Section, *Académie nationale de l'Air et de l'Espace*.

David Alan Rosenberg. B.A. (1970), American University; M.A. (1971), Ph.D. (1983), University of Chicago. Researcher for U.S. Navy and Smithsonian Institution (1966–71); consultant for Office, Chief of Naval Operations and private corporations (1974–78); instructor, University of Wisconsin–Milwaukee (1976–78); Associate Professor, Temple University.

Mark K. Wells. Colonel, USAF. B.S. (1975), USAF Academy; M.A. (1983), Texas Tech University; Ph.D. (1992), King's College, University of London. Commissioned in the USAF (1975); pilot training (1976). Served in a variety of flying and staff positions (1977–83); Assistant Professor of History, USAF Academy (1983–86). Student Air Command and Staff College (1987) and Army War College (1993). Military Assistant to the Supreme Allied Commander, Europe (1989–91). Permanent Professor and Head, Department of History, USAF Academy.

Kenneth R. Whiting. A.B. (1940), Boston University; M.A. (1941), UCLA; Ph.D. (1951), Harvard University. Student at Institute for the Study of the USSR, Munich (1957). Active duty, USAAF (1941–45). Lecturer at UCLA, Tufts University, and elsewhere. Chief, Documentary Research Division, Air University Library, Maxwell Air Force Base, Alabama. Died 1990.

Index

ABLE Day test bomb, 155
AC-47 gunship (US), 275
Ader, Clément, 32–33, 34
Advance Base Brigade, 53
Aerial photography, 232–33, 248; *see also* Intelligence work; Reconnaissance
  in Cuban missile crisis, 252–53
  RAF missions, 249–50
  U-2 flights, 249–52
Aero Club (France), 33
Aero Club (Great Britain), 7
Aeronautical Division of the Signal Corps (US), 49–50
Aeronautical Journal, 7
Aeronautical Manufacturing Service (France), 45
Aeroplane, 7, 60
Aichi, 79
Air Battalion of the Royal Engineers, 9
Air Board (Great Britain), 14
Airborne early warning and control aircraft, 258
Air combat, 2–3
Air Committee (Great Britain), 9–10
Air Corps IV (Germany), 104
Air Corps Tactical School (US), 4, 61, 264–66, 283
Aircraft
  early Japanese purchases, 74
  French requests of US, 56
  pusher *vs.* tractor, 51
  of Soviets in Spain, 88
  successful, 205
Aircraft carriers
  development in 1920s, 60–61
  early, 176
  early call for, 59
  in Gulf War, 298
  offensive power, 190
  role in US Navy, 196
  strategic bombing capability, 178
  use of steam catapults, 206
  world's first, 76
Aircraft production
  in France 1910–14, 34
  in France 1914–18, 45–47
  in France 1938–39, 25–26
  German flaws in, 124–25
  Goering's impact on, 128–29
  in Japan 1914–18, 75
  in Japan 1918–45, 79–80
  and politics in France, 45–46
  Roosevelt plans for 1940–41, 142
  Soviet figures for 1941–42, 96–97
  Soviet relocation in 1942, 96–97
  in US 1941–42, 68–69
  in USSR 1921–32, 85
  in USSR 1940–41, 92–93
  in USSR after 1945, 238
  during World War II, 20

Air defense, German view of, 126–27
Air division, 42–43
Air doctrine, 2–3; *see also* Douhet, Giulio; Mitchell, Gen. William
  air division, 42–43
  bombardment, 40–41
  critical year of 1948 (US), 163–68
  danger of role concentration, 260
  definition of, 175
  Douhet's theories, 24
  in early atomic era, 152–56, 156–63
  early British ideas, 7–9
  early US ideas, 52–54
  Eisenhower and US Navy, 194–96
  employment doctrine, 39–40
  at end of Gulf War, 308–11
  at end of Vietnam War, 283–84
  on eve of Vietnam, 266–67
  Flexible Response, 257–59;
  in Germany: concentration on military matters, 111–13; consequences of command traits, 117–30; low priority for technology, 115; narrow view of mission, 116–17; origin of modes of thinking, 130–36; overemphasis on offense, 115–16; overemphasis on tactics/operations, 114–15; preference for strategic bombing, 127–30;
  in Gulf War, 284;
  interaction with technology: faith in future of, 207–9; ignoring ballistic missiles, 213–15; in USAF, 205–17
  kamikaze operations, 80
  massive retaliation, 4–5
  need for openness, 5
  postwar US developments, 149–52
  relation to ground forces, 310
  Third Dimension of Warfare, 16
  in USAF Manual 1-8, 266
  USAF preparation for Vietnam, 264–67
  US developments in World War II, 143–46
  US experience 1919–37, 57–63
  US Navy: and atomic bomb, 178–80; changing role for carriers, 196–98; General Board study, 184–86; interwar/postwar years, 175–78; and Korea, 191–94; OP-55 study, 189–91; role/doctrine debate, 186–89; role/mission debate, 184; and strategic air offensive, 180–83
Air Force Historical Foundation (US), 219
Air Force Plans for Peace (Smith), 151
Air-ground cooperation, in German forces, 131
Air interdiction operations, 305
Airmindedness, 1
Air power
  analysis of Soviet World War II role, 103–6
  AWPD-1 document, 68
  balloons and dirigibles, 35–37
  in battle in World War I, 38–39
  change in focus, 1

Cold War: Berlin Airlift, 240–43; British cooperation, 249–51; NATO strategy, 243–47; stimulus to, 259; strategic environment, 237–40; U-2 flights, 251–52
  developments before 1918, 139–40
  diversity of development, 2–3
  early history in France, 31–34
  early history in Great Britain, 7–9
  early history of RAF, 9–10
  early Japanese development, 74–75
  early predictions about, 32–33
  early Soviet developments, 85–87
  early US Navy development, 52–53
  in Ethiopian War, 27
  in Europe in World War I, 20–23
  execution of Gulf War strategy, 298–301
  fighter tactics in World War I, 22
  fighter tactics in World War II, 22
  Great Britain in World War I, 13–16
  Gulf War planning, 290–98
  Gulf War retrospective, 308–11
  in-flight refueling, 253–54
  in interwar Europe, 23–27
  logistics in World War I, 44–47
  military ballooning, 32
  modern flexibility, 5
  political context, 1
  popular culture of, 1
  post–Cold War, xiii–xiv
  pre-1914 Europe, 19–20
  reconnaissance, 22–23
  results of Gulf War, 301–8
  social history, 2
  Soviet Cold War contribution, 254–56
  Soviet forces 1939–41, 90–93
  in Spanish Civil War, 27–28
  strategic bombardment, 240
  tactical vs. strategic, 25–26
  technical context, 1–2
  US developments 1903–11, 49–54
  US future employment, 147–48
  US in France 1917–18, 56–57
  US interwar expansion, 57–63
  US in World War II, 54–57
  US Navy in World War I, 57
  US on eve of World War II, 63–69
Air Power Conference, Freiburg, Germany, 2
Airships; see also Zeppelins
  German preferences for, 20
  of US Army, 50
Air-to-air missiles, 215
Air transport, Luftwaffe failures, 123
Air War Academy (Germany), 111–13, 114, 115, 116
AJ-1 Savage (US), 184, 192
Akagi aircraft carrier, 77
Akron dirigible, 60
Albatross D. II, D. III (France), 37
Alexander, Grand Duke, 19
Alksnis, Ya. I., 91
Allies, prewar ignorance of Japanese aviation, 73–74
American Expeditionary Force, 56
Ames, Dr. Joseph, 65
Anabuki, M/Sgt., 80

Andersen Air Force Base, Guam, 274
Anderson, Maj. Gen. Frederick L., 151
Andrews, Gen. Frank M., 62, 64, 141–42, 264
Anglo-American Combined Chiefs of Staff, 68
Annapolis, naval air training, 53
Antoinette airplanes (France), 33
Antonov An-2 Colt (USSR), 235
Arab-Israeli wars, 3
Arc Light bombing mission, 273, 275
Armed Forces Academy (Germany), 112
Armed Forces Special Weapons Project (US), 165–66
Armée de l'air, 25
Army Air Corps Act of 1926, 59
Army General Staff (Japan), 74
Army Group Center (Germany), 94
Army Group North (Germany), 94
Army Group South (Germany), 94
Army Reorganization Act of 1920, 58
Arnold, Gen. Henry H., 51, 59, 62, 65, 66, 67, 122, 152–53, 240, 264
  on air force capabilities, 149
  on need for independent air force, 150–51
  and public relations, 142
  on rapid demobilization, 157
  on Soviet threat, 239
  on US Navy, 149–50
Atlantic Charter, 67
Atlas program (US), 211
Atomic Energy Commission, 155, 162, 211, 246
Atomic weapons, 159, 160
  basis for NATO defenses, 247
  Bikini Atoll test, 181
  deployment in Europe, 246
  dropped on Japan, 265
  in Korean War era, 192
  limited-option position, 214
  and US Navy, 178–80
A-6 (US), 300
Aviatik (France), 38
Aviation Act of 1926 (US), 141
Aviation/Aeronautics; see also Air power
  airships, 20, 50
  balloon flights, 74, 76
  changes in 1930s, 62
  characteristics, 34–35
  dirigibles and balloons, 35–37
  earliest modern ideas, 7–8
  early French efforts, 31–34
  European developments 1914–17, 21
  first military appropriations, 51–52
  interwar era, 23–27
  pre–World War I era, 19–20
  rapid airplane development, 37–39
  sea-based, 176
  US developments 1918–41, 140–42
  in World War I, 20–23
  World War I improvements, 20–21
  Wright brothers, 49–51
Aviation Hall of Fame Air War College (US), 146
Aviation Industry Commissariat (USSR), 92
AWACs; see Airborne early warning and control aircraft
AWPD-1 document (US), 68, 265

Index

Baka guided missiles (Japan), 80
Baker, Newton, 62
Baker Board, 141
Balbo, Italo, 26
Baldwin, Hanson, 142
Baldwin, Stanley, 62
Balfour, Arthur, 10
Ballistic missiles
  attempt to ignore, 213–15
  development of, 209–13
Balloons
  Japanese experiments, 76
  Japanese flights 1877–1904, 74
  military flights, 32
  US flights, 49–50
  in World War I, 35–37
Baranov, P. I., 91
Barbarossa Operation (Germany), 3, 93
Barès, Maj. José, 36, 39, 40, 41
Barksdale Air Force Base, La., 62, 298
Barrel Roll operation, 278
Basic Doctrine Manual (USAF), 264, 266, 283
Basic National Security Policy (US), 195
Battle of Britain, 28, 67
  and purpose of Luftwaffe, 128
Battle of Fleurus, 32
Battle of Saint-Mihiel, 43, 56
Battle of the Ia Drang, 275
Battle of the Marne, 22–23, 38, 52
Battle of the Somme, 37, 42
Battle of Verdun, 37, 41
Battleships, 49
  Mitchell's views of, 58
B-17 bomber (US), 62, 64, 66, 67, 101, 213, 248
B-24 bomber (US), 101
B-25 bomber (US), 66
B-26 bomber (US), 66, 222–23
B-29 bomber (US), 62, 151–52, 154, 157, 158, 160, 162, 165, 184, 209, 223, 242, 243, 247–48, 248, 253–54, 255
B-35 bomber (US), 62
B-36 bomber (US), 157–58, 165, 206, 208–9
B-47 bomber (US), 158, 206, 214, 247–48
B-50 bomber (US), 158, 165, 206
B-52 bomber (US), 2, 208–9, 211, 247–48, 267, 270, 273, 274, 279, 282, 297, 298
B-58 bomber (US), 206, 211
B-60 bomber (US), 206
Bell, Alexander Graham, 50
Bell UH-1 helicopter (US), 274
Benson, Adm. William, 59
Berlin Airlift, 240–43
Berlin Blockade, 165–67, 168, 220
Bertram Collection, 132–33
Bevin, Ernest, 241
B-17 Flying Fortress (US), 142
Bialer, Uri, 1
Bikini Atoll, 155, 178
Bikini Evaluation Report, 181
B-36 investigation, 189
Black Hole. *see* Special Planning Group
Blanchard, Col. William H., 154–55, 167

Blandy, Adm., 189
Blériot, Louis, 33, 50
Blériot aircraft (France), 33, 34, 46
Blitzkrieg, 66, 67, 142
Blyukher, Gen. V. K., 88, 89
Bock, Gen. Fedor von, 94
Bock's Car, 265
Boeing 229, 62
Boeing Aircraft Company, 62
Boeing E-3 AWACs (US), 258
Boelcke, Oswald, 22
Bolling Mission, 56
Bolshevik Revolution, 23
Bombers
  acquired by US Army 1938–41, 64
  acquired by US Navy 1938–41, 64
  technology in 1930s, 62
Boog, Horst, 111
Boomer, Lt. Gen. Walter E., 307
Bradley, Gen. Omar N., 165, 167, 187, 237–38
  on strategic bombing, 244
Breguet XIV A2 (France), 389
Breguet XIV B2 (France), 389
Breguet XIV (France), 45
Brereton, Lt. Gen. Lewis, 155
Brewster Buffalo, 64
Bristol Jupiter, 86
"Britain and the Command of the Air," 8
British Chiefs of Staff, 245
British Royal Flying Corps, 52
Brodie, Bernard, 176, 180, 181
BROILER Plan, 164, 165
B-36-Super Carrier and Unification Hearings of 1949, 167
B-45 (UK), 249
Bundy, McGeorge, 270
Bureau of Aeronautics (US), 60
Burgess and Curtiss Factory, 53
Burke, Adm. Arleigh, 185, 187, 189, 196
Burke, Capt. C. J., 8, 9
Bush, George, 284, 293
  objectives in Gulf War, 291
Bush, Vannevar, 68, 210
C-7A Caribou (US), 274
Cambodia, 2
  bombing raids 1969–70, 277–78
  US bombing, 263
Canberra (UK), 252
Caproni, Count Gianni, 59
Caproni bombers (Italy), 57
Caquot, Capt. Albert H., 36
Carney, Adm. Robert B., 166, 195–96
Casablanca Conference, 143
Catapult scout planes, 60
Cate, James Lea, 55
Catton, Jack, 167
Caudron RX1 (France), 38, 45
Central Command (US), 290
Central Intelligence Agency (US), 148, 251
C-5 Galaxy (US), 274
Chamberlain, Neville, 64
Chambers, Capt. W. I., 52, 53
Chandler, Capt. Charles DeF., 51

Chang Hsüeh-liang, 88
CHARIOTEER Plan, 181–82
Checkmate division (USAF), 291
C-130 Hercules (US), 215, 274
Chief of Aeronautical Forces at the High Headquarters of the Armies (France), 38
China
  Communist takeover, 245
  Russia at war in, 28
  Soviet aid to, 89–90
Chinese Air Force, 77, 89
Chinese Eastern Railway, 88
Churchill, Winston, 4, 9, 67, 144–45, 250
Civil-military relations in France 1914–18, 45–46
Civil War, Russia, 23
Claude monoplanes, 77
Clausewitz, Carl von, 130, 131, 237
Clay, Gen. Lucius, 241
Clemenceau, Georges, 45
Clodfelter, Mark, 4, 5, 263
Close air support, 297–98, 301
Cold War, xiii
  Berlin Airlift, 240–43
  British cooperation, 249–51
  Cuban missile crisis, 252–53
  final stimulus to air power, 259
  flexible response and convergence, 257–59
  NATO strategy, 243–47
  role of reconnaissance, 247–49
  Soviet air power, 254–56
  strategic environment 1945, 237–40
Colson, Charles, 282
Combat support in Korea, 226–29
Combat wings, 43
Comet bomber (UK/US), 214
Commander in Chief, Far East (US), 193
Command of the Air (Douhet), 59
Commando Hunt operation, 278
Committee of Imperial Defence (Great Britain), 9
Condor Legion (Germany), 28, 63, 88
Conolly, Adm. Richard, 190
Consolidated B-24 (US), 66
Convair PB47-2 (US), 249
Conventional vs. guerrilla warfare, 267
Converse, Col. Elliott V., III, 314
Coolidge, Calvin, 58, 141
Coox, Alvin D., 73
Council of People's Commissars (USSR), 95
Cowdray, Lord, 10
Cowdray Board (Great Britain), 10
Crampton, J., 249
Cross-country flights, 33
CROSSROADS operation, 154, 155, 178, 180
C-7 transport (US), 277
C-47 transport (US), 275
C-54 transport (US), 241
C-97 transport (US), 242
C-123 transport (US), 274, 277
C-124 transport (US), 242
C-130 transport (US), 277
C-141 transport (US), 274
Cuban missile crisis, 252–53, 267

Cunningham, Lt. A. A., 53
Curtiss, Glenn, 50, 52
Curtiss JN-1, 51
Curtiss "Large America" flying boats, 57
Curtiss P-40 (US), 79
Curtiss seaplanes, 76
Curtiss Triad, 53
CVA-58 (US), 181, 182, 183, 184, 187, 190
Czechoslovakia, 164
Dargue, Gen. Herbert, 264
Davison, F. Trubee, 141
DBA long-range bomber (USSR), 94
DB-3-F long-range bomber (USSR), 92
DC-3 Dakota (US), 241
D-Day invasion, 132
Deane, Gen. John, 102, 143
Denfeld, Adm. Louis E., 165, 166–67, 183, 187, 188, 189, 191
Department of Aviation Research (Germany), 125
Department of Defense, Strategic Missiles Evaluating Committee (US), 211
Department of the Air Force, creation of, 178
Deputy Chief of the Air Staff for Research and Development (US), 153
Deputy CNO for Special Weapons (US), 178–79
Desert Shield, 290
Desert Storm, 260, 284; *see also* Gulf War
  air plan, 291–92
DH-4 reconnaissance bomber (Germany), 55, 56, 57
Directive No. 21 (Germany), 93
Directorate of Aeronautics (France), 45
Dirigibles, in World War I, 35–37
Dive bombers
  German, 25
  German emphasis on, 129
  in Spanish Civil War, 28
Douglas A4D Skyhawk (US), 196
Douglas B-18 (US), 63
Douglas 3D Skyraider (US), 196
Douglas SBD (US), 64
Douhet, Giulio, xiv, 8, 24, 25, 26, 59, 61, 64, 116, 131, 143, 145
Drachen balloon (Germany), 36, 41
Dreadnaught, 49
Dulles, John Foster, 195, 246–47
Dumesnil, Jacques-Louis, 45
Dunkirk evacuation, 67
Du Peuty, Maj., 39
Duval, Gen., 39, 41, 42–43, 45, 46, 47
Eaker, Gen. Ira C., 61, 65, 139–40, 143, 154, 219, 264
EARSHOT Plan, 161
Easter offensive of 1972, 279–80
EC-47 (US), 230–31
Education, of German officers, 131–33; *see also* Pilot training
Ehrhart, Maj. Robert C., xiii
Eighth Air Force (US), 5, 162
Eighth Army (US), 226, 229
8th Bomber Command (US), 143
8th Fighter-Bomber Group (US), 224, 230
8th Tactical Reconnaissance Squadron (US), 221, 222
Eighth United States Army Air Force, 2

Eisenhower, Dwight D., 143, 146, 157, 168, 177, 193, 194, 246, 263–64
  and U-2 flights, 251–52
  and US Navy, 194–96
Eisenhower administration, 211–12, 253
Ellyson, Lt. T. G., 52
Ely, Eugene, 52
Embick, Gen. Stanley, 64
Emme, Eugene M., xiv, 49
Emmons, Gen. Delos, 67
Employment doctrine, 39–40
Engineering journal, 8, 12
Enola Gay, 265
Escadrille, N., 3, 41
Essential elements of information, 231
Ethiopian War, 27
Europe; *see also* France; Germany; Great Britain; Italy; Russia
  air power in World War I, 20–23
  interservice rivalries 1918–36, 24
  interwar air power, 23–27
  pre-1914 air power, 19–20
  preference for tactical over strategic air power, 25–26
  pre–World War II conflicts, 27–29
  slow acceptance of strategic bombers, 24–25
Everest, Brig. Gen. Frank F., 158–59, 161
Executive Order 9877, 164
Fahd, King, 291
Fairchild, Maj. Muir, 65
Far East Air Force (US), 233
  pre–Korean War situation, 220, 221
  at start of Korean War, 222–24
Farman, Henry, 33
Farman aircraft (France), 33, 46, 74
Farman F 40 (France), 37
Farman F 50 (France), 45
Farman seaplanes, 76
Fascist air force, 26
F/A-18 (US), 300
F-80C (US), 235–36
F-6D reconnaissance plane (US), 232
Fechet, Maj. Gen. James E., 61
Federal Aviation Commission (US), 62
F-15E (US), 300
F-80 fighter (US), 216, 223, 224
F-82 fighter (US), 223
F-84 fighter (US), 224–25
F-111 fighter (US), 254
F-111F (US), 254, 299, 300
F2H-2B Banshee (US), 194
FH-4 reconnaissance bombers (US), 58
5th Air Force Training Center (US), 220
5th Air Force (US), 222–23, 224, 233
5th Bomb Wing (US), 162
58th Bomb Wing (US), 154
Fighter Command (Great Britain), 67
Fighter pilots, in France in World War I, 42
Fighter planes
  in France in World War I, 41–42
  World War II developments, 22
Fighter tactics, World War I, 22
Finland, 28, 216

Finletter, Thomas K., 163–63
Finnish Air Force, 91
First Aero Squadron (US)
  creation of, 51–52
  demise of, 54
First Air Cavalry Division (US), 275
First Air Division (France), 21
First Army (US), 56
First Aviation Objective (US), 67
First Flight Around the World, 140
1st Aero Squadron (US), 139
1st Air Division (Germany), 100
1st Battalion Welch Fusiliers, 242
Fisher, Col. William P., 153–54
509th Bomb Group (US), 154, 155, 160, 162
Five Year Plans (USSR), 85
Fleet Maritime Force (US), 61
Flexible Response doctrine, 257–59
Flight, 7–8
Flight Refueling Limited (UK), 254
Flight technology; *see* Aviation/Aeronautics
Flying aircraft carriers, 60
F-100 Misty Forward Air Controllers, 230
F-51 Mustang (US), 232, 233
Foch, Marshal Ferdinand, 56
Focke-Wulf Fw-190 (Germany), 101
Fokker, Anthony, 22
Fokker aircraft (Germany), 38
Fonck, René, 42
Formation flying, 22
Forrestal, James, 163, 164, 165–66, 178, 184, 187
Fort Leavenworth Signal Corps School, 50
Fort Myer, Va., 50, 51
46th Reconnaissance Squadron (US), 248
Foulois, Brig. Gen. Benjamin D., 50, 51, 54, 56, 58, 61, 62, 139, 141, 219
460th Tactical Reconnaissance Wing (US), 228
Fourth Air Fleet (Germany), 98, 122
F-5 photo reconnaissance plane (US), 232
France
  aircraft in battle 1914–18, 38–39
  aircraft production 1910–14, 34
  aircraft production 1938–39, 25–26
  air doctrine, 3
  air force structure in World War I, 21
  aircraft production in World War I, 20
  air power in World War I, 20–23, 34–39
  balloon flights, 32
  components of aviation in 1914, 35–36
  creation of independent air force, 24
  defense doctrine, 25–26
  dirigibles and balloons, 35–37
  early history of air power, 31–34
  early military aircraft, 33–34
  employment doctrine, 39–40
  fall to Germany 1940, 68
  fighter planes in World War I, 41–42
  first flying school, 33
  interservice rivalry 1914, 35
  logistics in World War I, 44–47
  losses in World War I, 31
  military advisers in Japan, 75

military aircraft in World War I, 37–38
multipurpose battle plan, 25, 27
origin of air division, 42–43
pilot losses in World War I, 44
pilot training 1910–14, 34
pilot training in World War I, 43–44
politics of aviation 1914–18, 45–46
pre-1914 air power, 19–20
pre–World War II air production, 2
purchase of US aircraft 1939–40, 65–66
request for US aircraft in World War I, 56
role in airplane development, xiv
strategic bombing in World War I, 40–41
withdrawal from NATO, 257
Franco, Gen. Francisco, 88–89
Franks, Lt. Gen. Frederick, 307
Frank, Maj. Walter, 61
Fredette, Raymond, 55
Free fire zones, 276
French Air Force, weakness in 1938, 64
French Army, first airplane orders, 33
Frey, Gen. H., 8
FROLIC Plan, 164, 165
F-86 Sabers (US), 235
F-117 stealth aircraft (US), 290, 291, 294, 298, 299, 309
F-87 Thunderbolt (US), 233
Fullerton, Col. J. D., 7, 8
Fundamental Field Manual No. 16 (Germany), 122, 127–28
Fundamental Order No. 1 (Germany), 135–36
F-4 (US), 279
F-16 (US), 300
F-51 (US), 227
F-80 (US), 233
F-84 (US), 233
F-101 (US), 236
Futrell, Dr. Frank, 59
Galland, Gen. Adolf, 127
Gallery, Rear Adm. Daniel V., 182, 183, 184, 187–88, 192
Gallieni, Joseph, 45
Gamelin, Gen. Maurice, 26
Garros deflector system, 22
Gates, Artemus, 178
Gavin, Lt. Gen. James, 226
Genda, Minoru, 81
General Directorate of Aeronautics (France), 45
General Headquarters Air Force (US), 62, 64, 66
authorization of, 141–42
George, Col. Henry, 68
George, Maj. Harold Lee, 4, 63, 66, 264, 265
German Air Force; see Luftwaffe
German Air Force High Command; see also Luftwaffe
air doctrine and traits, 111–17
controversy within, 117–21
emphasis on strategic bombing, 127–30
lack of controlling bodies, 122
low priority for intelligence work, 125–26
operational thinking, 120–24
organization of, 117–21
preference for offense, 126–27
principle of command, 133–35
socio-educational background, 131–33

technical blunders, 124–25
German Air Force League, 19
German Army
airships, 50
principle of command, 133–34
view of logistics, 131
German General Staff, intellectual limitations, 131
Germany
aircraft production 1930–45, 79
air defenses in World War I, 14
air doctrine, 3, 4
air doctrine in World War II, 111–17
air force structure in World War I, 21
aircraft production in World War I, 20
air power in World War I, 20–23
attack on USSR in 1941–42, 85, 93–98
in Battle of Verdun, 41
creation of independent air force, 24
dive bombers, 25
early World War II success, 28, 63–69
French air attacks against, 40–41
Gotha raids of 1917, 10–11
Japanese actions against 1914–18, 74
losses to US bombing in World War II, 144
medium bombers, 25
military condition in 1919, 116–17
pact with Soviets, 91
politicization of military, 135–36
popular culture of air power, 1
rocket bombs, 209–10
and Soviet counteroffensive, 98–102
submarines, 179
success of strategic bombing against, 4–5
traditional military attitudes, 130–31
World War II air production, 2
Zeppelin attacks on London, 53
GHQAF; see General Headquarters Air Force (US)
GIRDER project, 179, 185
Glosson, Brig. Gen. Buster, 292, 297
Goering, Herman, 98, 99, 106, 112, 115, 123, 135, 143;
see also Luftwaffe
committed to offense, 127
divide and conquer strategy, 118
impact on aircraft production, 128–29
as Luftwaffe commander, 117–21
quality of appointments, 124
relation to Supreme Command, 122
technical blunders, 125
Goldwater-Nichols legislation of 1986, 308–9
Golovanov, Gen. A. Ye., 97–98
Gorbachev, Mikhail, 259
Gotha bombers (Germany), 10–11, 55
GQG; see Grand Quartier Général des Armées
Graduated response, 214
Grand Quartier Général des Armées, 38, 39, 40, 41, 45
Great Britain
advisers to Japanese Navy, 77
after World War II, 238
air power in World War I, 13–16
creation of RAF, 1
early air power developments, 7–9
German bombing raids in 1917, 10–11

Index

interwar view of air power, 1
massive retaliation doctrine, 4–5
purchase of US aircraft 1939–40, 65–66
Great Patriotic War; *see* Red Army; Soviet air forces
Great Society, 268–69
Greenwood, John T., 149
Grey, Charles G., 7, 60
Gromyko, Andrei, 252
Grumman F6F Hellcat (US), 79
Grumman F4F Wildcat (US), 64, 79
Grumman TBF (US), 64
Guderian, Gen. Hans, 122
Guernica, Spain, 120
Guerrilla warfare, 267
Gulf War, xiii, 289–311; *see also* Desert Storm
   air planning stage, 290–98
   execution of US strategy, 298–301
   military objectives, 293
   political meddling, 309
   results, 301–8
   retrospective, 308–11
Gurevich, Mikhail, 92
Guynemer, George, 42
Haig, Gen. Sir Douglas, 9, 13
Haldeman, H. R., 279
Halder, Gen. Franz, 114
HALFMOON operation, 165
HALFMOON Plan, 166
Halsey, Adm. William, 145
Hansell, Haywood, 265
Harding, Warren, 58
Harmon, Lt. Gen. Hubert R., 187–88
Harriman, Averell, 132
Harris, Sir Arthur, 145
HARROW Plan, 165
Haushofer, Prof. Karl, 116
Heavy Bombardment Committee (US), 253–54
He-111 bomber (Germany), 98–99, 104
He-177 bomber (Germany), 124
He-162 fighter (Germany), 124
Heinrich, Prince or Prussia, 19
He-178 jet fighter (Germany), 125
Helms, Richard, 252–53
Henderson, Gen. G. F. R., 12
Hill-Norton, Adm. Peter, 238
Hindenburg, Field Marshal Paul von, 55
Hino, Capt., 74
Hiroshima bombing, 153, 265
Hirschauer, Gen., 45
Hispano-Suiza, 86
"History of Rockwell Field" (Arnold), 51–52
Hitachi, 79
Hitler, Adolf, 26, 57–58, 63, 64, 65, 66, 88, 111, 112, 116, 117, 120, 121, 123, 124, 131, 142, 143, 265
   and air power, 105–6
   and attack on USSR, 93, 94, 97, 104
   Barbarossa campaign, 3
   belief in offense, 127
   call for costly air support, 130
   control of ground strategy, 113
   illusions on Britain, Russia, and US, 126
   land-war mindedness, 127
   and leader principle, 135–36
   schedule of 1942 for USSR, 97
   during Soviet counteroffensive, 101
   view of air forces, 128
HMS *Prince of Wales,* 73
HMS *Repulse,* 73
Ho Chi Minh, 147, 267, 268, 269, 272–73
Ho Chi Minh Trail, 230, 272, 278
Hoeffding, Oleg, 273
Holley, Gen. Bill, 55
Hoover administration, 57
Hopkins, Harry, 65, 142
Horner, Lt. Gen. Charles A., 284, 292, 296, 305
Hosho aircraft carrier, 76
House Armed Services Committee
   B-36 investigation, 189
   Unification and Strategy Hearings, 189
Howell, Clark, 62
Howell commission, 62
H2S radar, 249–50
Humanistiche Gymnasium, 132
Humphreys, Lt. Frederic, 51
Hurley, Col. Alfred F., xiii, 55, 59
Hurricanes (Great Britain), 67
Hussein, Saddam, 284, 289, 291, 292, 295, 303–4
Hydro-airplane, 52
Hydrogen bomb, 168
I-15B (USSR), 94
ICBM. *see* Intercontinental ballistic missiles
Ideas and Weapons (Holley), 55
Identification Friend or Foe device, 215
I-15 fighter (USSR), 86, 89
I-16 fighter (USSR), 89
Il-28 bomber (USSR), 252
Il'hushin, Sergei V., 92
Il-4 long-range bomber (USSR), 92
Il-2 Shturmovik dive bomber (USSR), 105
Il'ya Muromets bomber (USSR), 23
Immelmann, Max, 22
Imperial Air Service (Russia), 23
Imperial All-Russia Aero Club, 19
Imperial Guards Division (Japan), 74
Independent Forces (Great Britain), 14
Independent Force (US), 56
Industrial targets, 14
Infantry Manual (US), 50
In-flight refueling, 253–54
Ingalls, David, 141
Instant Thunder bombing plan, 291, 293–94
Instruction of 13 December 1914 (France), 38
Intelligence work; *see also* Aerial photography; Reconnaissance
   in Cuban missile crisis, 252–53
   essential elements of information, 231
   in Europe, 231
   in Korean War, 230–31
   NATO reconnaissance, 247–49
   vs. operations in Luftwaffe, 125–26
   overflights of USSR, 249–51
   photography missions, 232–33
   in Vietnam, 230
Intercontinental ballistic missiles, 168, 209–13

Intercontinental cruise missiles, 216
Intern Allied Aviation Committee, 15
Internal Look 90 exercise, 290, 291
International Air Meet, France, 51
International Brigades, 88
Interservice rivalry
  in Europe 1918–36, 24
  in France in 1914, 35
  in Great Britain, 10, 12
  in US 1947, 164
  US Navy-Air Force 1952, 194
Iraqi armed forces, 284
Iraqi army; *see also* Republican Guard
  attrition rate, 306–7
  during bombing campaign, 305–6
Iraqi defenses, 301–2
Iraqi invasion of Kuwait, 291
Iraqi navy, 306
Iraqi nuclear weapons program, 302–3
Iraqi petroleum facilities, 304–5
Irvine, Clarence, 167
I-16 (USSR), 94
Italy
  air power failure in World War II, 2
  air power in 1930s, 26
  air tactics in World War I, 21
  bombing raids in 1911, 20
  creation of independent air force, 24
  Ethiopian War, 27
  medium bombers, 25
Japan
  aircraft production 1914–18, 75
  aircraft production 1918–45, 79–80
  airplane inventory in 1941, 79
  airplane losses in World War II, 80
  Allied prewar ignorance of, 73–74
  carrier production 1927–40, 77, 78
  expansionism in 1930s, 89
  homeland defenses, 82
  kamikaze attacks, 80
  political development 1868–1941, 73
  sinking of USS *Panay*, 63
  Soviet attack of 1945, 102–3
  Soviet confrontation in 1930s, 89–90
  US air power against, 145
  US urban attacks, 156
Japan Air Lines, 83
Japan Air Self Defense Force, 83
Japanese Army, invasion of China, 61
Japanese Army Air Forces
  combat experience, 75–76
  developments 1904–25, 74–75
  first combat experience, 74
  French advisers to, 75
  kamikaze attacks, 80
  pilot training, 78–79
  reasons for losses in World War II, 81–82
  at time of Soviet attack, 103
Japanese Naval Air Forces
  development 1877–1941, 76–78
  early losses in China 1937–39, 77
  kamikaze attacks, 80

losses in May 1945, 83
pilot training, 78–79
reasons for losses in World War II, 81–82
Japanese Navy, British advisers to, 77
Japan Logistical Command, 226, 229
Jenny training planes, 58
Jeschonnek, Gen. Hans, 106, 113, 114, 118, 119, 123, 131
Joffre, Gen. Joseph, 3, 34, 38, 39
Johnson, Gen. Leon, 225
Johnson, Louis A., 160, 192, 221
Johnson, Lt. Gen. Gerald W., 282
Johnson, Lyndon B., 147, 274, 276
  air war in Laos, 278
  air war in S. Vietnam, 276–77
  ambivalence about Vietnam, 268–69
  goals in Vietnam, 263, 277
  launches Rolling Thunder, 267–68
  mistrust of military, 270–71
  restrictions on bombing, 270
Joint Army and Navy Board (US), 60
Joint Chiefs of Staff (US)
  approval of Vietnam bombing, 274
  Bikini Evaluation Committee, 181
  emergency war plan of 1948, 244
  Joint Strategic Survey Committee (US), 177
  PINCHER studies, 158–59, 160
Joint Forces Air Component Commander, Gulf War, 284, 296
Joint Outline of the Long Range War Plan, 181–82
Joint Shuttle Bombing Mission, 143
Joint Strategic Survey Committee (US), 177
Joint Surveillance and Target Radar System aircraft (US), 309
Joint War Plans Committee (US), 158–59
Jones, R. V., 215
Journal of the Royal United Services Institution, 8
Junkers JU-88 bomber (Germany), 104
Junkers JU-87 Stuka dive bomber (Germany), 3, 66, 105
Jupiter missile, 212
JU-52 transport (Germany), 98–99, 105
Ju-86 transport (Germany), 98–99
Kaga aircraft carrier, 77
Kaganovich, M. M., 91
Kamikaze warriors, 80
Kanbara, Lt., 76
Kawanishi, 79
Kawasaki, 79
Kazushige, Gen. Ugaki, 75
KB-29 bomber (US), 165
KC-135 bomber (UK/US), 214
Keaney, Thomas A., xv, 4, 289
Kellogg-Briand Peace Treaty, 57
Kennedy, John F., 252
Kennedy, Robert F., 252
Kennedy administration, 212–13
Kenney, Gen. George C., 65, 67, 145, 167
Kesselring, Gen. Albert, 94, 122–23, 135
Kessner, August, 167
Key West Conference of 1948, 184, 189
Khalkhin-Gol Incident, 75, 90, 91, 92
Khe Sanh, Vietnam, 275, 276, 278
Khripin, V. V., 91

Index

Khrushchev, Nikita, 106, 252, 255–56
Kilner Board, 66
King, Adm. Ernest J., 176, 189
Kirkpatrick, Col. E. E., 158
Kissinger, Henry, 212, 279–80, 281, 282, 283
Klein, Cdr. Doyen, 179
Koller, Gen. Karl, 120, 127
Korean War, 3, 167–68, 168
  boundaries and sanctions, 235–36
  lessons from, 234–35
  USAF combat support, 226–29
  USAF losses and readiness, 231–34
  USAF prewar situation, 219–22
  US air power in, 146–47
  US air superiority, 224–25
  US forces at start of, 222–24
  US intelligence work, 229–31
  US Navy in, 191–94
Korten, Gen. Günther, 120
Kosovo, xiii
Kozhevnikov, Gen., 100
Krasovsky, Gen., 102
Krebs, Capt. Jean, 32
Kurds, 304
Kursk offensive of 1943, 100
Kuter, Capt. Laurence S., 63, 65, 219, 265
Kuwait; *see* Gulf War
Kwantung Army Air Force (China), 75
Kwantung Army (Japan), 103
Ky, Air Marshal Nguyen Cao, 274
La-5 fighter (USSR), 98
La-7 fighter (USSR), 103
La-9 fighter (USSR), 103
LaGG-3 fighter (USSR), 92, 98
Lahm, Lt. Frank, 51, 52
Lanchester, F. W., 7, 11, 13
Langley, Dr. Samuel, 49, 52
Langley Field, Va., 54, 62
Laos
  air war in, 278
  US bombing, 263
Lapchinskii, A. N., 26
Laser/electro-optically guided bombs, 280
Laser-guided bombs, 249, 300, 309
Lassiter Board, 141
Latham, Hubert, 33
La-11 (USSR), 236
Lavochkin La-7 (USSR), 101
Leader principle (Germany), 135–36
League of Nations, 88
Leahy, Adm. William D., 165
Leeb, Gen. Ritter von, 94
LeMay, Gen. Curtis, 64, 154–55, 167–68, 248, 249, 250, 254, 265, 266, 268, 283; *see also* Strategic Air Command
  and atomic-era air doctrine, 153
Lemp, Richard, 31
Le Prieur rockets, 38, 41
Lewis, Col. Isaac, 51
Liberty engines, 58
Libyan War of 1911, 20
Limited nuclear option, 214

Lincoln, Brig. Gen. George A., 160
Lindbergh, Charles A., 65, 66, 140
Linebacker II operation, 282–83
Linebacker operation of 1972 (US), 263, 280–81
Lloyd George, David, 1
Lockheed Electra, 215
Loerzer, Bruno, 120, 121
Logistics
  in France in World War I, 44–47
  German view of, 131
  prior to Korean War, 221
  USAF in Korea, 226–29
Loktionov, A. D., 91
London naval accord of 1930, 77
Long-range aerial navigation, 14
Long-Range Bombardment Aviation (USSR), 93–94
Long-range bombing, World War I, 13–14
Los Angeles dirigible, 60
Low-recoil machine gun, 51
Luck, Lt. Gen. Gary, 307
Ludendorff, Gen. Erich von, 130, 131
Luedecke, Col. Alvin R., 158–59
Luftwaffe, 58; *see also* German Air Force High Command
  analysis of role against Soviets, 104–6
  in attack on Poland, 66–67
  in attack on USSR, 93–96, 97
  in Battle of Britain, 67
  command thinking: concentration on military matters, 111–13; low priority for technology, 115; narrow view of mission, 116–17; overemphasis on offense, 115–16; overemphasis on tactical operations, 114–15
  consequences of air doctrine: in command structure, 117–21; dominance of offensive over defense, 126–27; low priority for intelligence, 125–26; operational thinking, 121–24; strategic bombing *vs.* tactical support, 127–30; technical blunders, 124–25
  on D-Day invasion, 132
  early World War II years, 63–69
  in late 1930s, 63
  line officers *vs.* engineers, 125
  losses at Stalingrad, 99
  neglect of air transport, 123
  operational failures, 122–23
  origins of modes of thinking: impact of "leader principle," 135–36; principle of command, 133–35; socio-educational background, 131–33; traditional military attitudes, 130–36
  in Poland 1939, 67
  pre–World War II, 26
  reliance on dive bombers, 129
  reliance on propaganda, 129
  during Soviet counteroffensive, 98–100
  in Spanish Civil War, 27–28, 129
  strategic bombing *vs.* tactical support, 127–30
  three assigned tasks, 127–28
Luke, Frank, 56
Lusitania sinking, 52
Lyautey, Louis, 45
Lyman, Deke, 142
MacArthur, Gen. Douglas, 57, 145, 220, 222, 224, 225
MacArthur-Pratt Agreement of 1931, 61
MacIsaac, Dr. David, xiii

Macon, 60
Maginot, André, 25
Magna Carta of British Air Power, 11
Mahan, Capt. Alfred Thayer, 49
MAKEFAST Plan, 160–61
Man and Superman (Shaw), 28–29
Manchukuo, 75
Manchurian Incident, 75
Manhattan Project, 153, 162, 209
Manstein, Gen. Erich von, 10, 101
Manual 1-8 (USAF), 266
Mao Tse-tung, 245
March Field, Calif., 62
Marshall, Gen. George C., 65, 66, 67, 142, 143, 145, 149–50, 152, 157, 241
Marshall Plan, 240
Martin B-10 bomber, 62
Martin MB-2 bombers (US), 58
Mason, R. A. "Tony," 5, 7, 237
Massive retaliation doctrine, 4–5, 195, 211–12, 247
Maxwell, Brig. Gen. Alfred R., 153–54
M-4 bomber (USSR), 255
McConnell, Gen. John P., 268, 274–75
McCracken, William, 141
McDonald, Maj. Gen. George C., 159
McDonnell-Douglas A-4 Skyhawk (US), 272
McDonnell-Douglas F-4 Phantom (US), 272
McKee, William F., 219
McNamara, Robert S., 268, 270, 271
McNarney, Gen. Joseph, 65, 166–67
Me-210 destroyer (Germany), 124
Medium bombers, 25
Me-109 fighter (Germany), 89, 92, 94
Me-262 fighter (Germany), 125, 209–10
Me-163 (Germany), 209–10
Me-109G (Germany), 101
Memorandum on Military Aviation (Ader), 32
Me-163 rocket fighter (Germany), 216
Metcalf, Dr. Arthur, 143
Meusnier, Lt., 32
Meyer, Gen. J. C., 282
Michelin, 34
MiG-23/27 bomber (USSR), 256
MiG-3 fighter (USSR), 92
MiG-15 fighter (USSR), 216, 233
MiG-29 (USSR), 259
MiG-31 (USSR), 259
MIKE Shot of 1952, 168
Mikoyan, Artem, 92
Milch, Field Marshal Erhard, 113, 118, 119, 120, 121, 125, 128, 135
Military Assistance Command, Vietnam, 274
Military Aviation, 32–33
Military ballooning, 32
Military Liaison Committee (US), 155
Millerand, Alexandre, 45
Millikan, Maj. Robert, 54
Milling, Lt. Thomas D., 52
Minoru, Gen. Genda, 77
Missile gap controversy, 253
Mitchell, Gen. William, xiv, 43, 140–41, 143, 145, 283
  conflict with armed services, 58–60

legacy of, 59
pre–World War I efforts, 54
and US preparation for Vietnam, 264–65
in World War I, 55–56
Mitchell, John, 279
Mitscher, Vice Adm. Marc, 185
Mitsubishi AGM Zero (Japan), 79
Mitsubishi bombers (Japan), 103
Mitsubishi Company, 79
Moffett, Adm. William, 60
Molotov, Vyacheslav M., 91
Moltke, Helmuth von (the elder), 130, 131
Momyer, Gen. William W., 274
Mongolia, Russia at war in, 28
Mongolian People's Republic, 75
Moon, Lt. Odas, 60
Moore, Lt. Gen. Joseph H., 268
Moorer, Adm. Thomas, 282
Morane-Saulnier airplane manufacturing, 34
Morrow Board, 58–59, 141
Mrozek, Donald J., 283
Munich Crisis of 1938, 64–65
Mussolini, Benito, 26, 88
NACA; *see* National Advisory Committee for Aeronautics
Nagasaki bombing, 153, 265
Naiden, Col. Earle, 66
Nakajima, 79
Napoleon Bonaparte, 32
National Academy of Sciences, 54
National Advisory Committee for Aeronautics (US), 59–60, 66, 68
National Council of Reconciliation and Concord (S. Vietnam), 281
National Defense Research Committee (US), 68
National Research Council (US), 54
National Security Act of 1947, 178
National Security Act of 1949, 188
National Security Council, 184
National Socialist ideology, 113, 135
Navaho missile (US), 211, 215
Naval Advisory Committee for Aeronautics (US), 53–54
Naval Aeronautics Station, Pensacola, 53
Naval Air Establishment (Japan), 77
Naval Appropriations Act of 1915, 53
Naval Appropriations Act of 1920, 60
Naval Expansion Act of 1938, 64
Naval War College (US), 49
Naval Wing of the Royal Flying Corps, 9
Navy Department (US), 149
Neumann, Capt., 9
Neumann, John von, 212
Neutrality Acts (US), 57–58
Neutrality Patrol (US), 66
New Economic Program, Lenin's, 85
New York Times, 64, 142
New York Tribune, 132
Nieuport XVII (France), 37
Nimitz, Adm. Chester W., 145, 159, 164, 177, 179, 180, 182, 183, 184, 189
94th Bombardment Squadron (US), 66
Nippon Hikoki, 79

Nishizawa, Warrant Officer, 80
Nitze, Paul, 245
Nivelle, Gen. Robert, 39
Nixon, Richard M.
  bombing N. Vietnam, 263
  and Easter offensive of 1972, 279–80
  expertise in foreign affairs, 280
  Vietnamization program, 277–78
  Vietnam peace negotiations, 281–82
No. 3 Wing of Royal Navy Air Service, 14
No. 2 Air Wing (Great Britain), 9
Nomonhan Incident; see Khalkin-Gol Incident
Nomura, Cdr. Ryosuke, 82
Norstad, Gen. Lauris, 150, 151–52, 164
North Atlantic Strategic Concept of 1949, 244–45
North Atlantic Treaty Organization, 229, 237, 2328
  Cold War strategy, 243–47
  flexible response and convergence, 257–59
  treaty, 193
  treaty signatories, 243–44
Northern Bombing Group (US), 57
North Korea, 5
North Vietnam, 5
  assault on south 1972, 278–80
  Christmas bombing 1972, 282–83
  destruction from Rolling Thunder, 273
  US bombing campaign, 263, 267–76, 280
Novikov, Gen. A. A., 97, 99
Nuclear power, 148
Nungesser, Charles, 42
Oberrealschule, 132
O'Donnell, Emmett, 167
Offensive warfare, German view, 115–16
Office of Scientific Research and Development (US), 68
Ofstie, Adm. Ralph, 183, 187, 188, 189
Okinawa campaign, 80
Olds, Capt. Robert, 63, 264
Operation IVY, 168
Operations Plan 1002-92 (US), 290, 291
Oppama naval air station (Japan), 76
OP-55 study, US Navy, 189–91
Ostfriesland battleship, 58, 60
Overy, Prof. Richard J., xiii, 1
Pacific Sea Frontier (US), 61
Painlevé, Paul, 45
Pan Am Goodwill Flight, 61, 140
Partridge, Maj. Gen. Earle E., 154, 219
Pathet Lao, 278
Patrick, Gen. Mason M., 56, 60
  court martial, 141
Pau flying school (France), 33
Paulus, Gen. Friedrich von, 98, 99
Pearl Harbor, 61
  Japanese attack on, 69, 78
Pearson, Drew, 182
Pe-8 bomber (USSR), 86–87
Pe-2 dive bomber (USSR), 92
Peripheral Basing Program (US), 157
Permanent Inspector of Aeronautics (France), 34
  abolished in 1914, 34–35
Perry, Robert, xiv, 205
Pershing, Gen. John J., 54, 55, 56, 139, 141

Pétain, Gen. Philippe, 25, 39, 46
Petlyakov, Vladimir M., 86, 92
Petlyakov Pe-2 (USSR), 101
P-36 fighter (US), 66
P-38 fighter (US), 66
P-39 fighter (US), 66
P-40 fighter (US), 66
P-47 fighter (US), 224
P-51 fighter (US), 208–9
Phony war, 66, 67
Pierced steel planking construction, 226
Pilot school
  Air Corps Tactical School, 264–66
  first, 33
  France, 43–44
Pilot training
  earliest, 33
  in France 1910–14, 34
  in France in World War I, 43–44
  at German Air War Academy, 111–12
  German failures, 123–24
  in Japan, 78–79
  United States Army Air Corps, 78
  in US before World War II, 66
  US loss rate after 1945, 221
  by US Navy, 53
  in USSR, 87
PINCHER studies, 158–59, 160, 179
Plocher, Gen. Hermann, 94
PO-2 biplane (N. Korea), 235
Poe, Gen. Bryce, II, 219
Poland, invasion of 1939, 66
Polaris missile, 212
Polikarpov, N. N., 86
Polish Air Force, 91
Position Emitter Location System, 233–34
Postwar strategic air force
  origin of, 149–52
  war planning 1945–48, 156–63
  year of change 1948, 163–68
Powell, Gen. Colin, 292
Power, Thomas, 167, 168
Powers, Francis Gary, 252, 253
Principle of command, Germany, 133–35
Principles and Applications of Naval Warfare, 180
Problems of Air Navigation (Douhet), 8
Proud Deep Alpha operation, 279
Prussia, pre-1914 air power, 19–20
Prussian Army, 131
Prussian General Staff, 19
Public relations, by US Army Air Corps, 142
Punitive Expedition to Mexico, 139
Push CAS system, 297–98, 301
Pusher aircraft, 51
P2V-3C Neptune (US), 184, 192
Question Mark, 61
Question Mark Flight, 140
Radford, Adm. Arthur, 177–78, 179, 189
Ramey, Roger, 167
RAND Corporation, 158
RB-45 (UK), 249, 252
RC-135 tanker (US), 254

Realgymnasium, 132
Reconnaissance; *see also* Aerial Photography; Intelligence work
   airmen, 22–23
   British cooperation in NATO, 249–51
   and Cuban missile crisis, 252–53
   by dirigible and balloon, 35–37
   role in Cold War, 247–49
   U-2 flights over USSR, 251–52
Red Air Force, 23; *see also* Soviet air forces pre–World War II, 26
Red Army (USSR)
   counterattack of 1941, 96
   in counteroffensive against Germany, 98–102
   early aviation developments, 85–87
   in Far East in 1930s, 87–88
   during German assault of 1941, 93–94
   Stalin's purges, 90–91
   in war against Japan, 102–3
   in Winter War, 91
Red Banner Far Eastern Army, 90
Reed, Lt. George, 52
Renard, Capt. Charles, 32
Republic F-105 Thunderchief fighter (US), 272
Rex (Italian liner), 64
RF-4C Phantom (US), 230–31, 231
RF-80 (US), 222–23, 227, 232
Ribot, Alexandre, 55–56
Richardson, Holden C., 53
Richthofen, Field Marshal Wolfram von, 98, 112, 124, 240
Rickenbacker, Eddie, 56
Risiko-Luftwaffe (risk air force), 129
Robineau, Gen. Lucien, xiv, 2, 3, 31
Rodgers, Lt. John, 53
Rogers, Col. Turner C., 155–56
Rogers, William P., 282
Rolling Thunder campaign of 1965–68, 263, 267–76, 280
Rommel, Gen. Erwin, 123, 131
Roosevelt, Franklin D., 1, 62, 142, 145, 147
   on air power, 65
   neutrality declaration 1939, 66
   pre–World War II years, 63–68
Roosevelt, Theodore, 49–50
Roques, Gen., 33, 34, 45
Royal Aircraft Factory, Farnborough, England, 9
Royal Air Force, 3, 55, 145
   in Berlin Airlift, 240–43
   creation of, 1, 10
   demobilization 1945, 238
   early history of, 9–10
   Fighter Command system, 67
   Gulf War planning, 293
   in NATO, 249–51
   rivalry with Royal Navy, 12
   and US basing requirements, 158
   weakness in 1938, 64
   in World War I, 13–16
Royal Air Force Bomber Command, 2, 161
   in World War II, 4–5
Royal Engineers Air Battalion, 9
Royal Flying Corps, 9–10, 13, 15
   Naval Wing, 9

Royal Navy, rivalry with RAF, 12
Royal Navy Air Service, 10, 57
   in World War I, 13–14
Royal Saudi Air Force, 293
Royal United Services Institution, 8
R-5 reconnaissance plane (USSR), 86
Rudenko, Gen. R. A., 102
Rundstedt, Gen. Gerd von, 94
Rusk, Dean, 268, 270
Russia; *see also* Soviet Union
   aircraft production in World War I, 20
   air power in World War I, 23
   German dive-bombing raids against, 3
   pre–World War I air power, 19–20
Russian I-16 aircraft, 77
Russo-Japanese Neutrality Pact, 102
Ryan, John D., 167
Rychagov, Gen. P., 90
Ryujo aircraft carrier, 77
Saburo, Sakai, 79
Saconney, M., 36
Salmson 2 A2 (France), 389
Salmson aircraft (France), 45, 75
SALT II negotiations, 148
SALT I negotiations, 280
Sanders, Wing Cmdr. R. S., 249
SANDSTONE operation, 165
SB-2 bombers (USSR), 89
Schlieffen, Field Marshal Alfred, 130, 131
Schlieffen Plan, 130–31
Schlight, John, 276
Schwarzkopf, Gen. H. Norman, 292, 295, 296, 297, 309
Scott, George C., 279
Scriven, Brig. Gen. George, 53
Scud missiles (Iraq), 293, 294, 299, 301
   air attacks on, 303
Sea-based aviation, 176
Seamans, Robert C., 278
Seaplane tender, 76
2nd Air Army (USSR), 102
2nd Air Division (US), 268, 270
2nd Air Fleet (Germany), 94
Seeckt, Gen. Hans von, 131
Selfridge, Lt. Thomas, 50
SE-5 pursuit planes (US), 58
Seventh Comintern Congress, 88
7th Air Force (US), 273, 274
Seventy Group Air Force (US), 151, 163
72nd Reconnaissance Squadron (US), 248
Seversky, Alexander de, 219
Shakurin, A. I., 91–92
Sharp, Adm. U. S. Grant, 268, 271, 274, 276
Shaw, George Bernard, 28–29
Shenandoah dirigible disaster, 58, 60
Sherman, Adm. Forrest P., 167, 179, 181, 191
Sherman, Gen. William, 264
Sherry, Michael S., 1
Shigemetsu, Mamorn, 89
Shi'ite Muslims, 304
Shiner, Col. John F. (Fred), 314, 135
Shinohara, Lt., 76, 80
Shoho aircraft carrier, 78

Index

Shokaku aircraft carrier, 78
Siberian Expedition of 1918–22, 74
Signal Corps. *see* United States Army Signal Corps
Sihanouk, Norodom, 277–78
Sikorski, Igor, 23
Silkworm antimissile ships (Iraq), 306
Silver Fish aircraft, 52–53
Sims, Adm. William S., 59
Sino-Japanese War, 117
Sino-Soviet split, 280
16th Air Army (USSR), 102
Sixth Army (Germany), 98, 99, 123
Sixth Fleet (US), 193
Smart munitions, 280
Smith, Perry McCoy, 151
Smushkevich, Ya. I., 91
Smuts, Gen. Jan C., 11–12
Snark missile (US), 211, 215
Society for the Promotion of Defense, Aviation and Chemical Warfare (USSR), 87
Sopwith fighter (Great Britain), 75, 77
Soryu aircraft carrier, 78
South Vietnamese Air Force, 274
Soviet Aircraft Industry, 86
Soviet air forces; *see also* Red Air Force
  during 1939–41, 90–93
  analysis of World War II role, 103–6
  combat experience in 1930s, 87–90
  counterattack of 1941, 96
  in counteroffensive against Germany, 98–102
  early aviation developments, 85–87
  in Far East 1939, 89–90
  in German attack of 1941–42, 93–98
  inventory in 1930s, 87
  lack of strategic bombing, 104–5
  restructuring in 1942, 97
  in Spanish Civil War, 88–89
  Stalin's purges, 90–91
  in war against Japan, 102–3
  in Winter War, 91
Soviet Far Eastern Air Force, 75–76
Soviet-German Non-Aggression Pact of 1939, 91
Soviet Rocket Forces, 256
Soviet Union
  air combat with Japan 1939, 76
  aircraft losses in 1941, 93–94
  aircraft production after 1945, 238
  aircraft production 1921–32, 85
  aircraft production 1940–41, 92–93
  aircraft production 1941–42, 96–97
  air defenses, 248
  air power in 1930s, 26
  Berlin Blockade, 165–67, 240–43
  building heavy bombers, 25
  confrontation with Japan 1930s, 89–90
  contribution to Cold War air power, 254–56
  and Cuban missile crisis, 252–53
  and Czechoslovakia, 164
  early military aviation, 85–87
  German attack in 1941–42, 85
  Germany attack in 1941–42, 93–98
  global intentions, 220
  missile defenses, 256
  and NATO, 243–47
  NATO overflights, 249–51
  popular culture of air power, 1
  pre–World War II wars, 28
  purges of 1930s, 26
  response to U-2 flights, 251–52
  split with China, 280
  submarines, 179
  surface-to-air missiles, 251–52
  and Truman, 168
  and US strategic plans 1945–48, 156–63
Spaatz, Gen. Carl, 65, 66, 67, 146, 151, 152, 158, 161, 164, 165, 177, 219, 237, 239, 240, 253–54, 264
Spaatz Board, 152–53, 156
Spad fighters (France), 75
Spad XII-cannon (France), 45
Spad XIII (France), 37, 45
Spanish Civil War, 63
  German forces in, 27–28, 129
  Soviet forces in, 88–89
Special Far Eastern Army (USSR), 87–88, 89
Special Planning Group (US), Gulf War, 292–93, 294–95
Speer, Albert, 68, 143–44
Spitfire (Great Britain), 64, 67, 92
Spruance, Adm. Raymond, 145
Sputnik, 212, 253
Squier, Maj. George O., 7–8
Stalin, Josef, 255, 256
  aid to China in 1930s, 89–90
  and air power, 106
  Five Year Plans, 85
  and German assault of 1941, 95–96
  and German threat in 1934–35, 88
  ill-advised offensive of 1942, 97
  nonbelligerent foreign policy, 87
  pact with Germany, 91
  purges of military, 90–91
  purges of 1930s, 26
  and Spanish Civil War, 88–89
  united front policy, 88
  and war against Japan, 102
Stalingrad, 123
  battle 1942–43, 97–98
  counteroffensive, 98–100
State Defense Committee (USSR), 95
Stavka VGK (USSR), 95
Steam catapults, 206
Stinger-class missiles, 299–300
Strategic Air Command, 154, 157, 158, 206
  atomic strike force, 161
  center of defense policy, 263–64
  emergency war plan of 1948, 244
  on eve of Vietnam, 266
  and flight refueling, 253–54
  and Korean War, 167–68
  LeMay in command of, 167–68
  responsibilities, 247–48
Strategic air offensive, US Navy, 180
Strategic air operations, 266
Strategic bombing
  Arc Light mission, 273, 275

Bradley's vision, 244
  in Cambodia 1969–70, 277–78
  from carriers, 178
  changes since 1970s, 3
  continuing controversy, 4
  critics of, 3
  effectiveness in Gulf War, 301–8
  extent in N. Vietnam, 263
  by France in World War I, 40–41
  by Germany in World War I, 21–22
  Gulf War strategy, 298–310
  Instant Thunder, 291, 293–94
  issue in 1945, 240
  lessons from World War II, 265
  Luftwaffe failure, 117
  mixed results in Vietnam, 275–76
  of N. Vietnam 1972, 279
  operation Linebacker, 280–81
  operation Linebacker II, 282–83
  by RAF in World War I, 14–16
  retrospect on Gulf War, 308–11
  Rolling Thunder campaign, 263, 267–76, 280
  shift away from, 3–4
  slow acceptance in Europe, 24–25
  Soviet inadequacy, 104–5
  Soviet potential in Cold War, 255–56
  Soviet views in 1930s, 26
  success in World War II, 4–5
  vs. tactical support in Germany, 127–30
  vital-targets issue, 4–5
Strategic Bombing Survey (US), 68, 144, 155, 176
Strategic Missiles Evaluating Committee (US), 211
Strategic reserves, German neglect of, 131
Stuka dive bomber (Germany), 3, 66, 105
Styer, Vice Adm. William, 164
Submarines, 179
SU-24 bomber (USSR), 256
SU-25 bomber (USSR), 259
SU-27 bomber (USSR), 259
SU-24 Fencer (NATO), 225
Suicide pilots, 80
Sukhoi 17/22 fighter (USSR), 256
Sullivan, John L., 179, 183
Supreme Allied Commander, Europe, 193
Supreme High Command (USSR), 95
Supreme War Council, 15
Surface-to-air missiles
  Iraq, 299–300
  USSR, 251–52
Surface-to-surface missiles, 240
Sweden, 24
Sweeney, Walter, 167
Sykes, Col. Frederick, 9, 14
Symington, W. Stuart, 163, 166
Synchronized machine gun, 22
Tachikawa, 79
Tactical air warfare, 3
Tactical Electronic Reconnaissance, 231
Tactical support, 127–30
Tactics
  formation flying, 22
  in Luftwaffe doctrine, 114–15

synchronized machine gun, 22
Taegu Air Base, Korea, 227
Taft, William Howard, 51
Tang, Truong Nhu, 275
Taylor, Gen. Maxwell, 212, 270
Taylor, John W. R., 23
TB-7 bomber (USSR), 86
TB-3 heavy bomber (USSR), 86, 89
Technology, 205–6
  application to military ends, 206
  vs. command in Germany, 133–35
  faith in future of, 207–09
  gap between invention and application, 215–16
  in German armed forces, 131–32
  interaction with air doctrine: ignoring ballistic missiles, 213–15; in USAF, 205–17
  Luftwaffe blunders, 124–25
  Luftwaffe view of, 115
  quality vs. quantity, 216–17
Tedder, Air Chief Marshal Sir Arthur, 158
Tet Offensive, 271, 278
Theodore von Karman Committee, 156–57
Thieu, Nguyen Van, 281
Third Dimension of Warfare, 16
Tho, Le Duc, 280, 281, 282
Thor missile, 212
Timoshenko, Gen. Semyen K., 91
Titan rocket, 212
Tiverton, Lt. Commander, 14
Tokugawa, Capt., 74
Tomahawk land-attack missiles, 291, 298, 300, 309
Torpedoes, aerial, 78
Towers, Adm. John, 53, 66
Tractor aircraft, 51
Treaty of Brussels, 243
Treaty of Versailles, 57–58, 117
Trenchard, Maj. Gen. Hugh, xiv, 4, 9, 13, 14–15, 55, 56, 59, 143, 145
Trench warfare, 42
Truman, Harry, 146, 149, 163, 164, 166, 168, 179, 181, 186, 192, 221, 250
  and Berlin Blockade, 241–42
  stops overflights of USSR, 249
Tsingtao, China, 74
TU-4 bomber (USSR), 255
TU-16 bomber (USSR), 255
TU-95 bomber (USSR), 255
Tukhachevsky, Gen. M. N., 90
TU-22M (USSR), 259
Tupolev, A. N., 86, 91
Tupolev 8B-2 fast bomber, 86
Tupolev Bureau (USSR), 255
Tupolev TB-3 bomber (USSR), 26
Tupolev TB-5 bomber (USSR), 25
TU-160 (USSR), 259
Twentieth Air Force (US), 156, 223
Twining, Nathan W., 219
287th Fighter Division (USSR), 98
Two-ocean navy concept (US), 64
Two-theater war proposal (US), 68–69
Type 1 fighter (Japan), 103
Type 97 fighter (Japan), 103

Index

U-Tapao Royal Thai Air Base, 274
Udet, Ernst, 22, 118, 124, 135
Undersecretariat of Aeronautics (France), 45
United front policy, Stalin's, 88
United States
 aircraft production 1930–45, 79
 air lessons from World War II, 145–46
 airplane development, xiv
 air power developments 1903–11, 49–54
 air power in Korea and Vietnam, 146–47
 air power in World War I, 54–57
 air power in World War II, 143–46
 air power on eve of World War II, 63–69
 and atomic weapons, 178–80
 aviation 1918–41, 140–42
 balloon flights, 49–50
 demobilization of 1919, 58
 early air doctrine, 52–54
 early air power, 52–53
 faith in future of technology, 207–9
 interservice rivalry 1947, 164
 interservice rivalry 1952, 194
 interwar air power, 57–63
 joint war plans after 1947, 186
 legacies of World War I air power, 139–40
 massive retaliation doctrine, 4–5, 95, 211–12, 247
 military buildup in 1980s, 259
 military objectives in Gulf War, 293
 nuclear-capable aircraft, 193
 origin of independent air force, 265–66
 origins of postwar air force, 149–52
 overseas expansion, 49
 rearmament 1938–41, 64
 Rolling Thunder aircraft, 272
 and strategic air combat, 180–83
United States Air Corps Tactical School; *see* Air Corps Tactical School (US)
United States Air Force, 164–65
 aircraft defects, 232
 air war in Laos, 278
 vs. Army control of B-52s, 274–75
 and atomic bomb, 152–56
 Basic Doctrine manual, 264, 266, 283
 in Berlin Blockade, 165–67
 Checkmate division, 291
 clash with Navy 1952, 194
 combat support in Korea, 226–29
 Electronic Intelligence, 248
 European basing requirements, 158
 European Theater, 157
 faith in future of technology, 207–9
 Gulf War: execution of strategy, 298–310; interdiction operations, 305; planning stage, 290–98; results of strategy, 301–8; retrospect on, 308–11
 interaction of technology and doctrine, 205–17
 Key West Conference of 1948, 164–65
 Korean War boundaries and sanctions, 235–36
 lessons of Korea, 234–35
 Linebacker II operation, 282–83
 Linebacker operation, 263, 280–81
 losses in Rolling Thunder, 272
 Manual 1-8, 266

 mission debate with Navy, 184
 organizational maturity, 175
 pilot training loss rate after 1945, 221
 plans for Desert Storm, 284
 pre-Korean War readiness, 219–22
 preparation for Vietnam, 264
 raids on Cambodia, 277–78
 readiness and losses in Korea, 231–34
 retention problems before Korea, 221–22
 strategic planning 1945–48, 156–63
 superiority in Korea, 224–25
 view of atomic weapons, 156
 year of change 1948, 163–68
United States Air Force Academy Eighth Military History Symposium, xiii
United States armed forces
 in Berlin Airlift, 240–43
 intelligence work in Korea, 229–31
 personnel problems in late 1940s, 220
 pre-Korean War readiness, 219–22
 spending cuts of 1949, 221
 at start of Korean War, 222–24
 strength in 1945, 238–39
United States Army
 adoption of Wright Flyer, 51
 air strength in April 1917, 54
 Board of Ordnance and Fortifications, 49
 creation of First Air Squadron, 51–52
 Infantry Manual, 50
 reorganization after 1900, 49
 vs. USAF control of B-52s, 274–75
 and Wright brothers, 49–50
United States Army Air Corps
 acquisitions 1938–41, 64
 air mail assignment, 62
 between 1918 and 1941, 140–42
 bomber technology in 1930s, 62
 decline after 1919, 58
 expansion in 1939, 65–66
 First Aviation Objective, 67
 in 1930s, 61
 pilot training, 78
 plans for World War II, 142
United States Army Air Corps Tactical School, 142
United States Army Air Force
 creation of, 68
 on eve of World War II, 1
 and Navy's strategic offensive, 180
 origins of postwar strategic force, 149–50
 two-theater war proposal, 68–69
 in World War II, 4–5
United States Army Signal Corps, 7
 Aeronautical Division, 49–50
 airships, 50
 Aviation Section, 51–52, 139
 first reconnaissance planes, 52
United States Army Signal Corps School, 50
United States Congress
 aircraft authorization 1940–42, 67
 aviation appropriation of 1911, 51
 expansion of Air Corps 1939, 65–66
 House Armed Services Committee, 189

Military Affairs and Appropriations Committee, 141
  naval aviation appropriation of 1911, 53
  Neutrality Acts, 57–58
  rejection of aeronautics budget 1908, 50
  World War I appropriations, 54
United States European Command, 228
United States Fleet Publication No. 1, 180
United States Historical Division, 129
United States Marine Corps
  aviators, 61
  at Khe Sanh, 277
  in Vietnam, 274
United States Military Academy, 58
United States Navy
  acquisitions 1938–41, 64
  air force challenge, 149–50
  air strength in 1917–18, 57
  air strength in April 1917, 54
  and atomic weapons, 181–83
  carrier force level, 187
  carriers in Gulf War, 298
  changing role for carriers, 196–98
  conflict with Mitchell, 58–60
  early flight training, 53
  Eisenhower's impact on, 194–96
  on eve of Pearl Harbor, 69
  General Board Study, 184–86
  impact of demobilization 1919, 59
  interwar/postwar air doctrine, 175–78
  and Key West Conference, 165
  Korean War developments, 191–94
  and massive retaliation doctrine, 195
  and Middle East oil, 159
  opposition to strategic air planning, 159–60
  OP-55 study, 189–91
  organizational maturity, 175
  role and mission debate, 184, 186–89
  Tomahawk missiles, 291, 298, 300, 309
  in Vietnam War, 273–74
United States Strategic Institute, 143
U-2 spy flights
  over Cuba, 252–53
  over USSR, 251–52
USS *Birmingham*, 52
USS *Constellation*, 279
USS *Coral Sea*, 279
USS *Enterprise*, 60
USS *Essex*-class carriers, 190
USS *Forrestal*, 192
USS *Forrestal*-class carriers, 194
USS *Hancock*, 279
USS *Kitty Hawk*, 279
USS *Langley*, 60, 61
USS *Lexington*, 60, 61
USS *Midway*-class carriers, 184, 190, 192
USS *Missouri*, 146
USS *Panay*, 63
USS *Pennsylvania*, 52
USS *Saratoga*, 60, 61
USS *United States* cancellation, 160, 187, 188, 190, 192
USS *Wasp*, 60
USS *Yorktown*, 60

Vandenberg, Gen. Hoyt S., 152–53, 163, 166, 167, 187, 188, 219
Vandenberg Resolution, 243
Vasilevsky, Gen. A. M., 100, 102
VC-5 nuclear attack squadron (US), 192
Very heavy bomber groups (US), 151–52, 163
Viet Cong, 267, 275–78
Vietnamization program, 277–78
Vietnam War, 3, 263–84
  air war in south, 276–77
  Arc Light bombing campaign, 273
  Commando Hunt operation, 278
  conflict over duration of bombing, 274–75
  free fire zones, 276
  intelligence work, 230
  Linebacker operation, 280–81
  Linebacker II operation, 282–83
  N. Vietnam assault of 1972, 278–80
  Nixon peace negotiations, 281–82
  Nixon's goals, 277–78
  Operation Barrel Roll, 278
  Operation Proud Deep Alpha, 279
  problem for US air leaders, 263–64
  Rolling Thunder campaign, 263, 267–76, 280
  Tet Offensive, 271
  USAF preparation for, 264–67
  US air power in, 146–47
  withdrawal of US troops, 279
Villa, Pancho, 54, 139
Villette, Giroud de, 32
Vincent, Daniel, 45
Vital-targets issue, 4–5
Vogt, Gen. John W., Jr., 279
Voisin LA 5 (France), 38
Volksschule, 132
Von Braun, Werner, 219
Von Hoeppner, Gen. Ernst, 43
Von Kluck, Gen. Alexander, 52
Voroshilov, Gen. K. E., 91
Vought F4U Corsair (US), 79
V-1 rocket (Germany), 209–10
V-2 rocket (Germany), 209–10
V-series bombers (US/UK), 214
VVS; *see* Soviet air forces
Wakamiya Maru seaplane tender, 76
Walker, Lt. Kenneth, 61, 63, 264–66, 265
War Ace, 42
Warden, Col. John A., 291, 292, 293
War Department (US), 149
  postwar air force planning, 151–52
War in the Air (Wells), 50–51
War Plans Division, United States Army Air Forces, 68
Warsaw Pact nations, 257–58
Warsaw Treaty, 237
Washington Conference of 1921–22, 77
Washington Naval Treaties 1921–22, 57, 60, 61
Washington Navy Yard, 53
Weapon System Evaluation Group (US), 194
Webb, James E., 163
Wedemeyer, Gen. Albert C., 164
Weiller, Lazare, 33
Wells, Col. Mark K., 2

Index

Wells, H. G., 50–51
West, Rebecca, 29
West Germany, 243
Westmoreland, Gen. William C., 274, 276
Westover, Gen. Oscar, 62, 65
Wever, Gen. Walter, 114, 115, 116, 128–29
Weyland, Maj. Gen. O. P., 163–63
Wheeler, Gen. Earle, 270
White, Thomas D., 219
Whiting, Kenneth R., 85
Wilhelm, Kaiser, 50
William I, Emperor, 131
Wilson, Maj. Donald, 63
Wilson, Woodrow, 52, 54
Winter War, 91, 216
Wohlstetter, Alfred, 158
Woodring, Harry, 64
World Disarmament Conference, 57
World War I
  airplane development, xiv
  air power in Europe, 20–23
  British air power, 13–16
  creation of British air power, 9–10
  creation of combat wings, 43
  France air power role, 34–39
  trench warfare, 42
  US air power, 54–57
  US Navy air power, 57
World War II
  air power in, xiv
  AWPD-1 bombing plan (US), 265
  dive bombers, 3
  legacies of air power, 139–40
  US air power, 143–46
Wright, Orville, 49–50
  flights at Fort Myer, 50–51
Wright, Wilbur, 33, 49–50
Wright brothers, 19, 32, 33, 139
Wright Company, 53
Wright Cyclone, 86
Wright Flyer, 49–50, 54
  adopted by US Army, 51
XB-19 bomber (US), 62
X-15 bomber (US), 62
X Corps (US), 226, 229
XF-10F fighter (US), 206
Yak-9DD fighter (USSR), 101
Yak-1 fighter (USSR), 92
Yak-3 fighter (USSR), 101
Yak-9 fighter (USSR), 98, 101
Yakovlev, Alexandr S., 92, 101, 105, 106
Yalta Conference, 102
Yamamoto, Adm. Isoroku, 77, 78
YB-17 aircraft, 62
YOKE test, 168
Yokosuka Naval Arsenal, 76
Zeppelin, Count Ferdinand von, 50
Zeppelins, 50
  attacks on London, 53
Zero fighter (Japan), 79
Zhigarev, Gen. Pavel, 97, 255
Zhukov, Gen. Georgi K., 75, 90, 91, 96, 98, 100
Zuiho aircraft carrier, 78
Zuikaku aircraft carrier, 78

Military History Symposium Series
of the
United States Air Force Academy

Vols. 1-6 (Carl W. Reddel, *Series Editor*)

An American Dilemma: Vietnam, 1964–1973 (1993)
Edited by Dennis E. Showalter and John G. Albert

A Revolutionary War: Korea and the Transformation of the Postwar World (1993)
Edited by William J. Williams

The Intelligence Revolution and Modern Warfare (1995)
Edited by James E. Dillard and Walter T. Hitchcock

Tooling for War: Military Transformation in the Industrial Age (1996)
Edited by Stephen D. Chiabotti

Forging the Sword: Selecting, Educating, and Training Cadets and Junior Officers in the Modern World (1998)
Edited by Elliott V. Converse III

Air Power: Promise and Reality (2000)
Edited by Mark K. Wells

Mark K. Wells, *Series Editor*

Future Wars: Coalition Operations and Global Strategy (forthcoming 2001)
Edited by Dennis E. Showalter

For further information on this series, visit the web site at
www.imprint-chicago.com